I0064795

Operations Research: Methodologies and Applications

Operations Research: Methodologies and Applications

Edited by **Justin Riggs**

CWILLFORD PRESS

New York

Published by Willford Press,
118-35 Queens Blvd., Suite 400,
Forest Hills, NY 11375, USA
www.willfordpress.com

Operations Research: Methodologies and Applications
Edited by Justin Riggs

© 2016 Willford Press

International Standard Book Number: 978-1-68285-024-4 (Hardback)

This book contains information obtained from authentic and highly regarded sources. Copyright for all individual chapters remain with the respective authors as indicated. All chapters are published with permission under the Creative Commons Attribution License or equivalent. A wide variety of references are listed. Permission and sources are indicated; for detailed attributions, please refer to the permissions page and list of contributors. Reasonable efforts have been made to publish reliable data and information, but the authors, editors and publisher cannot assume any responsibility for the validity of all materials or the consequences of their use.

The publisher's policy is to use permanent paper from mills that operate a sustainable forestry policy. Furthermore, the publisher ensures that the text paper and cover boards used have met acceptable environmental accreditation standards.

Trademark Notice: Registered trademark of products or corporate names are used only for explanation and identification without intent to infringe.

Printed in the United States of America.

Contents

Preface

Operations research is a multidisciplinary field of study which deals with various advanced analytical decision-making methods. This book is compiled in such a manner, that it will provide in-depth knowledge about concepts related to important topics such as decision analysis, manufacturing engineering, mathematical optimization, probability and statistics, and information management. The extensive content of this book provides the readers with a thorough understanding of the subject. It is an essential guide for both professionals and those who wish to pursue this discipline further.

The information contained in this book is the result of intensive hard work done by researchers in this field. All due efforts have been made to make this book serve as a complete guiding source for students and researchers. The topics in this book have been comprehensively explained to help readers understand the growing trends in the field.

I would like to thank the entire group of writers who made sincere efforts in this book and my family who supported me in my efforts of working on this book. I take this opportunity to thank all those who have been a guiding force throughout my life.

                                                                                          Editor

Analysis of *M/G/1* queueing model with state dependent arrival and vacation

Charan Jeet Singh[1*], Madhu Jain[2] and Binay Kumar[3]

Abstract

This investigation deals with single server queueing system wherein the arrival of the units follow Poisson process with varying arrival rates in different states and the service time of the units is arbitrary (general) distributed. The server may take a vacation of a fixed duration or may continue to be available in the system for next service. Using the probability argument, we construct the set of steady state equations by introducing the supplementary variable corresponding to elapsed service time. Then, we obtain the probability generating function of the units present in the system. Various performance indices, such as expected number of units in the queue and in the system, average waiting time, etc., are obtained explicitly. Some special cases are also deduced by setting the appropriate parameter values. The numerical illustrations are provided to carry out the sensitivity analysis in order to explore the effect of different parameters on the system performance measures.

Keywords: State dependent, Queue, Arbitrary service time, Vacation, Supplementary variable, Average queue length

Background

In some daily life congestion problems, the service time of the units may not follow exponential distribution. Such situations can be noticed in the clinics performing X-rays and blood test, etc. of patients and in bank at cash counters and many other places. In queueing systems with arbitrary service time distribution, the number of units in the system at time t and the length of time for which the unit is in service (if any) are sufficient to determine the future stochastic properties of these variables. Several researchers have contributed in the direction of general distributed service time queueing system. To mention a few notable works of researchers in this area, we refer Baba (1986), Doshi (1990) and Medhi (1997) and references cited therein. The queueing system under the special consideration with respect to idle period (i.e., vacation) is not new. Levy and Yechiali (1975) have considered such model under the assumption that the server takes a sequence of vacations until it finds at least one unit is waiting in the system. The

analysis of *M/G/1* queue by using the method of supplementary variables has been done by Takagi (1991).

Kimura (1981) assumed general service time distribution function and used a diffusion approximation technique to determine the optimal policy. The queueing model with setup and vacation was considered by Choudhury (2000). Further, Choudhury (2002) studied a queueing system with two different vacation times under multiple vacation policy. Single server queueing system with time homogeneous breakdowns and deterministic repair times was analyzed by Madan (2003). Wang (2004) worked on the *M/G/1* queueing system with second optional service and server breakdown. The multiple vacations system was considered by Wu and Takagi (2006). Choudhury (2008) discussed the queue size distribution of a queue with a random set-up time and Bernoulli vacation schedule under a restricted admissibility policy.

Recently, Maraghi et al. (2009) have studied batch arrival queueing system with random breakdowns and Bernoulli schedule random vacations having general vacation time. They have obtained steady state results in terms of probability generating functions for the number of customers in the queue. Choudhury and

* Correspondence: cjsmath@gmail.com
[1]Department of Mathematics Guru Nanak Dev University, Amritsar, Punjab, 143005, India
Full list of author information is available at the end of the article

Kalita (2009) studied the steady state behavior of a model with repeated attempts and Bernoulli vacation schedule, which was a generalization of the classical model. Banik (2010) analyzed a queueing system with a single working vacation to obtain the performance measures using the embedded Markov chain. Thangraj and Vanitha (2010) discussed the single server model with two stages of heterogeneous service with different service time distributions subject to random breakdowns and compulsory service vacations with arbitrary vacation periods.

The queueing system with server vacations can be used to model a system wherein the server stops working during a vacation. Such system has wide applicability in analyzing the performance of various real life traffic situations of day-to-day as well as industrial queues. In some systems, the arrival of the units occurs according to a Poisson process or general distributed fashion with different arrival rates depending on the server's status.

Madan (1999) discussed the steady state behavior of an arbitrary service time queue with deterministic service vacation. In his investigation, he has considered that the customers arrive at the system with uniform arrival rates. In many congestion scenarios, the arrival rates of the customers are influenced by the status of the server. The queueing models with variable arrival rates of the units can be observed in health care systems. Besides this, it is applicable in the banks, at the checkout counters in the supermarket, etc. In some industrial scenario, the arrival rate of the units may also be dependent upon the states of the system, especially in production and manufacturing systems, wherein the management may optimize the cost of inventory by controlling the arrival rate of new units.

The present investigation is the extension of vacation model for single server general distributed service time studied by Madan (1999) and addresses the analysis of $M/G/1$ queueing system with deterministic server vacation in which the arrival rate of the units are state dependent. The layout of the investigation is as follows: The model description, by stating the requisite assumptions and notations, is given in the 'Model description' section. In 'The steady state equations' section, we construct the set of steady state equations by introducing the supplementary variable corresponding to elapsed service time. In 'The analysis' section, we obtain the probability generating function of the queue size distribution in different states. Some system characteristics of the model are presented in 'Performance measures' section. By selecting appropriate parameter values, some special cases are deduced in 'Special cases' section. In 'Numerical illustration' section, numerical illustration is provided to explore the effect of different parameters on the performance measures. In the 'Conclusions' section, the

noble features and future scope of the present model are highlighted.

Model description

Consider $M/G/1$ queueing system with deterministic server vacations under the following assumptions:

- The server may decide to take a vacation of fixed length d (>0) at the completion of each service with probability p or may continue to be available in the system for the next service with probability $1 - p$.
- The units arrive in the system according to Poisson fashion with state dependent rates.
- The service of the units is rendered according to the general (arbitrary) distribution.
- The FCFS service discipline is followed to select the customer for the service.

The notations used in the formulation of the model are as follows:

λ_1:   mean arrival rate of the units in idle state

$\lambda_2(\lambda_3)$: mean arrival rate of the units in busy (vacation) state

$B(v)$:   distribution function of the service time

$b(v)$:   density function of service time

$\bar{b}(.)$:   Laplace transform of $b(.)$

x: elapsed service time

$\mu(x)dx$: hazard rate of completion of the service of the unit during the interval $(x, x + dx)$ with elapsed time x

K_r: probability of r arrivals during a vacation period

n: number of units in the queue, excluding the unit, which is in service (≥ 0)

$W_n(t,x)$: probability of n units in the system at time t when the server is busy in rendering service to the unit with elapsed service time lying between x and $x + dx$

$V_n(t)$:   probability of n units in the queue at time t when the server is on vacation

$Q(t)$:   probability that there are no units in the system and the server is in idle state at time t

$P_q(z)$:   probability generating function of the queue length whether the server is on vacation or available in the system

$P(z)$:   probability generating function of the number of units in the system

The hazard rate $\mu(x)$ is given by Equation 1:

$$\mu(x) = \frac{b(x)}{1 - B(x)} \tag{1}$$

where,

$$b(v) = \mu(v) \exp. \left[-\int_0^v \mu(x)dx \right] \tag{2}$$

In steady state, we have Equation 3:

$$W_n(x) = \lim_{t \to \infty} W_n(t, x); \quad V_n = \lim_{t \to \infty} V_n(t); \tag{3}$$

$$Q = \lim_{t \to \infty} Q(t)$$

and Equation 4

$$K_r = \frac{e^{-\lambda_3 d}(\lambda_3 d)^r}{r!}, \quad r = 0, 1, 2 \ldots \tag{4}$$

The steady state equations
In this section, we formulate the set of governing equations of the system using the appropriate rates as follows:

$$\frac{d}{dx} W_n(x) + (\lambda_2 + \mu(x)) W_n(x) = \lambda_2 W_{n-1}(x); \tag{5}$$
$$n \geq 1$$

$$\frac{d}{dx} W_0(x) + (\lambda_2 + \mu(x)) W_0(x) = 0 \tag{6}$$

$$\lambda_1 Q = (1 - p) \int_0^\infty W_0(x)\mu(x)dx + V_0 K_0 \tag{7}$$

$$V_n = p \int_0^\infty W_n(x)\mu(x)dx; n \geq 0 \tag{8}$$

The above equations are to be solved subject to the following boundary conditions:

$$W_n(0) = (1 - p) \int_0^\infty W_{n+1}(x)\mu(x)dx + V_0 K_{n+1}$$
$$+ V_1 K_n + \ldots + V_{n+1} K_0; \ n \geq 1 \tag{9}$$

$$W_0(0) = (1 - p) \int_0^\infty W_1(x)\mu(x)dx + V_0 K_1$$
$$+ V_1 K_0 + \lambda_1 Q \tag{10}$$

We define the following probability generating functions:

$$W(x, z) = \sum_{n=0}^\infty W_n(x)z^n; \quad W(z) = \sum_{n=0}^\infty W_n z^n; \tag{11}$$

$$V(z) = \sum_{n=0}^\infty V_n z^n$$

The analysis
In order to derive various performance indices, we obtain the probability generating function of the number of units in the system using the above set of equations as follows: on multiplying Equations 5 and 6 by z^n, summing over n and using Equation 11, we have Equation 12:

$$\frac{d}{dx} W(x, z) + (\lambda_2 - \lambda_2 z + \mu(x)) W(x, z) = 0 \tag{12}$$

Similarly, multiplying Equation 8 by z^n, summing over n and using Equation 11, we have Equation 13:

$$V(z) = p \int_0^\infty W(x, z)\mu(x)dx \tag{13}$$

Now, we obtain

$$\sum_{n=0}^\infty K_n z^n = \sum_{n=0}^\infty \frac{e^{-\lambda_3 d}(\lambda_3 d)^n}{n!} z^n = e^{-\lambda_3 d(1-z)}$$

Using Equations 9 to 11, we derive Equation 14:

$$zW(0, z) = (1 - p) \int_0^\infty W(x, z)\mu(x)dx$$
$$+ V(z)e^{-\lambda_3 d(1-z)} - K_0 V_0 + \lambda_1 Q z \tag{14}$$
$$- (1 - p) \int_0^\infty W_0(x)\mu(x)dx$$

With the help of Equation 7, the above Equation can be written as

$$W(0, z) = \frac{(1 - p) \int_0^\infty W(x, z)\mu(x)dx}{z}$$
$$\frac{+ V(z)e^{-\lambda_3 d(1-z)} + \lambda_1 Q(z - 1)}{z} \tag{15}$$

From Equation 12, we get Equation 16:

$$W(x, z) = W(0, z)e^{-\lambda_2(1-z)x - \int_0^x \mu(t)dt} \tag{16}$$

Using Equation 16, we have Equation 17:

$$W(z) = W(0, z)\left[\frac{1 - \bar{b}(\lambda_2 - \lambda_2 z)}{(\lambda_2 - \lambda_2 z)}\right] \tag{17}$$

where

$$\bar{b}(\lambda_2 - \lambda_2 z) = \int_0^\infty e^{-\lambda_2(1-z)x} b(x)\,dx.$$

Theorem 1 *The probability generating function of the queue length, whether the server is on vacation or available in the system is*

$$P_q(z) = \frac{\left[\frac{\lambda_1}{\lambda_2}\{\bar{b}(\lambda_2 - \lambda_2 z) - 1\} + \{p\lambda_1(z-1)\bar{b}(\lambda_2 - \lambda_2 z)\}\right]\left[1 - \frac{\lambda_1(1+p\mu)}{\mu(1-\lambda_3 pd)+\lambda_1 p\mu+\lambda_1-\lambda_2}\right]}{z - \bar{b}(\lambda_2 - \lambda_2 z) + p\bar{b}(\lambda_2 - \lambda_2 z)(1 - e^{-\lambda_3 d(1-z)})} \tag{18}$$

Proof: For proof, see *Proof of Theorem 1*[a] in the 'Endnotes' section.

Theorem 2 *The probability generating function of the number of units in the system is*

$$P(z) = \frac{\left[\left(1 - \frac{\lambda_1}{\lambda_2}\right)z + \bar{b}(\lambda_2 - \lambda_2 z)\left\{p\left(1 - e^{-\lambda_3 d(1-z)}\right) - 1 + zp\lambda_1(z-1) + \frac{\lambda_1}{\lambda_2}z\right\}\right]}{z - \bar{b}(\lambda_2 - \lambda_2 z) + p\bar{b}(\lambda_2 - \lambda_2 z)(1 - e^{-\lambda_3 d(1-z)})}$$
$$\times \left[1 - \frac{\lambda_1(1+p\mu)}{\mu(1-\lambda_3 pd)+\lambda_1 p\mu+\lambda_1-\lambda_2}\right] \tag{19}$$

Proof: For proof, see *Proof of Theorem 2*[b] in the 'Endnotes' section.

Performance measures

Now, we shall establish various performance measures using the probability generating function of the queue length as follows:

Theorem 3 *The expected number of units in the queue is*

$$L_q = \frac{\begin{aligned}&\left[\lambda_1\lambda_2 E(v^2)(1 - \lambda_3 pd + \lambda_2 p) + \frac{2\lambda_1\lambda_2 p}{\mu}\left(1 - \frac{\lambda_2}{\mu}\right) + \frac{2\lambda_1\lambda_2\lambda_3 pd}{\mu^2} + p^2\lambda_1\lambda_3{}^2 d^2 + \frac{p\lambda_3{}^2\lambda_1 d^2}{\mu}\right]\\&\times\left[1 - \frac{\lambda_1(1+\mu p)}{\mu(1-\lambda_3 pd)+\lambda_1(1+\mu p)-\lambda_2}\right]\end{aligned}}{2\left[1 - \frac{\lambda_2}{\mu} - \lambda_3 pd\right]^2} \tag{20}$$

Proof: The expected number of units in the queue (L_q) is obtained using

$$L_q = \lim_{z\to 1}\frac{d}{dz}P_q(z)$$

For detailed proof, see *Proof of Theorem 3*[c] in the 'Endnotes' section.

The expected number of units in the system can be obtained as

$$L = L_q + \rho \tag{21}$$

The expected waiting time in the queue is given by Little's formula

$$W_q = \frac{L_q}{\lambda_{eff}};\ \lambda_{eff} = \lambda_1 Q + \lambda_2 W(1) + \lambda_3 V(1) \tag{22}$$

Special Cases

It is worthwhile to establish the performance measures in some special cases by setting appropriate parameters to tally our results with some existing results. When $p = 0$, Equation 20 gives

$$L_q = \frac{[\lambda_1\lambda_2 E(v^2)]}{2\left(1 - \frac{\lambda_2}{\mu}\right)^2}\left[\frac{\mu - \lambda_2}{\mu + \lambda_1 - \lambda_2}\right] \tag{23}$$

This provides the average queue length of $M/G/1$ model with state dependent rates. For no server vacation model, when units arrive according to Poisson fashion with homogeneous rate (λ) in all states by setting $\lambda = \lambda_1 = \lambda_2 = \lambda_3$, we have Equation 24:

$$L_q = \frac{\lambda^2 E(v^2)}{2\left(1 - \frac{\lambda}{\mu}\right)} \tag{24}$$

Equation 24 provides the well-known result of $M/G/1$ queue (see Gross and Harris 2003).

In $M/E_k/1$ deterministic vacation queueing model, we put

$$E(v^2) = \frac{k+1}{k\mu^2},$$

Equation 20 reduces to

$$L_q = \begin{bmatrix}\dfrac{\lambda_1\lambda_2\left(\frac{k+1}{k\mu^2}\right)(1 - \lambda_3 pd + \lambda_2 p) + \frac{2\lambda_1\lambda_2 p}{\mu}\left(1 - \frac{\lambda_2}{\mu}\right) + \frac{2\lambda_1\lambda_2\lambda_3}{\mu^2}pd + p^2\lambda_1\lambda_3{}^2 d^2 + \frac{p\lambda_3{}^2\lambda_1 d^2}{\mu}}{2\left(1 - \frac{\lambda_2}{\mu} - \lambda_3 pd\right)^2}\\ \times\left[1 - \frac{\lambda_1(1+\mu p)}{\mu(1-\lambda_3 pd)+\lambda_1(1+\mu p)-\lambda_2}\right]\end{bmatrix} \tag{25}$$

In $M/M/1$ deterministic vacation queueing model, we set

$$E(v^2) = \frac{2}{\mu^2}$$

in Equation 20 so that

$$
L_q = \left[\frac{\left[\frac{2\lambda_1\lambda_2}{\mu^2}(1 - \lambda_3 pd + \lambda_2 p) + \frac{2\lambda_1\lambda_2 p}{\mu}\left(1 - \frac{\lambda_2}{\mu}\right) + \frac{2\lambda_1\lambda_2\lambda_3 pd}{\mu^2} + p^2\lambda_1\lambda_3^2 d^2 + \frac{p\lambda_3^2\lambda_1 d^2}{\mu}\right]}{2\left(1 - \frac{\lambda_2}{\mu} - \lambda_3 pd\right)^2} \right]
$$
$$
\times \left[1 - \frac{\lambda_1(1 + \mu p)}{\mu(1 - \lambda_3 pd) + \lambda_1(1 + \mu p) - \lambda_2} \right]
$$

$$(26)$$

Numerical Illustration

In this section, we present the numerical illustration to evaluate the queue size distribution for the single server deterministic vacation model using the analytical results derived in previous section. The effects of variation of service rate (μ), vacation time (d) and vacation probability (p) on the queue length are displayed in Tables 1 and 2. The service time assumed to be generally distributed, therefore, second moment of service time $E(v^2)$ for different distributions are as follows:

- $M/E_k/1$ deterministic vacation model: $E(v^2) = \frac{k+1}{k\mu^2}$
- $M/M/1$ deterministic vacation model: $E(v^2) = \frac{2}{\mu^2}$
- $M/D/1$ deterministic vacation model: $E(v^2) = \frac{1}{\mu^2}$.

For computation purposes, we set default parameters as $\lambda_1 = 1.0\lambda$, $\lambda_2 = 0.9\lambda$, $\lambda_3 = 0.7\lambda$, $p = 0.02$ and $d = 3$. From Tables 1 and 2, we observe that the queue length (L_q) decreases with the increase in service time. It is also noticed that the queue length decreases with the increase in the number of service phases (k). As far as the effects of parameters d and p are concerned, we see that the L_q increases significantly with the increment of both d and p.

Figure 1a,b respectively reveals that the L_q increases with the increase in arrival rate (λ). However, the effect is more prevalent for higher values of λ. Figure 2a,b exhibits the queue length L_q for $M/M/1$ and $M/E_5/1$ models by varying p. It is observed that the L_q increases with the increase in λ as well as p; the increment is more significant for larger values of λ and p. Figure 3a,b reveal that the L_q decreases with the increase in the number of phases (k).

Table 1 L_q by varying μ and d for $p = 0.02$, $\lambda_1 = 2$, $\lambda_2 = 1.5$, $\lambda_3 = 1$

d	2		3		4		5		6	
μ	$k = 1$	$k = 5$	$k = 1$	$k = 5$	$k = 1$	$k = 5$	$k = 1$	$k = 5$	$k = 1$	$k = 5$
2.1	2.46	1.58	2.90	1.94	3.52	2.47	4.37	3.22	5.56	4.26
2.2	2.01	1.29	2.35	1.58	2.83	1.99	3.47	2.57	4.35	3.36
2.3	1.68	1.08	1.96	1.32	2.34	1.66	2.86	2.13	3.55	2.76
2.4	1.43	0.93	1.67	1.13	1.99	1.41	2.42	1.81	2.99	2.33
2.5	1.24	0.80	1.44	0.98	1.72	1.23	2.09	1.57	2.57	2.01

μ, service rate; d, vacation time; k, service phases.

Table 2 L_q by varying μ and p for $d = 3$, $\lambda_1 = 2$, $\lambda_2 = 1.5$, $\lambda_3 = 1$

p	0.025		0.030		0.035		0.040		0.045	
μ	$k = 1$	$k = 5$	$k = 1$	$k = 5$	$k = 1$	$k = 5$	$k = 1$	$k = 5$	$k = 1$	$k = 5$
2.1	3.24	2.22	3.64	2.54	4.10	2.91	4.65	3.36	5.31	3.90
2.2	2.61	1.79	2.90	2.03	3.23	2.31	3.62	2.63	4.08	3.01
2.3	2.16	1.49	2.39	1.68	2.65	1.90	2.95	2.15	3.29	2.44
2.4	1.84	1.27	2.02	1.43	2.23	1.61	2.47	1.81	2.74	2.04
2.5	1.59	1.10	1.74	1.24	1.92	1.39	2.11	1.56	2.34	1.75

μ, service rate; p, vacation probability; k, service phases.

Results and discussion

Finally, we conclude that when arrival rate (service rate) increases, the queue length increases (decreases); this situation matches with our expectation in various real life congestion problems. The queue length decreases with the increase in the number of service phases. The queue length increases slightly with the increase in the vacation time as well as vacation probability for low traffic condition, but as traffic increases, there is remarkable increase in it. Therefore, in case of heavy traffic, the frequent vacation of the server has adverse effect on the grade of service and it should be avoided as much as possible.

Conclusions

For the real life congestion situations, where the arrival of units depends on the status of the server, our study may be very helpful in the design and development phases of the concerned systems. The fields of distributed computer system, wireless communications, production and manufacturing system, etc. have the major sources of motivation for the growth and creation of such queueing models.

Methods

In this investigation, we have studied the steady state behavior of a single server queueing model with vacation and varying arrival rates. The supplementary variable method is used to determine the probability generating function of the queue size which is further employed to evaluate other performance measures in explicit form. The sensitivity analysis is carried out which demonstrates the computational tractability and validity of the analytical results.

Endnotes

[a]*Proof of Theorem 1* From Equations 2 and 16, we get Equation 27

$$
\int_0^\infty W(x, z)\mu(x)dx = W(0, z)\,\bar{b}(\lambda_2 - \lambda_2 z)
$$

$$(27)$$

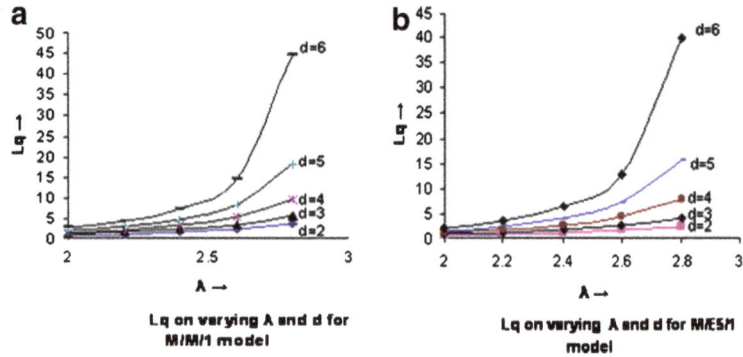

Figure 1 L_q for (a) $M/M/1$ and (b) $M/E_5/1$ models on varying arrival rate λ and vacation time d.

On using Equation 27 in Equation 15, we have

$$W(0,z) = \frac{V(z)e^{-\lambda_3 d(1-z)} + \lambda_1 Q(z-1)}{\left[z - \bar{b}(\lambda_2 - \lambda_2 z)(1-p)\right]} \qquad (28)$$

On using Equation 28, Equation 17 gives,

$$W(z) = \left[\frac{1 - \bar{b}(\lambda_2 - \lambda_2 z)}{(\lambda_2 - \lambda_2 z)}\right] \\ \times \left[\frac{V(z)e^{-\lambda_3 d(1-z)} + \lambda_1 Q(z-1)}{\left[z - \bar{b}(\lambda_2 - \lambda_2 z)(1-p)\right]}\right] \qquad (29)$$

With the help of Equation 27, Equation 13 becomes

$$V(z) = p(W(0,z))\bar{b}(\lambda_2 - \lambda_2 z) \qquad (30)$$

On using Equation 28, from Equation 30 we have

$$V(z) = p\left[\frac{V(z)e^{-\lambda_3 d(1-z)} + \lambda_1 Q(z-1)}{\left[z - \bar{b}(\lambda_2 - \lambda_2 z)(1-p)\right]}\right] \\ \times \bar{b}(\lambda_2 - \lambda_2 z) \qquad (31)$$

On simplifying Equation 31, we have

$$V(z) = \frac{p(\lambda_1 Q(z-1))\bar{b}(\lambda_2 - \lambda_2 z)}{\left[\begin{array}{c}z - \bar{b}(\lambda_2 - \lambda_2 z) \\ +p\bar{b}(\lambda_2 - \lambda_2 z)\left(1 - e^{-\lambda_3 d(1-z)}\right)\end{array}\right]} \qquad (32)$$

From Equations 29 and 32, we get

$$W(z) = \frac{\lambda_1\left[\bar{b}(\lambda_2 - \lambda_2 z) - 1\right]Q}{\left[\lambda_2\left[z - \bar{b}(\lambda_2 - \lambda_2 z) + p\bar{b}(\lambda_2 - \lambda_2 z)\left(1 - e^{-\lambda_3 d(1-z)}\right)\right]\right]} \qquad (33)$$

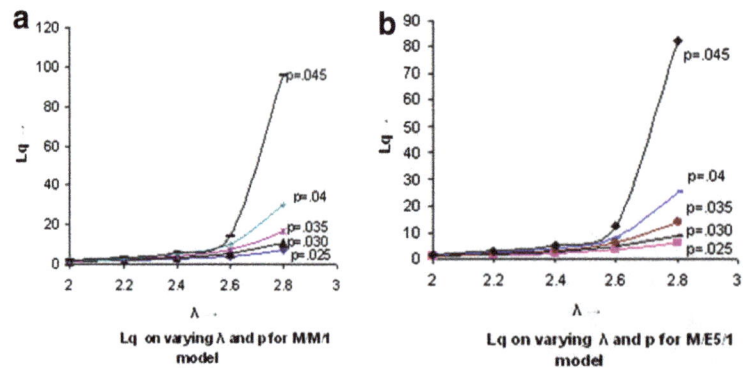

Figure 2 L_q for (a) $M/M/1$ and (b) $M/E_5/1$ models on varying arrival rate λ and optional vacation probability p.

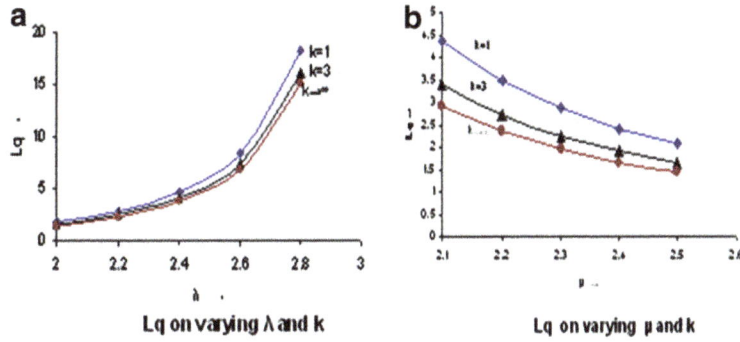

Figure 3 (a) L_q on varying arrival rate λ, service phases k and (b) L_q on varying service rate μ and service phases k.

Since $\bar{b}(0) = 1$; $-\bar{b}'(0) = \frac{1}{\mu}$ and $\bar{b}''(0) = E(v^2)$, we have

$$V(1) = \lim_{z \to 1} V(z) = \frac{\lambda_1 \mu p Q}{\mu - \lambda_2 - \lambda_3 \mu p d} \tag{34}$$

and

$$W(1) = \lim_{z \to 1} W(z) = \frac{\lambda_1 Q}{\mu - \lambda_2 - \lambda_3 \mu p d} \tag{35}$$

The normalizing condition $Q + V(1) + W(1) = 1$ gives the unknown constant Q as

$$Q = 1 - \frac{\lambda_1(1 + p\mu)}{\mu(1 - \lambda_3 p d) + \lambda_1 p \mu + \lambda_1 - \lambda_2}; \tag{36}$$
$$\lambda_2 < \mu(1 - \lambda_3 p d)$$

From Equation 36 we get

$$\rho = 1 - Q = \frac{\lambda_1(1 + p\mu)}{\left[\begin{array}{c}\mu(1 - \lambda_3 p d) \\ +\lambda_1 p \mu + \lambda_1 - \lambda_2\end{array}\right]} < 1 \tag{37}$$

On using Equation 36 in Equations 32 and 33, we have

$$V(z) = \frac{\left[\lambda_1 p \bar{b}(\lambda_2 - \lambda_2 z)(z - 1)\right] \times \left[1 - \dfrac{\lambda_1(1 + p\mu)}{\mu(1 - \lambda_3 p d) + \lambda_1 p \mu + \lambda_1 - \lambda_2}\right]}{\left[\begin{array}{c} z - \bar{b}(\lambda_2 - \lambda_2 z) \\ +p\bar{b}(\lambda_2 - \lambda_2 z)\left(1 - e^{-\lambda_3 d(1-z)}\right)\end{array}\right]} \tag{38}$$

$$W(z) = \frac{\lambda_1 \left[\bar{b}(\lambda_2 - \lambda_2 z) - 1\right] \times \left[1 - \dfrac{\lambda_1(1 + p\mu)}{\mu(1 - \lambda_3 p d) + \lambda_1 p \mu + \lambda_1 - \lambda_2}\right]}{\left[\lambda_2\left[z - \bar{b}(\lambda_2 - \lambda_2 z) + p\bar{b}(\lambda_2 - \lambda_2 z)\left(1 - e^{-\lambda_3 d(1-z)}\right)\right]\right]} \tag{39}$$

On adding Equations 38 and 39, we have

$$P_q(z) = V(z) + W(z) = \frac{\left[\begin{array}{c}\frac{\lambda_1}{\lambda_2}\{\bar{b}(\lambda_2 - \lambda_2 z) - 1\} \\ +\{p\lambda_1(z-1)\bar{b}(\lambda_2 - \lambda_2 z)\}\end{array}\right] \times \left[1 - \dfrac{\lambda_1(1 + p\mu)}{\mu(1 - \lambda_3 p d) + \lambda_1 p \mu + \lambda_1 - \lambda_2}\right]}{\left[\begin{array}{c}z - \bar{b}(\lambda_2 - \lambda_2 z) \\ +p\bar{b}(\lambda_2 - \lambda_2 z)\left(1 - e^{-\lambda_3 d(1-z)}\right)\end{array}\right]} \tag{40}$$

[b]*Proof of Theorem 2* By using Equations 36 and 40, we get

$$P(z) = Q + zP_q(z)$$
$$= \frac{\left[\left(1 - \frac{\lambda_1}{\lambda_2}\right)z + \bar{b}(\lambda_2 - \lambda_2 z)\left\{p\left(1 - e^{-\lambda_3 d(1-z)}\right) - 1 + zp\lambda_1(z-1) + \frac{\lambda_1}{\lambda_2}z\right\}\right]}{z - \bar{b}(\lambda_2 - \lambda_2 z) + p\bar{b}(\lambda_2 - \lambda_2 z)(1 - e^{-\lambda_3 d(1-z)})}$$
$$\times \left[1 - \frac{\lambda_1(1 + p\mu)}{\mu(1 - \lambda_3 p d) + \lambda_1 p \mu + \lambda_1 - \lambda_2}\right] \tag{41}$$

[c]*Proof of Theorem 3* From Equation 40, we have

$$L_q = \lim_{z \to 1} \frac{d}{dz} P_q(z) = P'(1) = \lim_{z \to 1} \frac{D'(z)N''(z) - N'(z)D''(z)}{2[D'(z)]^2}$$
$$= \frac{D'(1)N''(1) - N'(1)D''(1)}{2[D'(1)]^2} \tag{42}$$

where

$$N'(1) = \left[\lambda_1\left(p + \frac{1}{\mu}\right)\right]\left[1 - \frac{\lambda_1(1 + \mu p)}{\mu(1 - \lambda_3 p d) + \lambda_1(1 + \mu p) - \lambda_2}\right],$$

$$N''(1) = \left[\frac{2p\lambda_1\lambda_2}{\mu} + \lambda_1\lambda_2 E(v^2)\right]\left[1 - \frac{\lambda_1(1 + \mu p)}{\mu(1 - \lambda_3 p d) + \lambda_1(1 + \mu p) - \lambda_2}\right],$$

$$D^{'}(1) = 1 - \frac{\lambda_2}{\mu} - \lambda_3 pd \text{ and}$$

$$D^{''}(1) = -\left[\lambda_2^2 E(v^2) + \frac{2\lambda_2\lambda_3 pd}{\mu} + p\lambda_3^2 d^2\right].$$

On using above values, Equation 42 gives the result as given in Equation 20.

Competing interests

The authors declare that they have no competing interests.

Acknowledgments

The authors are thankful to the learned referee and editor for their valuable comments and suggestions for the improvement of the paper. First author is also thankful to the University Grants Commission, New Delhi for the financial support under major research project no. 37-191/2009 (SR).

Author details

[1]Department of Mathematics Guru Nanak Dev University, Amritsar, Punjab, 143005, India. [2]Department of Mathematics, Indian Institute of Technology Roorkee, Roorkee, Uttarakhand, 247667, India. [3]Department of Mathematics M.L.U. DAV College, Phagwara, Punjab, 144402, India.

Authors' contributions

CJS has worked on the modeling and analysis of non-markovian $M/G/1$ model. The queue size distribution and various performance measures via. queue theoretic approach based on supplementary variable and generating function method have been obtained by MJ. BK has performed numerical experiment and carried out sensitivity analysis by taking an illustration. All authors read and approved the final manuscript.

Authors' information

Charan Jeet Singh is Associate Professor in the department of Mathematics, Guru Nanak Dev University, Amritsar (India). Madhu Jain is a faculty member in the department of Mathematics, I.I.T. Roorkee (India). Binay Kumar is Lecturer in the department of Mathematics, M.L.U. DAV College, Phagwara (India).

References

Baba Y (1986) On the $M^x/G/1$ queue with vacation time. Operation Res Letter 5:93–98

Banik AD (2010) Analysis of single working vacation in $GI/M/1/N$ and $GI/M/1/\infty$ queueing system. Int J Operational Res 7(3):314–333

Choudhury G (2000) An $M^x/G/1$ queueing system with a setup period and vacation period. QUESTA 36:23–28

Choudhury G (2002) Some aspects of $M/G/M$ queue with two different vacation times under multiple vacation policy. Stoch Anal Appl 20(5):901–909

Choudhury G (2008) A note on the $M^x/G/1$ queue with a random set-up time under a restricted admissibility policy with a Bernoulli vacation schedule. Statistical Methodology 5:21–29

Choudhury G, Kalita S (2009) A two-phase queueing system with repeated attempts and Bernoulli vacation schedule. Int J Operational Res 5(4):392–407

Doshi BT (1990) Single server queues with vacations. In: Takagi H (ed) Stochastic analysis of computer and communication systems. Elsevier, North-Holland, Amsterdam, pp 217–265

Gross D, Harris CM (2003) Fundamentals of queueing theory, 3rd edn. Wiley, New York

Kimura T (1981) Optimal control of an $M/G/M$ queueing system with removable server via diffusion approximation. Eur J Oper Res 8:390–398

Levy Y, Yechiali U (1975) Utilization of idle time in an $M/G/M$ queueing system. Manag Sci 22:202–211

Madan KC (1999) An $M/G/M$ queue with optional deterministic server vacations. Metron, LVII 57(3–4):83–95

Madan KC (2003) An $M/G/M$ type queue with time homogeneous breakdowns and deterministic repair times. Soochow J Mathematics 29(1):103–110

Maraghi FA, Madan KC, Dowman KD (2009) Batch arrival queueing system with random breakdowns and Bernoulli schedule server vacations having general time distribution. Int J Inf Manag Sci 20:55–70

Medhi J (1997) Single server queueing system with poisson input: a review of some recent development. In: Balakrishnann N (ed) Advances in combinatorial method and applications in probability and statistics. Birkhauser

Takagi H (1991) Queueing analysis. A foundation of performance evaluation, vacation and priority systems, 1st edn., Elsevier, Amsterdam

Thangraj V, Vanitha S (2010) $M/G/M$ queue with two stage heterogeneous service compulsory server vacation and random breakdowns. Int J Contemporary Mathematical Sci 5(7):307–322

Wang J (2004) An $M/G/M$ queue with second optional service and server breakdowns. Computers and Mathematics with Applications 47:1713–1723

Wu DA, Takagi H (2006) $M/G/M$ queue with multiple working vacations. Perform Eval 63:54–681

2

A seller-buyer supply chain model with exponential distribution lead time

Mehrab Bahri[1] and Mohammad Jafar Tarokh[2*]

Abstract

Supply chain is an accepted way of remaining in the competition in today's rapidly changing market. This paper presents a coordinated seller-buyer supply chain model in two stages, which is called Joint Economic Lot Sizing (JELS) in literature. The delivery activities in the supply chain consist of a single raw material. We assume that the delivery lead time is stochastic and follows an exponential distribution. Also, the shortage during the lead time is permitted and completely back-ordered for the buyer. With these assumptions, the annual cost function of JELS is minimized. At the end, a numerical example is presented to show that the integrated approach considerably improves the costs in comparison with the independent decisions by seller and buyer.

Keywords: Integrated inventory model, Stochastic lead time, Supply chain coordination, Cost, Optimization

Background

Supply chain takes on an importance because of the rapid market changes which is the result of the explosion of product varieties with short life cycles in today's global market (Ben-Daya et al. 2008). The effective collaboration of partners and coordination of all activities within the supply chain are prerequisites in such competitive and dynamic market conditions (Soroor et al. 2009b; Tarantilis 2008). So in the recent years, supply chain was dealt with from many points of view such as pricing problem in Hu et al. (2010), Chen and Kang (2010), and Huang et al. (2010), or the fuzzy conditions in supply chain elements in Xu and Zhai (2010) and many others.

One major subject in this topic is managing the inventory across the whole supply chain to reduce the costs for customers (Soroor et al. 2009a). To deal with this problem, some researchers follow real-time data processing such as radio frequency identification, global positioning system, flow control sensors, cellular telephones, navigation systems, and satellite positioning systems (Tarantilis 2008; Soroor et al. 2009c). There are some other parallel researches that try to elicit a mathematical model from the integrated supply chain. In order to construct such a model, it is necessary to use some simplifying assumptions. The first integrated supply chain model which was introduced by Goyal (1977) and contained only one seller and one buyer was called Joint Economic Lot Sizing (JELS) problem. It was under the most simplifying and deterministic conditions. Goyal (1977) presented a solution to the problem under the assumption of the seller's infinite production rate and lot-for-lot policy for the shipments from the seller to the buyer. In this policy, before shipment, the entire production lot should be ready and each production lot is sent to the buyer as a single shipment. Banerjee (1986) eliminated the infinite production rate assumption, but retained lot-for-lot policy. Then the lot-for-lot policy was relaxed by Goyal (1988) in an effort to generalize the problem. By constructing a model which allowed shipments to take place during production, Lu (1995) decreased the assumption of completing a batch before starting shipments. Banerjee and Kim (1995), Ha and Kim (1997), and Kim and Ha (2003) also considered JELS model with equal-shipment policies. Viswanathan (1998) proposed an optimal policy for a particular model relating to problem parameters. Hill (1997) further took the geometric growth factor as a decision variable and so generalized the model of Goyal (1995). Goyal and Nebebe (2000) suggested another simple geometric policy to produce acceptable results. This was a model in which a small shipment is followed by a series of larger and equal-sized shipments. Another generalization made

* Correspondence: mjtarokh@kntu.ac.ir

[2]Department of Industrial Engineering, K.N. Toosi University of Technology, Tehran 1439955471, Iran

Full list of author information is available at the end of the article

by Hill (1999) found the optimal solution and proposed an exact iterative algorithm for solving the problem. The structure of the optimal policy presented on the basis of geometric series was followed by equal-sized shipments. Hill and Omar (2006) and Zhou and Wang (2007) relaxed the assumption of holding costs. The resultant assumption is that the successive shipment sizes are increased by a fixed factor when the vendor's holding cost is larger than the buyer's. The unreliability of the process on JELS is considered by Ben-Daya and Zamin (2002a) and Huang (2004). Other extensions of this problem are considered as stochastic parameters. Ben-Daya and Zamin (2002b) considered a JELS problem under equal-shipment policy with stochastic demand.

Ouyang et al. (2004) assumed the lead time to be stochastic and controllable. He also permitted a shortage during the lead time. In this paper we relax the assumption of deterministic lead time of transporting products from the seller to the buyer and assume lead time as a stochastic variable with exponential distribution. This paper is organized in the following form. In Section Definition of the problem, we define the problem, and also introduce notations and assumptions. The Model formulation Section gives a discussion on independent and integrated policies for seller and buyer. Solution for separate and joint models Section deals with the optimal solution of the independent and integrated model. Numerical results Section presents some numerical examples to compare two models. The conclusion of the paper is given in the Conclusions Section.

Definition of the problem
Model notation
The model is a supply chain which represented a seller with a constant produce rate of 'p', and his product is sent to a buyer by shipment equal size of 'Q'. The demand for the buyer inventory is constant value of 'D', when the inventory drops to r; buyer makes a new order in the size of 'Q'. The notations below are used in the model:

D demand rate for buyer inventory
p production rate of the seller
Q shipment size
r buyer's reorder point
A_v seller's setup cost
A_b buyer's ordering cost
h_v holding cost for the seller
h_b holding cost for the buyer
π shortage cost for the buyer per unit per unit time
n number of shipments
T buyer's cycle time (time between two successive orders)
L lead time to replenish the buyer's order
$TC_b(r,Q)$ buyer's expected total cost per unit time
$tc_b(r,Q,L)$ buyer total cost of one time cycle in term of L

$tc_b(r,Q)$ expected total cost in one cycle for given r and Q
TI_v seller's total inventory
TC_v seller's expected total cost
$STC(r,Q,n)$ separate total cost for buyer and seller
$JTC(r,Q,n)$ joint total cost for buyer and seller

Model assumptions
The assumptions made in the paper are as follow:

(1) Product is manufactured with a finite product rate p.
(2) The final demand for product is deterministic and constant D, where $p > D$.
(3) The lots delivered to buyer by seller in equal size batches, Q.
(4) The buyer makes an order of Q-size as soon as the inventory drops to r.
(5) In each setup, seller manufactures nQ product to reduce the average setup cost.
(6) Shortage is acceptable and completely back ordered for the buyer.
(7) The policy of shipment is non-delayed i.e., as soon as receiving an order, the seller delivers it, if available; otherwise, it manufactures product of nQ-size and delivers orders during manufacturing phase and after that.
(8) The lead time to deliver the shipment from seller to buyer follows an exponential distribution with parameter λ, i.e., L approximately $\exp(\lambda)$.
(9) Time horizon is infinite.

Model formulation
To obtain the buyer's expected total cost per unit time, $TC_b(r,Q)$, it is considered that the orders are received in a sequence which are not necessarily the same as they have been made. Some extra simplifying assumptions are introduced here to avoid facing intractable problem.

(1) The orders do not cross in time (Hadly and Whitin 1963).
(2) At the start of each time cycle, the net stock is considered to be r.

Figure 1 presents different possible cases in each cycle for the buyer. Considering the demand rate as D, from the time of ordering in which the inventory position is r until the time r/D current cycle inventory vanishes and after this time if a new lot reaches, shortage cost should be paid because the shipment size is Q. If the lot is delivered to the buyer after time of Q/D, the ordering of new lot is given before reaching the previous lot.

Figure 1, case 1 shows the condition in which ordered lot ships the buyer before consuming the recent stock ($L \le r/D$). In this condition shortage cost does not exist. In case 2, the lead time is more than r/D, but

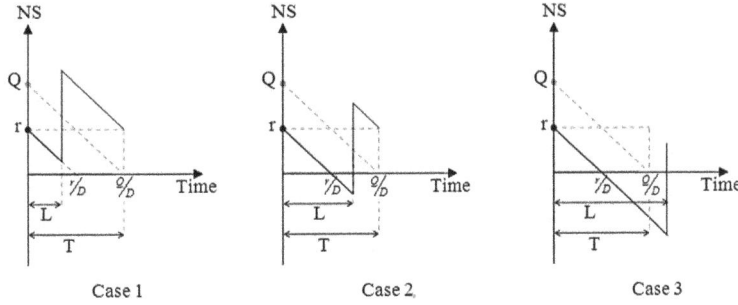

Figure 1 Net stock vs. time for the buyer.

in order that the outstanding order would be delivered before the time of Q/D, it is possible to send next order at the time in which the inventory position decrease to r. But in case 3 in which the lead time exceeds Q/D, the next order is released before receiving the outstanding order.

The buyer total cost of single cycle in terms of L, $tc_b(r,Q,L)$, can be written as follows:

$$
tc_b(r,Q,L) = \begin{cases}
A_b + h_b\left(\dfrac{Q^2}{2D} + \dfrac{rQ}{D} - QL\right), L \leq \left. r \middle/ D \right. \\[2ex]
A_b + h_b\dfrac{(r+Q-LD)^2}{2D} \\[1ex]
\quad + \dfrac{\pi}{2D}(LD-r)^2, \left. r \middle/ D \right. < L \leq \left. Q \middle/ D \right. \\[2ex]
A_b + h_b\dfrac{r^2}{2D} + \dfrac{\pi}{2D}(LD-r)^2, \\[2ex]
\quad \left. Q \middle/ D \right. < L,
\end{cases} \tag{1}
$$

and the expected total cost in one cycle for given r and Q, $tc_b(r,Q)$ is as follows:

$$
tc_b(r,Q) = A_b + \int_0^{\frac{r}{D}} \left(h_b\left(\frac{Q^2}{2D} + \frac{rQ}{D} - Ql\right)\right) f_L(l)dL
$$
$$
+ \int_{\frac{r}{D}}^{\frac{Q}{D}} \left(h_b\frac{(r+Q-lD)^2}{2D} + \frac{\pi}{2D}(lD-r)^2\right) f_L(l)dL
$$
$$
+ \int_{\frac{Q}{D}}^{\infty} \left(h_b\frac{r^2}{2D} + \frac{\pi}{2D}(lD-r)^2\right) f_L(l)dL \tag{2}
$$

Substituting $f_L(l) = \lambda e^{-\lambda l}$ into the above expression and by dividing the length of the buyer order cycle, Q/D, we obtain the buyer's expected total cost per unit time as the following expression,

$$
TC_b(r,Q) = \frac{DA_b}{Q} + h_b\left(r + \frac{Q}{2} - \frac{D}{\lambda}\right) + \frac{D^2(h_b + \pi)}{\lambda^2 Q}e^{-\frac{\lambda}{D}r}
$$
$$
+ \frac{Dh_b}{Q}\left(\frac{r}{\lambda} - \frac{D}{\lambda^2}\right)e^{-\frac{\lambda}{D}Q}. \tag{3}
$$

The way of obtaining the seller's total inventory is presented in such literatures as Lee (2005), Wee and Chung

(2007), Lin (2008), Ouyang et al. (2007), Chang et al. (2006), and Ouyang et al. (2004). We present a simple manner in Figure 2 for calculating the seller's total inventory.

By using S as area of a surface, we can write:

$$
TI_v = S_{ABCE} - S_{ADE} - \left[\frac{Q^2}{D} + 2\frac{Q^2}{D} + \ldots + (n-1)\frac{Q^2}{D}\right]
$$
$$
= nQ\left(\frac{Q}{p} + (n-1)\frac{Q}{D}\right) - \frac{n^2Q^2}{2p}
$$
$$
- \left[\frac{Q^2}{D}(1 + 2 + \ldots + (n-1))\right]
$$
$$
= \frac{nQ^2}{2D}\left((n-1)\left(1 - \frac{D}{p}\right) + \frac{D}{p}\right) \tag{4}
$$

and since orders are received by the seller at known intervals $T = \frac{Q}{D}$, the seller's expected total cost (TC_v) is as follows:

$$
TC_v(n) = \frac{1}{nT}(A_v + h_v.AI_v)
$$
$$
= \frac{DA_v}{nQ} + h_v\frac{Q}{2}\left((n-1)\left(1 - \frac{D}{p}\right) + \frac{D}{p}\right) \tag{5}
$$

We have proved (see Appendix 2) that $TC_v(n)$ is convex in n, and optimal solution (n^*), satisfies the condition:

$$
n^*(n^* - 1) \leq \frac{2DA_v}{h_vQ^2\left(1 - \frac{D}{p}\right)} \leq n^*(n^* + 1) \tag{6}
$$

Solution for separate and joint models

First, we consider solution for the case in which the seller and buyer optimize their total cost functions separately. The optimal solution will be denoted by (r_s^*, Q_s^*, n_s^*) in this case, and we show sum of costs of seller and buyer with

$$
STC(r_s^*, Q_s^*, n_s^*) = TC_b(r_s^*, Q_s^*) + TC_v(n_s^*) \tag{7}
$$

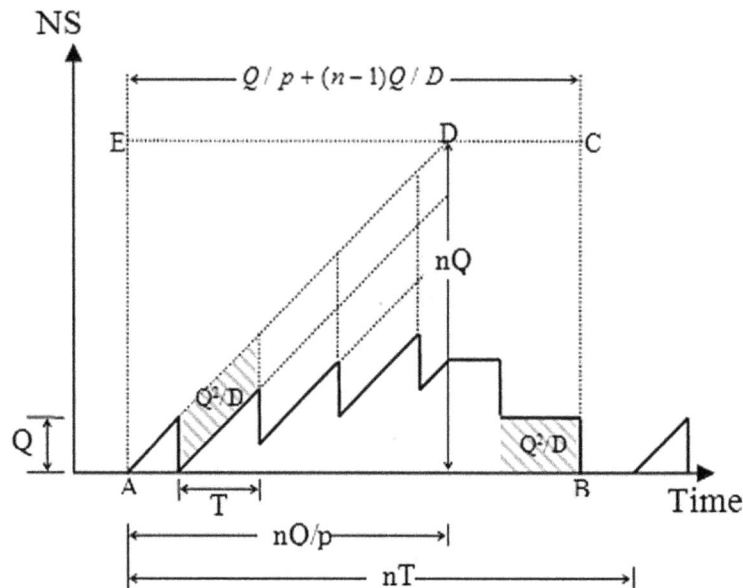

Figure 2 Net stock vs. time for the seller.

In the other case, joint total cost (JTC) function is optimized as a whole. In this case we sum the costs together to obtain joint total cost

$$\text{JTC}(r, Q, n) = \text{TC}_b(r, Q) + \text{TC}_v(n)$$

$$= \frac{D(nA_b + A_v)}{nQ} + h_b \left(r + \frac{Q}{2} - \frac{D}{\lambda} \right)$$

$$+ \frac{D^2(h_b + \pi)}{\lambda^2 Q} e^{-\frac{\lambda}{D}r} + \frac{Dh_b}{Q} \left(\frac{r}{\lambda} - \frac{D}{\lambda^2} \right) e^{-\frac{\lambda}{D}Q}$$

$$+ h_v \frac{Q}{2} \left((n-1)\left(1 - \frac{D}{p}\right) + \frac{D}{p} \right) \qquad (8)$$

and show its optimal solution with (r_j^*, Q_j^*, n_j^*). Because of convexity of TC_b (see Appendix 1) and TC_v, the obtained separate solution is general. On the other hand, it can be shown that in spite of convexity of JTC on each of r, Q, or n alone, it is not a convex function generally. It can be observed that $\frac{\partial \text{JTC}}{\partial n} = \frac{\partial \text{TC}_v}{\partial n}$ and $\frac{\partial^2 \text{JTC}}{\partial n^2} = \frac{\partial^2 \text{TC}_v}{\partial n^2}$ and hence, the optimality condition on n (Equation 6) is valid here. Thus in the procedure of solving JTC, we set $n = i$, with start of $i = 1$, and calculate the optimal values of r and Q, and increase i one unit in each step until the optimality in condition (Equation 6) is satisfied. It is not necessary to focus on the method of finding optimal solution of JTC (when n is fixed), TC_b and TC_v because general methods like derivative-free method in many types of software such as MATLAB (MathWorks, Natick, MA, USA) can solve this problem easily.

Results and discussion
Numerical results
In this part an example with the following data will be presented as follows: $D = 1,000/\text{years}$, $A_b = \$25/\text{order}$, $h_v = \$4/\text{unit/year}$, $h_b = \$5/\text{unit/year}$, $\pi = \$30/\text{unit/year}$; and we change p by the values 5,000, 7,000, and 9,000, and $1/\lambda$ by the values 10, 15, 20, 25, 30, 35, 40, 45, and 50 to explore variation effects of lead time to the percentage of saving in individually optimized total cost over integrated inventory policy. As mentioned by Goyal (1977) and Ouyang et al. (2004), the total benefit under integrated optimization should be shared by both parties to encourage them to cooperate together. It can be done as follows:

$$\text{JTC}_v = \frac{\text{TC}_v\left(n_s^*\right)}{\text{STC}\left(r_s^*, Q_s^*, n_s^*\right)} \cdot \text{JTC}\left(r_j^*, Q_j^*, n_j^*\right) \qquad (9)$$

$$\text{JTC}_b = \frac{\text{TC}_b\left(r_s^*, Q_s^*\right)}{\text{STC}\left(r_s^*, Q_s^*, n_s^*\right)} \cdot \text{JTC}\left(r_j^*, Q_j^*, n_j^*\right) \qquad (10)$$

Referring to the existing literature, we consider percentage saving 'ps' as (STC-JTC)/STC*100. Figure 3 shows ps increases for more variable lead time, which is the characteristic of many cases in unpredictable real environment. It also shows that the percentage of improvement increases when there is a rise in the production rate. For instance, the percentage of improvement is 3.7 % for $p = 5,000$ (increase from 1.6 for $1/\lambda = 5$ days to 5.3 for $1/\lambda = 50$ days) and where it is 6.2 % for $p = 9,000$ (increase from 2.4 for $1/\lambda = 5$ days

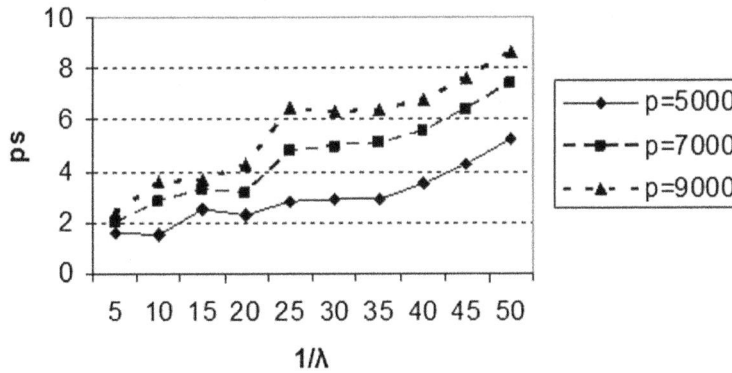

Figure 3 Effect of lead time variability on percentage saving.

Table 1 Optimal solution of non-integrated model vs. joint model

Parameters		Non-integrated optimization						Integrated optimization						Ps
P	$1/\lambda$	r_s^*	Q_s^*	n_s^*	TC_b	TC_v	STC	r_j^*	Q_j^*	n_j^*	JTC_b	JTC_v	JTC	
5,000	5	−2.4	114.6	4	492.4	1,468.5	1,960.9	−7.6	166.8	3	484.4	1,444.4	1,928.8	1.6
	10	10.4	130.9	4	570.4	1,444.6	2,015.0	2.9	172.7	3	561.8	1,422.8	1,984.6	1.5
	15	26.7	149.0	3	674.9	1,431.4	2,106.3	6.2	247.6	2	658.1	1,395.6	2,053.7	2.5
	20	44.1	169.1	3	794.2	1,397.3	2,191.5	22.1	255.9	2	776.0	1,365.3	2,141.4	2.3
	25	61.5	191.1	3	922.9	1,385.7	2,308.6	40.3	264.6	2	897.0	1,346.8	2,243.9	2.8
	30	78.7	214.8	2	1,058.0	1,360.6	2,418.6	18.3	459.9	1	1,027.8	1,321.8	2,349.6	2.9
	35	95.6	239.7	2	1,197.5	1,313.7	2,511.2	33.2	473.9	1	1,162.4	1,275.2	2,437.6	2.9
	40	112.3	265.5	2	1,340.2	1,284.3	2,624.4	49.2	488.2	1	1,293.6	1,239.6	2,533.2	3.5
	45	128.8	292.0	2	1,485.2	1,268.9	2,754.1	66.1	502.8	1	1,421.1	1,214.1	2,635.2	4.3
	50	145.1	319.0	2	1,632.0	1,265.0	2,897.0	83.6	517.9	1	1,545.2	1,197.7	2,742.9	5.3
7,000	5	−2.4	114.6	4	492.4	1,494.7	1,987.1	−7.5	165.2	3	482.7	1,465.1	1,947.8	2.0
	10	10.4	130.9	4	570.4	1,474.5	2,044.9	−6.1	239.3	2	554.2	1,432.6	1,986.8	2.8
	15	26.7	149.0	3	674.9	1,448.4	2,123.3	6.2	247.6	2	652.8	1,400.9	2,053.7	3.3
	20	44.1	169.1	3	794.2	1,416.6	2,210.8	22.1	255.9	2	769.3	1,372.1	2,141.4	3.1
	25	61.5	191.1	3	922.9	1,407.5	2,330.4	3.3	456.8	1	878.9	1,340.5	2,219.4	4.8
	30	78.7	214.8	2	1,058.0	1,360.6	2,418.6	16.4	471.0	1	1,004.5	1,291.9	2,296.5	5.0
	35	95.6	239.7	2	1,197.5	1,313.7	2,511.2	30.9	485.6	1	1,136.2	1,246.5	2,382.8	5.1
	40	112.3	265.5	2	1,340.2	1,284.3	2,624.4	46.5	500.5	1	1,264.7	1,212.0	2,476.7	5.6
	45	128.8	292.0	2	1,485.2	1,268.9	2,754.1	63.0	515.8	1	1,389.7	1,187.3	2,577.0	6.4
	50	145.1	319.0	2	1,632.0	1,265.0	2,897.0	80.1	531.5	1	1,511.4	1,171.5	2,682.9	7.4
9,000	5	−2.4	114.6	4	492.4	1,509.2	2,001.6	−12.1	231.3	2	480.5	1,472.8	1,953.3	2.4
	10	10.4	130.9	4	570.4	1,491.2	2,061.6	−6.1	239.3	2	549.7	1,437.1	1,986.8	3.6
	15	26.7	149.0	3	674.9	1,457.8	2,132.8	6.2	247.6	2	649.9	1,403.8	2,053.7	3.7
	20	44.1	169.1	3	794.2	1,427.3	2,221.6	−8.6	448.6	1	759.9	1,365.7	2,125.7	4.3
	25	61.5	191.1	3	922.9	1,419.7	2,342.6	2.4	462.9	1	862.9	1,327.3	2,190.2	6.5
	30	78.7	214.8	2	1,058.0	1,360.6	2,418.6	15.3	477.5	1	991.4	1,275.0	2,266.3	6.3
	35	95.6	239.7	2	1,197.5	1,313.7	2,511.2	29.6	492.5	1	1,121.4	1,230.3	2,351.7	6.4
	40	112.3	265.5	2	1,340.2	1,284.3	2,624.4	45.0	507.8	1	1,248.4	1,196.3	2,444.7	6.8
	45	128.8	292.0	2	1,485.2	1,268.9	2,754.1	61.2	523.4	1	1,371.9	1,172.1	2,544.0	7.6
	50	145.1	319.0	2	1,632.0	1,265.0	2,897.0	78.1	539.6	1	1,492.3	1,156.6	2,648.9	8.6

to 8.6 for $1/\lambda = 50$ days). By joining and coordinating, both seller and buyer can control the variability effect of lead time by decreasing the number of shipments and increasing batch sizes for higher level of production rate. However, this does not happen when they work separately and cannot react to the variation of shipment lead time (Table 1).

Conclusions

An integrated model for JELS in which lead time of delivering the shipment is not deterministic, which follows an exponential distribution, was presented in this article. We showed that integrated inventory policy increases profit of both buyer and seller, if they can compromise to divide the gained benefit of coordination. This policy is completely executable in a unit system containing two parts, but may not be suitable for separate systems when percentage of improvement is low. A numerical example showed that the cooperation between two supply chain partners in the integrated situation is more useful in unreliable purchasing environments in terms of lead times of shipments. Stochastic lead time with an exponential distribution is the difference of the current research with the previous ones. Authors are in the process of decreasing. simplifying or adding assumptions of this model ((1) the orders don not cross in time and (2) at the start of cycle time, the net stock is considered to be r) to conform it more to the real system. The model also can be extended to situations, such as general distributions for lead times, multi vendor case, stochastic demand, and stochastic price. We hope that this extension will be helpful to researchers who are interested in integrating decisions on supply chain.

Methods

The used method in this article is mathematical modeling. This model is an expansion of previous version with a change in delivery lead time assumption. Like all such models, to confirm the performance of model, a computer simulation experiment was conducted.

Appendix 1

Here we survey convexity of TC_b. By calculating the second partial derivations of TC_b, we get

$$h_{11} = \frac{\partial^2 (TC_b)}{\partial r^2} = \frac{h_b + \pi}{Q} e^{-\frac{\lambda}{D}r},$$

$$h_{12} = h_{21} = \frac{\partial^2 (TC_b)}{\partial Q \partial r} = \frac{D(h_b + \pi)}{\lambda Q^2} e^{-\frac{\lambda}{D}r}$$
$$- \frac{h}{Q}\left(\frac{D}{\lambda Q} + 1\right) e^{-\frac{\lambda}{D}Q}$$

$$h_{22} = \frac{\partial^2 (TC_b)}{\partial Q^2} = \frac{2A_b D}{Q^3} + \frac{2D^2(h_b + \pi)}{\lambda^2 Q^3} e^{-\frac{\lambda}{D}r}$$
$$+ \frac{2Dh_b}{\lambda Q}\left(r - \frac{D}{\lambda}\right)\left[\frac{1}{Q^2} + \frac{\lambda}{DQ} + \frac{\lambda^2}{2D^2}\right] e^{-\frac{\lambda}{D}Q}$$

Since $h_{11} > 0$, to prove H is positive definite and therefore TC_b is convex, it should be only shown that Hessian determinant i.e., $|H| = \begin{vmatrix} h_{11} & h_{12} \\ h_{21} & h_{22} \end{vmatrix}$ is positive. To avoid the complexity of calculation in general cases for all values of parameters, we construct some conditions that stated problems in literature that were satisfied, and so they are logically acceptable for the application purpose. These conditions are $0 < \frac{\lambda}{D} < 1, r \geq 1$ and $\pi > h$. Also, we define $\alpha = \frac{\lambda}{D}$.

$$|H| = \frac{2A_b D(h_b + \pi)}{Q^4} e^{-\alpha r} + \frac{(h_b + \pi)^2}{\alpha^2 Q^4} e^{-2\alpha r}$$
$$- \frac{h_b^2}{Q^2}\left(\frac{1}{\alpha Q} + 1\right)^2 e^{-2\alpha Q} + \frac{2h_b(h_b + \pi)r}{\alpha Q^2}$$
$$\times \left(\frac{1}{Q^2} + \frac{\alpha}{Q} + \frac{\alpha^2}{2}\right) e^{-\alpha(r+Q)} - \frac{h_b(h_b + \pi)}{Q^2} e^{-\alpha(r+Q)}$$

The first sentence is positive. The sum of next two sentences is also positive since $r < Q$, thus $e^{-2\alpha r} > e^{-2\alpha Q}$, and with substitution, it will be to gain

$$\frac{(h_b + \pi)^2}{\alpha^2 Q^4} e^{-2\alpha r} - \frac{h_b^2}{Q^2}\left(\frac{1}{\alpha Q} + 1\right)^2 e^{-2\alpha Q} > \frac{1}{\alpha^2 Q^4}$$
$$\times \left[(h_b + \pi)^2 - h_b^2(1 + \alpha Q)^2\right] e^{-2\alpha Q} = \frac{1}{\alpha^2 Q^4}$$
$$\times \left[((h_b + \pi) + h_b(1 + \alpha Q))((h_b + \pi) - h_b(1 + \alpha Q))\right]$$
$$e^{-2\alpha Q} = \frac{1}{\alpha^2 Q^4}\left[(2h_b + \pi + \alpha Q)(\pi - h_b \alpha Q)\right] e^{-2\alpha Q}$$

In the last expression, all elements including second parenthesis are positive, since

$$\alpha Q = \frac{\lambda}{D} Q = \lambda \frac{D}{Q} < 1 \Rightarrow -h_b \alpha Q > h_b \Rightarrow \pi - h_b \alpha Q$$
$$> \pi - h_b > 0.$$

On the other hand, the sum of the last two sentences of $|H|$ is also positive, because

$$\frac{2h_b(h_b + \pi)r}{\alpha Q^2}\left(\frac{1}{Q^2} + \frac{\alpha}{Q} + \frac{\alpha^2}{2}\right) e^{-\alpha(r+Q)} - \frac{h_b(h_b + \pi)}{Q^2} e^{-\alpha(r+Q)}$$
$$> \frac{h_b(h_b + \pi)}{Q^2}\left[\frac{1 + (1 + \alpha Q)^2 - \alpha Q}{\alpha Q^2}\right] e^{-\alpha(r+Q)} > 0$$

Consequently TC_b is strictly convex.

Appendix 2

In equating the first derivation of TC_v with zero, we obtain

$$\frac{d}{dn}TC_v(n) = -\frac{DA_v}{n^2Q} + h_v\frac{Q}{2}\left(1 - \frac{D}{p}\right) = 0$$

$$n(n-1) \leq n^2 = \frac{2DA_v}{h_vQ^2\left(1 - \dfrac{D}{p}\right)} \leq n(n+1)$$

On the other hand, derivation of the second order of TC_v is $\frac{d^2}{dn^2}TC_v(n) = \frac{2DA_v}{n^3Q} > 0$, and hence, TC_v is strictly convex on n.

Competing interests

The authors declare that they have no competing interests.

Authors' contributions

MB carried out the literature review and constructed the proposal model and drafted the manuscript. MJT supervised the research and guided him to do correction. He also revised and improved the model, then designed the simulation for validation of the model. Both authors read and approved the final manuscript.

Author's information

Mehrab Bahri is a PhD student from the Department of Industrial Engineering, Science and Research Branch in the Islamic Azad University in Tehran.

Author details

[1]Department of Industrial Engineering, Science and Research Branch, Islamic Azad University, Tehran 14778, Iran. [2]Department of Industrial Engineering, K. N. Toosi University of Technology, Tehran 1439955471, Iran.

References

Banerjee A (1986) A joint economic-lot-size model for purchaser and vendor. Decis Sci 17:292–311

Banerjee A, Kim SL (1995) An integrated JIT inventory model. Intern J Oper & Production Management 15(9):237–244

Ben-Daya M, Zamin SA (2002a) Effect of preventive maintenance on the joint economic lot sizing problem with imperfect processes. Systems Engineering Department, King Fahd University of Petroleum and Minerals, Dhahran, Saudi Arabia, Technical Report

Ben-Daya M, Zamin SA (2002b) Joint economic lot sizing problem with stochastic demand. Systems Engineering Department, King Fahd University of Petroleum and Minerals, Dhahran, Saudi Arabia, Technical Report

Ben-Daya M, Darwish M, Ertogral K (2008) The joint economic lot sizing problem: review and extensions. Eur J Oper Res 185:726–742

Chang HC, Ouyang LY, Wu KS, Ho CH (2006) Integrated vendor-buyer cooperative inventory models with controllable lead time and ordering cost reduction. Eur J Oper Res 170:481–495

Chen LH, Kang FS (2010) Integrated inventory models considering permissible delay in payment and variant pricing strategy. Appl Math Model 34(1):36–46

Goyal SK (1977) An integrated inventory model for a single supplier-singl e customer problem. Int J Prod Res 15(1):107–111

Goyal SK (1988) A joint economic-lot-size model for purchaser and vendor: a comment. Decis Sci 19:236–241

Goyal SK (1995) A one-vendor multi-buyer integrated inventory model: a comment. Eur J Oper Res 82:209–210

Goyal SK, Nebebe F (2000) Determination of economic production-shipment policy for a single-vendor single-buyer system. Eur J Oper Res 121:175–178

Ha D, Kim SL (1997) Implementation of JIT purchasing: an integrated approach. Production Planning & Control 8:152–157

Hadley G, Whitin TM (1963) Analysis of inventory systems. Wiley, New York

Hill RM (1997) The single-vendor single-buyer integrated production-inventory model with a generalized policy. Eur J Oper Res 97:493–499

Hill RM (1999) The optimal production and shipment policy for the single-vendor single-buyer integrated production-inventory model. Int J Prod Res 37:2463–2475

Hill R, Omar M (2006) Another look at the single-vendor single-buyer integrated production-inventory problem. Int J Prod Res 44(4):791–800

Hu O, Wei Y, Xia Y (2010) Revenue management for a supply chain with two streams of customers. Eur J Oper Res 200(2):582–598

Huang CK (2004) An optimal policy for a single-vendor single-buyer integrated production inventory problem with process unreliability consideration. Int J Prod Econ 91:91–98

Huang CK, Tsai DM, Wu JC, Chung KJ (2010) An integrated vendor-buyer inventory model with order-processing cost reduction and permissible delay in payments. Eur J Oper Res 202(2):473–478

Kim SL, Ha D (2003) A JIT lot-splitting model for supply chain management: enhancing buyer–supplier linkage. Int J Prod Econ 86:1–10

Lee W (2005) A joint economic lot size model for raw material ordering, manufacturing setup, and finished goods delivering. Omega 33:163–174

Lin YJ (2008) An integrated vendor-buyer inventory model with backorder price discount. Computers &Industrial Engineering. doi:10.1016/j.cie.2008.10.009

Lu L (1995) A one-vendor multi-buyer integrated inventory model. Eur J Oper Res 81:312–323

Ouyang LY, Wu KS, Hu CH (2004) Integrated vendor-buyer cooperative models with stochastic demand in controllable lead time. Int J Prod Econ 92(3):255–266

Ouyang LY, Wu KS, Ho CH (2007) An integrated vendor-buyer inventory model with quality improvement and lead time reduction. Int J Production Economics 108:349–358

Soroor J, Tarokh MJ, Keshtgary M (2009a) Preventing failure in IT-enabled systems for supply chain management. Int J Prod Res 47(23):6543–6557

Soroor J, Tarokh MJ, Shemshadi A (2009b) Theoretical and practical study of supply chain coordination. J Bus Ind Mark 24(2):131–142

Soroor J, Tarokh MJ, Shemshadi A (2009c) Initiating an state of the art system for real time supply chain coordination. Eur J Oper Res, Elsevier BV 196(2):635–650

Tarantilis CD (2008) Topics in real-time supply chain management. Comput Oper Res 35:3393–3396

Viswanathan S (1998) Optimal strategy for the integrated vendor-buyer inventory model. Eur J Oper Res 105:38–42

Wee HM, Chung CJ (2007) A note on the economic lot size of the integrated vendor-buyer inventory system derived without derivatives. Eur J Oper Res 177:1289–1293

Xu R, Zhai X (2010) Analysis of supply chain coordination under fuzzy demand in a two-stage supply chain. Appl Math Model 34(1):129–139

Zhou Y, Wang S (2007) Optimal production and shipment models for a single-vendor-single-buyer integrated system. Eur J Oper Res 180(1):309–328

3

Modeling the operational risk in Iranian commercial banks: case study of a private bank

Omid Momen[1,2*], Alimohammad Kimiagari[2] and Eaman Noorbakhsh[1,3]

Abstract

The Basel Committee on Banking Supervision from the Bank for International Settlement classifies banking risks into three main categories including credit risk, market risk, and operational risk. The focus of this study is on the operational risk measurement in Iranian banks. Therefore, issues arising when trying to implement operational risk models in Iran are discussed, and then, some solutions are recommended. Moreover, all steps of operational risk measurement based on Loss Distribution Approach with Iran's specific modifications are presented. We employed the approach of this study to model the operational risk of an Iranian private bank. The results are quite reasonable, comparing the scale of bank and other risk categories.

Keywords: Operational risk, Copula, Loss distribution approach, Bank

Background

Nowadays, risk management becomes an important module of every industry. However, its magnitude in banking industry is much more obvious because, usually, the profit of every bank is directly related to the amount of risk it takes. It means that the more risk it takes, the more profit it can earn. However, this huge amount of risk should be carefully managed in order to reduce the possibility of loss or bankruptcy. Therefore, the Bank for International Settlements (BIS) has founded the Basel Committee on Banking Supervision (hereafter Basel Committee), which has developed several documents containing basic standards, guidelines, and consultative papers for risk management and banking supervision. One of the most recent and well-known documents of BIS is Basel II accord. It includes the most popular and trusted guidelines in banking supervision and risk management, which are generally acquiesced by central banks all over the world including the Central Bank of Iran.[a]

Basel II accord has classified major banking risks into three different types: credit risk, market risk, and operational risk. *Credit risk* is an investor's risk of loss arising from a borrower who does not make payments as promised. *Market risk* is the risk that the value of a portfolio, either an investment portfolio or a trading portfolio, will decrease due to the change in value of the market risk factors. The four standard market risk factors are stock prices, interest rates, foreign exchange rates, and commodity prices. *Operational risk*, which is the main focus of this study based on Basel II accord, has been defined as the risk of loss resulting from inadequate or failed internal processes, people and systems, or from external events. This definition includes legal risk, but excludes strategic and reputational risk (Basel Committee on Banking Supervision 2006).

In the last two decades, a significant number of financial institutions have experienced loss or bankruptcy due to the mismanagement of operational risks. Some famous instances are as follows: First, *Societe Generale Bank*, alleged fraud by a trader, lost 4.9 billion € in 2008. Second, Former currency trader was accused of hiding US \$691 million in losses at *Allfirst Bank of Baltimore* in 2002. Third, UK's *Barings Bank* collapsed after trader Nick Leeson lost £860 million (US \$1.28 billion at the time) on futures trades in 1995 (BBC News 2008). For more related cases, go to Gallati (2003). In the case of Iran, most of the banks have been state-owned up to a few years ago, and the government has prevented them from insolvency. However, emerging of private banks, along with service development of both private and state-owned banks in recent years, led in to a more competitive market, which encounters banks with more complex operational risks that need to be

* Correspondence: omid.momen@aut.ac.ir
[1]Karafarin Bank, Tehran, Iran
[2]Amirkabir University of Technology, Tehran, Iran
Full list of author information is available at the end of the article

considered. Since the operational risk has greatly affected a large number of banks globally, as seen in non-Iranian cases above, and due to the lack of attention to the subject in Iranian banks and legislators, a new trend of research in this area is indispensable.

Measuring is one of the main steps in operational risk management. Basel II accord introduces three different ways for measuring operational risk in financial institution: the first method is Basic Indicator Approach (BIA), which calculates Capital-at-Risk (CaR) as a fraction of the bank's gross income; the second proposed method is called Standardized Approach (SA), which divides the institution into eight specified business lines and, in each one, computes the business line-specific CaRs as a percentage of their relevant gross incomes then adds these eight CaRs to obtain the bank's total CaR; and finally, Basel II suggests Advanced Measurement Approaches (AMA) in which banks are permitted to develop their own methodology to assess yearly operational risk exposure within a confidence interval of 99.9% or more. The first two methods are easy to apply but undesirable among banks because, as a consequence of their conceptual simplicity, BIA and SA models do not provide any insights into drivers of ethods in Iran, refer to Karafarin Bank (2009) and Erfanian and Sharbatoghli (2006). However, the third category of methods (i.e., AMA) has not been implemented in any bank in Iran, which is much more sensitive to risk; therefore, it is recommended by Basel Committee and widely applied by international banks.

Among the eligible variants of AMA, over the last few years, a statistical model widely used in the insurance sector and often referred to as the Loss Distribution Approach (LDA) has become a standard in the banking industry around the world (for two examples see Chapelle et al. (2007) and Aue and Kalkbrener (2006)). Anyway, to our knowledge, it is not employed by any bank in Iran. When applying the LDA in Iranian banking circumstance, some issues arise: First, operational loss events have not been recorded thoroughly, so available loss data are rare and inferences of their related distributions need special concern. Second, because there is no bankruptcy reported, there are no data available for extreme losses. Third, the previous methods implemented in Iran (BIA and SA) do not explicitly account for dependence structure of risks. Therefore, the objective of this study is to present the comprehensive LDA framework for the measurement of operational risk of banks in Iran, whereas we try to provide recommendations to resolve Iran's specific issues by utilizing available statistical and mathematical techniques.

The methodology of this study has been applied in Karafarin Bank, which is an Iranian private bank. For more information about Karafarin Bank, visit its home page (Karafarin Bank 2010).

This paper is organized as follows: in 'Methodology', a comprehensive methodology of measuring operational risk is discussed; then in 'Empirical analysis', we apply the methodology to loss data of Karafarin Bank, and results are reported. Finally, concluding remarks will be presented in the last section.

Case description
Methodology
The Basel Committee encourages banks to use Advanced Measurement Approaches for modeling operational risk. Although AMA includes a wide range of proprietary models, the most popular one is by far the Loss Distribution Approach (Chapelle et al. 2007). LDA is a parametric technique that estimates two separate distributions for frequency and severity of operational losses and then combines them through n-convolution[b] (see Frachot et al. (2001) for details). However, as mentioned before, the basic LDA encounters some problems when applied to Iranian banks, which suffer from loss data unavailability, unreported large losses, and lack of attention to dependence structure of operational risks.

The Basel Committee has provided a basic framework that banks should use to classify their operational loss data. This framework includes seven operational risk event categories and eight banking business lines. In order to comply with Basel II, it is necessary to consider this classification as presented in Table 1. For more definitions and instances about the categories, see 'Annexes 8 and 9' of Basel Committee on Banking Supervision (2006).

LDA should be applied in all cells of Table 1 separately, and then, the resulting loss distributions will be integrated considering the dependence structure. In order to keep the integrity of the methodology in the following sections ('Frequency distribution', 'Severity distribution', 'Loss distribution for a specified risk category', and 'Loss distribution for bank as a whole'), the comprehensive methodology of measuring operational risk in a commercial bank will be described as presented in Figure 1.

Frequency distribution
In LDA, occurrence of operational losses of a specified bank is modeled by a so-called frequency distribution. This distribution is discrete and, for short periods of time, usually estimated either by Poisson or by negative binomial distributions (Aue and Kalkbrener 2006). The difference between these two distributions is that the intensity parameter is deterministic in the first case and stochastic in the second. More precisely, if the intensity of a Poisson process follows a gamma distribution, the negative binomial distribution arises (Embrechts et al. 2003).

In this study, a score-based approach (see Panjer (2006) and Klugman et al. (2004)) has been used for selecting between the Poisson and negative binomial

Table 1 Business Line Event Type mapping according to Basel II framework

	Internal fraud	External fraud	Employment practices and workplace safety	Clients, products, and business practices	Damage to physical assets	Business disruption and system failures	Execution, delivery, and process management
Corporate finance							
Trading and sales							
Retail banking							
Commercial banking							
Payment and settlement							
Agency services							
Asset management							
Retail brokerage							

distributions. In order to implement the score-based approach, three different statistic hypothesis tests have been utilized including Cramer-von Mises (Anderson 1962), Kolmogorov-Smirnov (Stephens 1974), and likelihood ratio (McGee 2002).

Severity distribution

Modeling severity distribution for economical impact of operational losses is not as straightforward as modeling frequency of losses. Some studies like Chapelle et al. (2007) and de Fontnouvelle et al. (2004) indicate that classical distributions are unable to fit the entire range of observations for modeling the severity of operational losses. Hence, as in Alexander (2003), Chapelle et al. (2007), de Fontnouvelle et al. (2004), and King (2001), in this study, the discrimination between ordinary (i.e., high frequency/low impact) and large (i.e., low frequency/high impact) losses has been considered, as presented in Figure 2. The 'ordinary distribution' includes all losses in a limited range denoted $[L; U]$ (L being the collection threshold used by the bank), while the 'extreme distribution' generates all the losses above the cut-off threshold U. The severity distribution will then be defined as a mixture of the corresponding mutually exclusive distributions.

For modeling ordinary losses, distribution such as the exponential, Weibull, gamma, or lognormal distribution that is strictly positive continuous distribution can be employed. More precisely, let $f(x;\theta)$ be the chosen

parametric density function, where θ denotes the vector of parameters, and let $F(x;\theta)$ be the cumulative distribution function (cdf) associated with $f(x;\theta)$. Then, the density function $f^*(x;\theta)$ of the losses in $[L; U]$ can be expressed as

$$f^*(x;\theta) = \frac{f(x;\theta)}{F(U;\theta) - F(L;\theta)} \qquad (1)$$

The corresponding log-likelihood function is:

$$\ell(x;\theta) = \sum_{i=1}^{N} \ln\left(\frac{f_i(x_i;\theta)}{F(U;\theta) - F(L;\theta)}\right) \qquad (2)$$

where $(x_1, \ldots, x_N)$ is the sample of observed ordinary losses. It should be maximized in order to be estimated (Chapelle et al. 2007).

In Iranian banks, due to lack of recorded operational loss data, modeling distribution of large losses is not as clear-cut as ordinary losses because there are not enough observations available for severe operational losses (Momen 2008). For such samples, classical maximum likelihood methods yield inappropriate distributions for estimating the occurrence probability of exceptional losses because the resulting distributions are not sufficiently heavy tailed. To resolve this issue, a procedure developed by Chapelle et al. (2007) has been used. This procedure is built upon the results of Balkema and de Haan (1974) and Pickands (1975), which state that, for a broad class of distributions, the values of the random variables above a

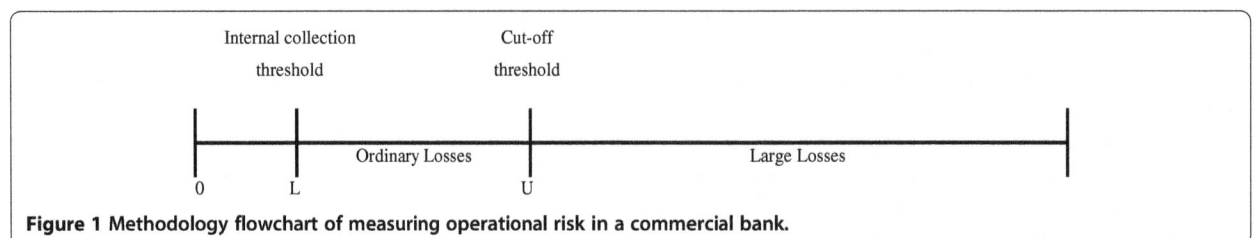

Figure 1 Methodology flowchart of measuring operational risk in a commercial bank.

2. Methodology

Initiation and Termination	2.1 Frequency Distribution	2.2 Severity Distribution	2.3 Loss Distribution for a Specified Risk Category	2.4 Loss Distribution for Bank as a Whole

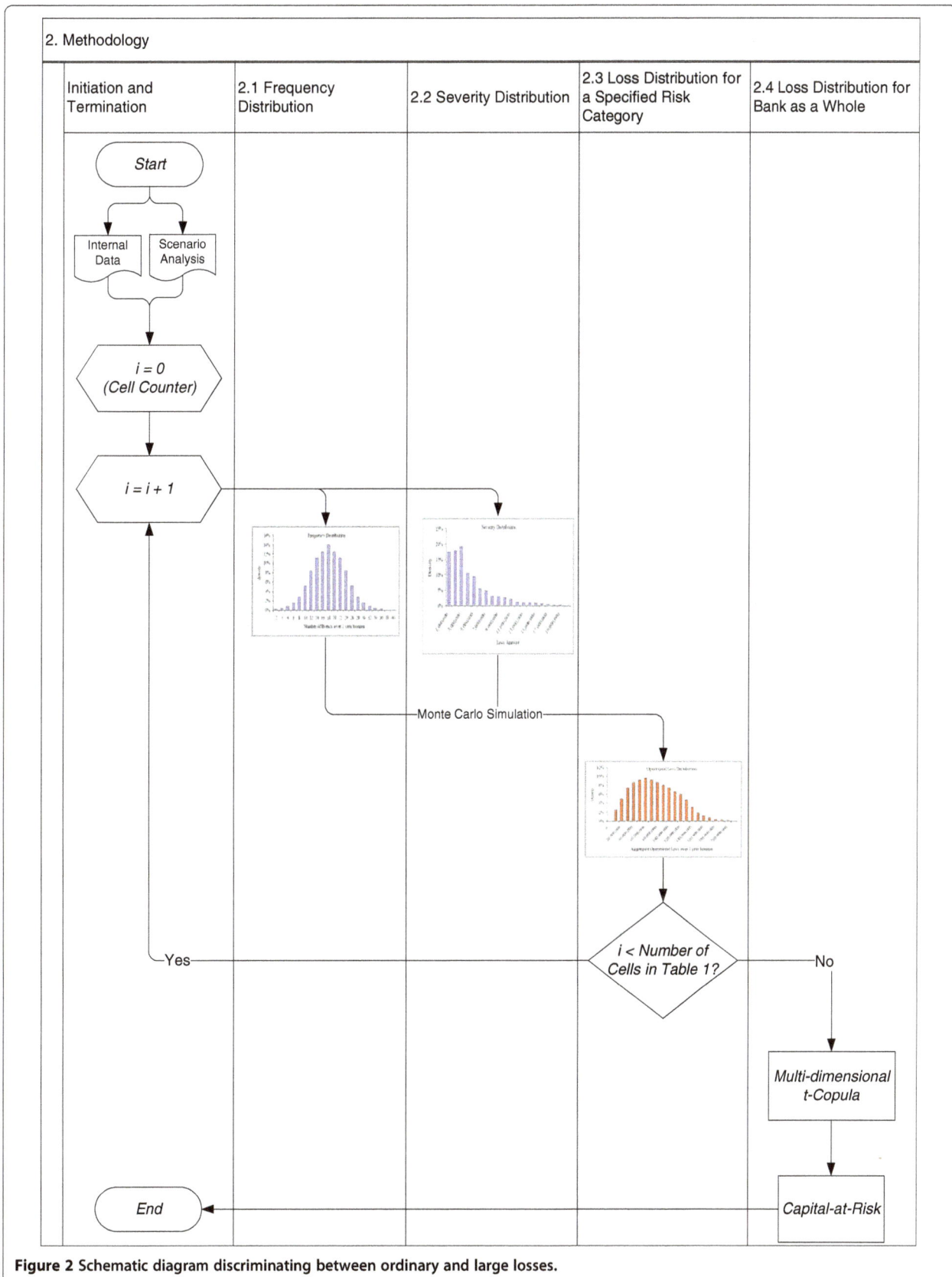

Figure 2 Schematic diagram discriminating between ordinary and large losses.

sufficiently high threshold U follow a generalized Pareto distribution (GPD) with parameters ξ (shape index or tail parameter), β (the scale index), and U (the location index). The GPD can thus be thought of as the conditional distribution of X given $X > U$ (Embrechts et al. 1997).

As indicated before, another problem with operational risk modeling in Iranian banks is that large catastrophic losses like bankruptcy of a bank have not been reported. Hence, tail of the severity distribution cannot be modeled precisely. This problem is somehow specific to Iran and other developing countries. North American and European banks have access to operational risk databases like Algorithmics® and ORX®. Since Iranian banks do not have access to such databases, the well-known method of scaling external data (see Shih et al. (2000) for a review and refer to Aue and Kalkbrener (2006), Chapelle et al. (2007), and Moscadelli (2005) for some applications) is not applicable to them. Therefore, we used a scenario analysis approach to enrich operational loss database with catastrophic losses (and ordinary losses if needed).

In scenario analysis approach, banking experts are asked to provide following information about operational risks:

- Scenario configuration (which event or combination of events)
- Impact assessment (how much can it cost)
- Frequency of occurrences (how many times can it happen)

Loss distribution for a specified risk category
In LDA, the loss for the business line i and the event type j between times t and $t + \tau$ is:

$$\vartheta(i,j) = \sum_{n=1}^{N(i,j)} \xi_n(i,j) \tag{3}$$

where i and j are indices of Table 1, and $\xi(i,j)$ is the random variable that represents the amount of one loss event for the business line i and the event type j (which follows severity distribution). The loss severity distribution of $\xi(i,j)$ is denoted by $F_{i,j}$. $N(i,j)$ is a random variable indicating the number of events between times t and $t + \tau$, which has a probability function $p_{i,j}$ (frequency distribution).

Let $G_{i,j}$ be the distribution of $\vartheta(i,j)$. $G_{i,j}$ is then a compound distribution:

$$G_{i,j}(x) = \begin{cases} \sum_{n=1}^{\infty} p_{i,j}(n) F_{i,j}^{n*}(x), x > 0 \\ p_{i,j}(0), x = 0 \end{cases} \tag{4}$$

where $*$ is the *convolution* operator on distribution functions, and F^{n*} is the n-fold convolution of F with itself (Frachot et al. 2001).

In general, there is no analytical expression of the compound distribution (Feller 1968; Frachot et al. 2001). Therefore, computing the loss distribution requires using a numerical algorithm. The most widely used algorithms are the Monte Carlo method (Fishman 1996; Panjer 2006), Panjer's recursive approach (Panjer 1981), and the inverse of the characteristic function (Heckman and Meyers 1983; Robertson 1992).

In this study, the Monte Carlo method is used for computing the loss distribution for each cell in Table 1 (Fishman 1996). This method includes the following steps:

1. One random draw from the frequency distribution is taken (n).
2. n random draws from the severity distribution are taken (for example: first draw US $5,000,000, second draw US $1,200,000, ..., the nth draw US $12,500,000).
3. The US dollar value of losses is summed (for example: US $45,000,000 result is one observation in aggregate loss distribution).
4. The above steps should be repeated m times (for example: 1,000,000 times).

These m observations are used to model the loss distribution for an individual cell of Table 1.

Loss distribution for bank as a whole
The methodology outlined in the sections 'Frequency distribution', 'Severity distribution', and 'Loss distribution for a specified risk category' is applicable to a specified category of operational loss data (i.e., one cell of Table 1 which is calculated in one loop of Figure 1.) However, in order to comply with Basel II, one should consider all 56 categories of risks according to Table 1. For this purpose, Basel Committee recommends calculating the total capital charge of the bank by simple summation of the capital charges of all 56 risk categories; by this proposal, Basel Committee has assumed a perfect positive dependence between the risks implicitly. In spite of this, banks are interested in considering the dependence structure by other appropriate techniques because the basic assumption of Basel Committee will result in large requirements of capital; therefore, banks will have an unacceptable high level of opportunity costs (Aue and Kalkbrener 2006; Chapelle et al. 2007; Moscadelli 2005). Traditionally, correlation is used to model dependence between variables (risk categories here), but recent studies show the superiority of copula over correlation for modeling dependence due to higher flexibility of the copula compared to conventional correlation. Another important reason to choose copula instead of correlation is that the latter is unable to model dependence between extreme events (Kole et al. 2007), which are the main

concern in operational risk modeling. Therefore, in this study, the dependence among aggregate losses will be modeled by copulas in order to combine the marginal distributions of different risk categories into a single joint distribution. A brief definition of copula follows:

Copula. A copula is a multivariate joint distribution defined on the n-dimensional unit cube $[0,1]^n$ in a way that every marginal distribution is uniform in the interval $[0,1]$. Specifically, $C : [0,1]^n \rightarrow [0.1]$ is an n-dimensional copula (briefly, n-copula) if

1. $C(u) = 0$ whenever $u \in [0,1]^n$ has at least one component equal to 0.
2. $C(u) = u_i$ whenever $u \in [0,1]^n$ has all the components equal to 1 except the ith one, which is equal to u_i.
3. $C(u)$ is n-increasing, i.e., for each hyper rectangle
4.

$$B = x_{i=1}^n [x_i, y_i] \subseteq [0,1]^n : V_C(B):$$
$$= \sum_{z \in x_{i=1}^n [x_i, y_i]} (-1)^{N(z)} C(z) \geq 0 \tag{5}$$

5. where $N(z) = \text{card}\{k/z_k = x_k\}$. $V_C(B)$ is the so-called C-volume of B (for more details see Cherubini et al. (2004), Genest and McKay (1986), Nelsen (1999), and Panjer (2006)).

There are various types of copulas in the literature. In this study, we decide to employ a multivariate copula, which is more applicable in practice; the traditional candidate for modeling dependence is Gaussian copula (for application of this copula in a real bank see Aue and Kalkbrener (2006)). However, due to the following four reasons, we preferred a multidimensional t-copula over it: First, operational loss distributions share some similar characteristics with asset portfolio (like skewness, heavy tails, and tail dependence); according to the findings of Kole et al. (2007) for asset portfolio, their procedure provides clear evidence against Gaussian copula but does not reject the t-copula. Second, t-copula assigns more probability to tail events than the Gaussian copula, which makes it appropriate in operational risk modeling where extreme losses are a subject of more concern for banks. Third, t-copula exhibits tail dependence, which is appealing in operational risk modeling. And finally, t-

copula is capable of modeling dependence in the tail without giving up the flexibility to model dependence in the center (Kole et al. 2007); it means that this copula fits well in the entire range of observations. However, to our knowledge, the usage of this copula in a real bank has not been reported anywhere in the world.

t-copula. Multivariate t-copula (MTC) is defined as follows:

$$T_{R,v}(u_1, u_2, \ldots, u_n) = t_{R,v}\left(t_v^{-1}(u_1), t_v^{-1}(u_2), \ldots, t_v^{-1}(u_n)\right) \tag{6}$$

where R is a symmetric, positive definite matrix with $\text{diag}(R) = (1,1,\ldots,1)^T$, and $t_{R,v}$ is the standardized multivariate Student's t distribution with correlation matrix R and v degrees of freedom. $t_v - 1$ is the inverse of the univariate cdf of Student's t distribution with v degrees of freedom. Using the canonical representation, it turns out that the copula density for the MTC is

$$C_{R,v}(u_1, u_2, \ldots, u_n) = |R|^{-\frac{1}{2}} \frac{\Gamma\left(\frac{v+n}{2}\right)}{\Gamma\left(\frac{v}{2}\right)} \left(\frac{\Gamma\left(\frac{v}{2}\right)}{\Gamma\left(\frac{v+1}{2}\right)}\right)^n \tag{7}$$

$$\frac{\left(1 + \frac{1}{v}\varsigma^T R^{-1}\varsigma\right)^{-\frac{v+n}{2}}}{\prod_{j=1}^n \left(1 + \frac{\varsigma_j^2}{v}\right)^{-\frac{v+1}{2}}}$$

where $\varsigma_j = t_v^{-1}(u_j)$ (Cherubini et al. 2004). Using t-copula, we can now calculate the Capital-at-Risk of the bank.

Capital-at-Risk. With LDA, the capital charge (or the Capital-at-Risk) is a Value-at-Risk measure of risk, which is defined as follows:

Given some confidence level $\alpha \in (0,1)$, the Value-at-Risk (VaR) at the confidence level α is given by the smallest number l in a way that the probability that the loss L exceeds l is not larger than $(1 - \alpha)$ (McNeil et al. 2005):

$$\text{VaR}_a = \inf\{l \in R : P(L > l) \leq 1 - a\}$$
$$= \inf\{l \in R : F_L \geq a\} \tag{8}$$

The left equality is a definition of VaR. The right equality assumes an underlying probability distribution, which makes it true only for the parametric VaR. The left equality means that we are $100(1 - \alpha)\%$ confident

Table 2 BLET table for Karafarin bank

	Business disruption and system failures	Execution, delivery, and process management
Retail banking	1,1	2,1
Commercial banking	2,1	2,2

BLET, Business Line Event Type.

Table 3 Distribution of loss data

	1,1	1,2	2,1	2,2
Frequency (%)	61	28	6	4
Severity (%)	4	53	1	42

Table 4 Frequency distribution results

	Distribution	Parameter(s)	Cramer-von Mises	Kolmogorov-Smirnov	Log-likelihood
1,1	Negative binomial	(0.0034,4.9739)	9.3202	0.1786	218.99
	Poisson	1440.6	9.3224	0.5058	3498.65
2,1	Negative binomial	(0.0023,2.0416)	6.989	0.1569	160.88
	Poisson	882	6.9997	0.6666	542.3
1,2	Negative binomial	(0.0247,6.734)	4.9528	0.1822	90.16
	Poisson	266.4	4.8982	0.5325	366.86
2,2	Negative binomial	(0.0059,1.7478)	3.3195	0.1494	66.1
	Poisson	294.9	3.3179	0.4999	740.59

that the loss in the related period will not be larger than the VaR.

Discussion and evaluation
Empirical analysis
Operational loss data of Karafarin Bank have been identified and categorized according to the Basel II Event Types (ETs) as follows:

1. Internal fraud
2. External fraud
3. Employment practices and workplace safety
4. Clients, products, and business practices
5. Damage to physical assets
6. Business disruption and system failures
7. Execution, delivery, and process management

For more detailed classification, definition, and examples of these risk categories, please see 'Annex 9' of Basel Committee on Banking Supervision (2006). Related data of each of the above risk categories have been gathered in eight Basel II defined Business Lines (BLs) as follows:

1. Corporate finance
2. Trading and sales
3. Retail banking
4. Commercial banking
5. Payment and settlement
6. Agency services
7. Asset management
8. Retail brokerage

Please see 'Annex 8' of Basel Committee on Banking Supervision (2006) for more information about activity groups and principle for business line mapping.

A combination of seven ETs and eight BLs provides a 56-cell matrix (Business Line Event Type) as presented in Table 1. Data of this matrix are used for all operational risk calculations.

Karafarin Bank, in line with other banks (for example, Deutsche Bank (Aue and Kalkbrener 2006), National Bank of Belgium (Chapelle et al. 2007), and Bank of Italy (Moscadelli 2005)) considers its operational loss data as confidential; however, main operational risk events related to this study are software and hardware failures, disruption in telecommunication, data entry error, accounting error, collateral management failure, inaccurate reports, incomplete legal documents, unauthorized access to accounts, damage to client assets, and vendor disputes. The methodology mentioned in the section 'Case description' (Figure 1) was applied to Karafarin Bank data.

All data in 56 loss categories have been considered and collected, but due to the scarcity of data, only four cells were used for modeling in this study, as presented in Table 2. Distribution of loss data in the four concerning cells is presented in Table 3. For the sake of confidentiality, all data have been multiplied by a constant scalar and then used in calculations.

In order to estimate frequency distributions, we employed the methodology presented in the section 'Frequency distribution'. Three different goodness-of-fit tests were used including Cramer-von Mises, Kolmogorov-Smirnov, and log-likelihood. In order to provide a reliable

Table 5 Scenario analysis summary table

Basel II classification	Business disruption and system failure		Execution, delivery, and process management										
	Systems		TCEM		M & R		CI		CAM		TC		V & S	
	F	S	F	S	F	S	F	S	F	S	F	S	F	S
Retail banking														
Commercial banking														

CAM, customer/client account management; CI, customer intake and documentation; F, number of occurrences in 1 year; M & R, monitoring and reporting; S, total amount of loss in Iranian rials in 1 year; TC, trade counterparties; TCEM, transaction capture, execution, and maintenance; V & S, vendors and suppliers.

Table 6 Severity distribution results for cell (1,1)

1,1	Distribution	Parameter(s)	Anderson-Darling	Cramer-von Mises	Kolmogorov-Smirnov	Log-likelihood
Ordinary losses	Exponential	403400	153.56	128.3329	0.4566	5553.93
	Extreme value	(800600,1016000)	199.22	128.3333	0.501	6518.25
	Gamma	(0.3053,1321100)	4.9381	128.3215	0.1423	5165.82
	Generalized extreme value	(3.4248,32722,9523.2)	52.2085	128.3323	0.3679	5345.78
	Generalized Pareto	(2.5217,22679)	26.762	128.3307	0.06731	5135.49
	Lognormal	(10.6559,3.4317)	19.1277	128.3295	0.098	5122.04
	Weibull	(186000,0.4277)	11.8198	128.3244	0.0893	5124.39
Large losses	Generalized Pareto	(0.2167,3546300,3809600)				

selection, these three tests were used together, while Aue and Kalkbrener (2006) used one. Another point here is that weighted sum method (Triantaphyllou 2002) was employed in order to guarantee the convergence of several individual goodness-of-fit tests to one best-fitted distribution (see Momen (2008) for details). Analysis of this study like Chapelle et al. (2007), approved the selection of negative binomial distribution for all risk categories, which was confirmed by dispersion analysis (i.e., variance of frequencies are greater than their mean) as shown in Table 4.

With the intention of estimating severity distribution, the methodology presented in the section 'Severity distribution' was followed. By using this method, the first barrier for Iranian banks in measuring operational risk (i.e., the effect of insufficient recorded losses) was resolved.

We tested the fitness of exponential, Weibull, lognormal, gamma, extreme value, generalized extreme value, and generalized Pareto distributions for modeling of economic impact of ordinary operational losses. In order to select among the above mentioned distributions, Anderson-Darling, Cramer-von Mises, Kolmogorov-Smirnov, and log-likelihood goodness-of-fit tests were employed. This diversification among distributions and tests increases the reliability of distribution selection procedure.

In the present work, in order to resolve the second problem of Iranian banks in calculating operational risk (i.e., unreported immense losses), additional examples and descriptions of real large loss events, as recommended by Basel Committee, have been provided (Momen 2008). Therefore, bank experts have been asked to provide scenarios about frequency and severity of large losses in 1 year. These scenarios have been summarized in spreadsheets, like Table 5, and added to the database of operational losses of the bank in order to enrich it with enough large losses.

This type of scenario analysis is more explainable to the management and adds benefits of expert ideas to the quantitative calculations, while previous works in Iran like Erfanian and Sharbatoghli (2006) has missed experts' ideas and only relayed on data of gross income. Tables 6, 7, 8, and 9 show the results of fitting severity distribution for Karafarin Bank's data. Aggregate operational losses for each cell are estimated using Monte Carlo simulation, and approximate distributions are presented in Table 10.

According to the section 'Loss distribution for bank as a whole', for the aim of integrating different aggregate distributions, we decided to model operational loss of banks using t-copula, which solves the third problem for Iranian banks (i.e., missed dependence structure of risk categories). To our knowledge, this copula is globally

Table 7 Severity distribution results for cell (1,2)

1,2	Distribution	Parameter(s)	Anderson-Darling	Cramer-von Mises	Kolmogorov-Smirnov	Log-likelihood
Ordinary losses	Exponential	2114400	281.3302	64.333324	0.5919	3413.78
	Extreme value	(6098600,1.1668000)	47.877	64.333332	0.4831	3907.8
	Gamma	(0.1961,10784000)	6.1354	64.319297	0.1563	2883.33
	Generalized extreme value	(3.0985,30968,9944)	18.1442	64.333007	0.3678	3019.36
	Generalized Pareto	(3.5933,7802.2)	4.2116	64.330906	0.116	2890.69
	Lognormal	(10.7++3,3.4547)	0.5228	64.326428	0.0574	2859.47
	Weibull	(270030,0.3154)	0.8667	64.321569	0.0836	2860.82
Large losses	Generalized Pareto	(0.3571,90367000,50082000)				

Table 8 Severity distribution results for cell (2,1)

2,1	Distribution	Parameter(s)	Anderson-Darling	Cramer-von Mises	Kolmogorov-Smirnov	Log-likelihood
Ordinary losses	Exponential	376370	19.3114	18.999946	0.4809	825.58
	Extreme value	(595510,457840)	8.17	18.999992	0.3981	915.56
	Gamma	(0.4791,785640)	0.7217	18.999279	0.1956	777.48
	Generalized extreme value	(1.1516,136860,87175)	7.852	18.999882	0.368	807.17
	Generalized Pareto	(0.2377,294460)	2.1162	18.999679	0.1348	775.66
	Lognormal	(11.5036,2.7569)	1.4252	18.99966	0.1063	774.32
	Weibull	(288120,0.6192)	0.9279	18.999382	0.1502	774.25
Large losses	Generalized Pareto	(0.7,760080,2190300)				

new in application of operational risk with data of real bank in the published works.

According to the definition of Capital-at-Risk and the confidentiality factor (multiplied by all raw loss data), Capital-at-Risk of Karafarin Bank modeled with t-copula and 99.9% confidence level is equal to (Momen 2008):

$$\text{CaR} = 746286286124 \cong 7.4 \times 10^{11} (\text{IRR})^c$$

This means that with a 99.9% of confidence, the operational loss of Karafarin Bank will not be greater than $7.4 \times 10^{11} (\text{IRR})$. This result quite satisfied the management of Karafarin Bank because it is in tune with their presumptions of operational risk, and it is reasonable compared to the scale of market and credit risk exposures. Moreover, as presented in Table 11, it requires much less capital compared to other approaches that are provided by Basel II. Therefore, by using the present model, banks have the opportunity to use the extra unallocated capital for creating further income within a controlled level of operational risk.

Conclusions

In this paper, a comprehensive methodology of operational risk assessment was addressed. To our knowledge, there are no published works to model operational risk of an Iranian commercial bank appropriately; the main reason is the existence of some inconveniences to measure operational risk using available methods. Therefore, our main objective was to propose a practical framework for Iranian bankers. In this regard, we presented the most important issues facing operational risk analysts and suggested solutions for them through an all-inclusive methodology. The first issue was lack of recorded operational loss, and the second problem was unreported large losses in Iranian banking system. We suggested dividing the severity distribution to different ranges and to deal with each range separately. Moreover, scenario analysis was used to enrich the loss database, provide examples of magnificent losses, and exploit the opinion of experts. The third issue discussed in this study was the dependency structure of operational loss categories, where we proposed t-copula for modeling. The presented methodology was employed to calculate the operational risk of Karafarin Bank, and then, the successive steps of calculations and modeling in Karafarin Bank were reported.

This framework could be applied to other Iranian commercial banks for modeling operational risk and for calculating the Capital-at-Risk of the bank. All aspects of this research could be extended in various ways, provided more complete and robust operational database (i.e., cells of Table 1). Moreover, some researches could be conducted using other multivariate copulas and compare the results with the present work. Another interesting and unexplored area is modeling operational risk with other advanced measurement approaches rather than Loss Distribution Approach (presented here), like Bayesian approaches, Neural networks, Fuzzy modeling, and so on.

Table 9 Severity distribution results for cell (2,2)

2,2	Distribution	Parameter(s)	Anderson-Darling	Cramer-von Mises	Kolmogorov-Smirnov	Log-likelihood
Ordinary losses	Exponential	166620	56.0893	11.66666629	0.5592	671.1794
	Extreme value	(250170,166420)	10.6922	11.66666665	0.4688	758.4963
	Gamma	(0.4890,340770)	1.458	11.66660707	0.1817	583.895
	Generalized extreme value	(1.5483,53833,29959)	4.4841	11.66665921	0.3679	607.7055
	Generalized Pareto	(0.2395,210330)	1.3356	11.66661431	0.1627	585.8375
	Lognormal	(107199,2.7984)	0.3886	11.66660781	0.1255	579.2755
	Weibull	(131650,0.6311)	0.2868	11.66660625	0.0922	579.04
Large losses	Generalized Pareto	(0.4857,35066000,22345000)				

Table 10 Aggregated distributions

	Business disruption and system failures	Execution, delivery, and process management
Retail banking	Gamma	Gamma
	(4.9427,1495600000)	(1.7654,84917000000)
Commercial banking	Gamma	Weibull
	(5.8349,104890000)	(35733000000,1.4442)

Table 11 CaR in different methodologies and their capital requirements (IRR)

	Capital-at-Risk (CaR)	Percentage of Karafarin Bank capital required (%)
Basic Indicator Approach	1,288,019,455,849	64
Standardized Approach	1,264,472,491,808	63
This study	746,286,286,124	37
Karafarin Bank Capital (2010)	2,000,000,000,000	

Endnotes

[a]Bank Markazi.

[b]n is a random variable which follows the frequency distribution.

[c]1 USD $\cong 10,600$ Iranianrials (IRR).

Competing interests
The authors declare that they have no competing interests.

Authors' contributions
OM led the research, carried out the operational risk studies, modified the methodology, participated in the case calculations, and drafted the manuscript. AK participated in the literature review and worked on the statistical models. EN led the data gathering, reviewed the Basel documents, and edited the first draft. All authors read and approved the final manuscript.

Acknowledgments
The authors wish to thank Dr. Parviz Aghili Kermani, the former CEO of Karafarin Bank, and Dr. Mani Sharifi for their helpful comments. We are also grateful to the personnel of Karafarin Bank for providing us the operational loss data.

Author details
[1]Karafarin Bank, Tehran, Iran. [2]Amirkabir University of Technology, Tehran, Iran. [3]Insead Business School, Paris, France.

References
Alexander C (2003) Operational risk: regulation, analysis and management. FT Prentice Hall, London
Anderson TW (1962) On the distribution of the two-sample Cramer-von Mises criterion. Annals of Mathematical Statistics 33(6):1148–1159
Aue F, Kalkbrener M (2006) LDA at work: Deutsche Bank's approach to quantifying operational risk. Journal of Operational Risk 4:49–93
Balkema AA, de Haan L (1974) Residual life time at great age. Annals of Probability 2:792–804
Basel Committee on Banking Supervision (2006) International convergence of capital measurement and capital standards: a revised framework -
comprehensive version (Basel II Framework). Bank for International Settlements, Basel
BBC News (2008) Rogue trader to cost SocGen $7bn. http://news.bbc.co.uk/2/hi/business/7206270. Accessed 12 Feb 2009
Chapelle A et al (2007) Practical methods for measuring and managing operational risk in the financial sector: A clinical study. Journal of Banking & Finance 32:1049–1061
Cherubini U, Luciano E, Vecchiato W (2004) Copula Methods in finance. Wiley, Chichester
de Fontnouvelle P, Rosengren E, Jordan J (2004) Implications of alternative operational risk modeling techniques. Working Paper. Federal Reserve Bank of Boston, Boston
Erfanian A, Sharbatoghli A (2006) Motale Tatbighi va ejraye modelhaye riske amaliati mosavabe komite bal dar banke Sanat o Madan. Faslname elmi o pajooheshie sharif (Issue 34):59–68
Embrechts P, Hansjorg F, Kaufman R (2003) Quantifying regulatory capital for operational risk. RiskLab, Zurich
Embrechts P, Kluppelberg C, Mikosch T (1997) Modelling extrenal events for insurance and finance. Springer, Berlin
Feller W (1968) An introduction to probability theory and its applications, vol 1, 3rd edn. Wiley series in probability and mathematical statistics. John Wiley, New York
Fishman GS (1996) Monte Carlo: concepts, algorithms and applications. Springer, New York
Frachot A, Georges P, Roncalli T (2001) Loss distribution approach for operational risk. Groupe de Recherche Op'erationnelle, Cr'edit Lyonnais, France
Gallati R (2003) Risk management and capital adequacy. McGraw-Hill, New York
Genest C, McKay J (1986) The joy of copulas: bivariate distributions with uniform variables. The American Statistician 40:280–283
Heckman PE, Meyers GG (1983) The calculation of aggregate loss distributions from claim severity and claim count distributions. In Proceedings of the Casualty Actuarial Society, vol 71., Balmar, Arlington, pp 22–61
Karafarin Bank (2010) Karafarin Bank official website. http://www.karafarinbank.com/MainE.asp. Accessed 05 Jan 2010
Bank K (2009) Karafarin Bank report and financial statements for the year ended March 20, 2009. Karafarin Bank, Tehran
King JL (2001) Operational risk, measurement and modelling. Wiley, New York
Kole E, Koedijk K, Verbeek M (2007) Selecting copulas for risk management. Journal of Banking & Finance 8(31):2405–2423
Klugman SA, Panjer HH, Willmot GE (2004) Loss models: from data to decisions. Wiley, New Jersey
McGee S (2002) Simplifying likelihood ratios. Journal of General Internal Medicine 17(8):646–649
McNeil AJ, Frey R, Embrechts P (2005) Quantitative Risk Management: Concepts Techniques and Tools. Princeton University Press, New Jersey
Momen O (2008) Developing a model for measuring operational risk and capital adequacy for a commercial bank: case study for Karafarin Bank. Industrial Engineering Department, Amirkabir University of Technology, MSc Thesis
Moscadelli M (2005) Operational risk: practical approaches to implementation. RiskBooks, London
Nelsen RB (1999) An introduction to copulas. Springer, New York
Panjer HH (1981) Recursive evaluation of compound distributions. Astin Bulletin 12:22–26
Panjer HH (2006) Operational risk modeling analytics. Wiley, New Jersey
Pickands J (1975) Statistical inference using extreme order statistics. Annals of Statistics 3:119–131
Robertson JP (1992) The computation of aggregate loss distributions. In Proceedings of the Casualty Actuarial Society, vol 79., Balmar, Arlington, pp pp 57–133
Shih J, Samad-Khan AH, Medapa P (2000) Is size of operational risk related to firm size? Operational Risk. http://www.risk.net/operational-risk-and-regulation/feature/1508327/is-the-size-of-an-operational-loss-related-to-firm-size
Stephens MA (1974) EDF statistics for goodness of fit and some comparisons. Journal of the American Statistical Association 69:730–737
Triantaphyllou E (2002) Multi-criteria decision making methods: a comparative study. Kluwer, Norwell

Risk determinants of small and medium-sized manufacturing enterprises (SMEs) - an exploratory study in New Zealand

Ariful Islam[1*] and Des Tedford[2]

Abstract

The smooth running of small and medium-sized manufacturing enterprises (SMEs) presents a significant challenge irrespective of the technological and human resources they may have at their disposal. SMEs continuously encounter daily internal and external undesirable events and unwanted setbacks to their operations that detract from their business performance. These are referred to as 'disturbances' in our research study. Among the disturbances, some are likely to create risks to the enterprises in terms of loss of production, manufacturing capability, human resource, market share, and, of course, economic losses. These are finally referred to as 'risk determinant' on the basis of their correlation with some risk indicators, which are linked to operational, occupational, and economic risks. To deal with these risk determinants effectively, SMEs need a systematic method of approach to identify and treat their potential effects along with an appropriate set of tools. However, initially, a strategic approach is required to identify typical risk determinants and their linkage with potential business risks. In this connection, we conducted this study to explore the answer to the research question: what are the typical risk determinants encountered by SMEs? We carried out an empirical investigation with a multi-method research approach (a combination of a questionnaire-based mail survey involving 212 SMEs and five in-depth case studies) in New Zealand. This paper presents a set of typical internal and external risk determinants, which need special attention to be dealt with to minimize operational risks of an SME.

Keywords: SMEs, disturbance, Risk, Risk determinants, Strategic risk management

Background

In the dynamic and highly competitive business environment, manufacturing industries are under tremendous pressure due to the free market economy, rapid technological development, and continuous changes in customer demands (Islam et al. 2006). To cope with the current business trends, the demands on modern manufacturing systems have required increased flexibility, higher quality standards, and higher innovative capacities (Monica and John 1999). 'These demands emphasize the need for high levels of overall system reliability that include the reliability of all human elements, machines, equipment, material handling systems and other value added processes and management functions throughout the manufacturing system' (Islam et al. 2006). Whatever the resources they possess, the manufacturing organizations encounter undesirable events and unwanted setbacks such as machine breakdowns, material shortages, accidents, and absenteeism that make the system unreliable and inconsistent (Monica and John 1999; Islam 2008; Islam et al. 2008; Mitala and Pennathurb 2004; Monostori et al. 1998; Toulouse 2002). In fact, undesirable events and unwanted setbacks (internal and external) in day-to-day operations are common in small and medium-sized manufacturing enterprises (SMEs; Islam 2008). The authors of this paper chose the word 'disturbance' to represent any of these undesirable events and setbacks. They define the disturbance as 'an undesirable or unplanned event that causes the deviation of system performance in such a way that it incurs a loss,' and the definition is published by the authors elsewhere (Islam et al. 2006; Islam 2008). This research adopts the

* Correspondence: arifbd@yahoo.com
[1]Department of Industrial and Production Engineering, Shahjalal University of Science and Technology, Sylhet 3114, Bangladesh
Full list of author information is available at the end of the article

definition of disturbance. As a disturbance creates undesirable consequences that are obviously detrimental to a business performance, we finally refer to a disturbance as a 'risk determinant' on the basis of its significant presence in the system and its consequential negative impact on business and operational performance. Disturbances are linked to undesirable consequences which may originate from different circumstances (Monostori et al. 1998). 'Whatever the sources of disturbances, the consequences resulting from them could be; difficulties to continue work, decreased productivity, reduced production rate, increased defective products, unplanned rework, delayed delivery to market, unexpected downtime, human loss, etc.' (Islam et al. 2006; Islam 2008). In practice, there is a financial loss due to any consequential effects of disturbances. The combined effect of different disturbances could effectively cripple an SME's business performance which may ultimately put it at risk of complete failure (Islam et al. 2006). The risks can, in general, be categorized into three groups: operational, occupational, or economic. The first category of risks involves the loss of production and the loss of production capability that includes productivity losses, quality-related losses, interrelated activity losses, and asset losses. The second category comprises the risks associated with employees' health, safety, and well-being, while the third category encompasses business risks associated with the financial penalties resulting from either of the first two categories as well as compensation claims and damage to reputation. While dealing with risks, the term 'hazard' automatically comes into the scenario; thus, the definition of a hazard can play an important role when dealing with risks in the industrial context. A hazard is a condition that can cause harm, injury, death, damage, or loss of equipment or personnel (Bahr 1997) and can exist without anything actually failing within the enterprise. There are four types of hazards, namely *catastrophic* (death or serious personnel injury or loss of a complete system), *critical* (severe injury or loss of valuable equipment), *minor* (minor injury or minor system damage), and *negligible* (no resulting significant injury or system damage). While examining the definitions of a hazard, it can be noticed that a hazard ultimately represents a situation or condition that has the potential to harm people, property, or the environment. However, a question now presents itself, that if there is no chance to harm any of these three elements (people, property, environment), can we classify the situation as a hazard? For an example, the absence of a key machine operator may have no impact on any of these three elements, but it has the potential to develop financial risk to the organization in terms of loss of production; however, the impact might be severe for a small business if the absence is prolonged. There

might be some debate as to whether absenteeism should be included in the hazard category or not, but most people would agree to recognize it as a potential operational disturbance which could have serious consequences for an SME. Operational disturbances can be seen from different perspectives and can also be described with various words such as disruptions, failures, errors, defects, losses, and waste (Islam 2008). However, all potential disturbances and their consequential losses should be considered in the risk management of SMEs because they can be both time-consuming and costly. We believe that this type of disturbance should be studied under the umbrella of risk management. Consequently, while studying risk management in SMEs, we prefer to use the term 'disturbance' instead of hazard. According to our definition, therefore, a disturbance represents all types of hazards as well any other unwanted setback that can produce uncertainty or a loss for an organization.

The focus of our research was to identify typical risk determinants of SMEs that need to be considered in developing an integrated risk management approach which should include strategic, operational, occupational, financial, and technology-oriented risks. The research is, therefore, built in a specific research question - what are the typical risk determinants of manufacturing SMEs?

Based on the findings related to the question, we have identified a set of key internal and external operational disturbances, which are eventually highlighted as 'risk determinants' based on their occurrence and consequential effects on the business performance of SMEs. This paper presents the identified risk determinants and describes a methodology to identify them.

SMEs are viewed as a source of flexibility and innovation, and they make significant contributions to the economies of many countries, both in terms of the number of SMEs and the proportion of the labor force employed by them (Hoffman et al. 1998; Ministry of Economic Development 2004). However, SMEs are perceived as high-risk ventures, and the entry and exit rates support this perception (Zacharakis et al. 1999). Previous research has indicated that there is little difference between small business failure rates in developed and developing economies, and it is estimated that 50% of all start-ups fail in their first year, while 75% to 80% fail within the first 3 to 5 years in the USA (Anderson and Dunkelberg 1990). It has also been shown that up to 50% of the small businesses started in South Africa eventually failed (Watson and Vuuren 2002). In New Zealand, 40% to 50% of small businesses failed within the first 10 years, and a negative correlation was found between a firm's total full-time employment and its failure rate (Ministry of Economic Development 2004). Business failure is often caused by a lack of knowledge, misplaced overconfidence, lack of financial performance strategies, or a lack of internal management planning (Gibson and Cassar 2005; Hartcher et al. 2003). In spite of high failure rates, however, small businesses continue to be an essential component of the economy of many countries as they account for a significant percentage of all entities and collectively employ large numbers of the workforce. Generally, SMEs depend on financial factors such as profit or sales when considering business risks (Waring and Glendon 1998). However, monetary factors alone may ignore many issues affecting the long-term reputation of the SME and its staff. A recent research study has suggested that risk management is less well developed within SMEs where the strong enterprise culture sometimes mitigates against managing risks in a professional structured way (Virdi 2005). According to the study, the SMEs are reluctant to adopt a formal risk management strategy despite having the evidence that businesses that adopt risk management strategies are more likely to survive and grow. Zacharakis et al. (1999) identify some reasons for failures of small businesses that include both internal and external causes. The internal causes of failure include poor management, lack of risk management planning, and failure to adopt a risk limit threshold. The external causes included government policies, the vulnerability resulting from small size, competition from larger businesses, civil strife, natural disasters, and general economic downturns. It was also found that 'overconfidence' could often drive small business operators to devalue the importance of fundamental risk assessment that ultimately caused their failure. Although there are some other causes for failure that are highlighted in this section, our research is not intended to investigate the reasons behind the absolute failures of SMEs. Rather, it deals with identifying the potential risks existing when operating SMEs within their current infrastructures so that they can avoid potential failures by implementing a strategic risk management approach. Because manufacturing involves a complicated mix of people, systems, processes, and equipment, an effective research strategy needs to be multidisciplinary in its approach to establishing a risk management framework (Islam 2008). Because of some infrastructural, technological, financial, and human resource-related limitations, SMEs may keep themselves away from adopting a positive approach towards strategic risk management (Islam et al. 2006; Islam 2008; Hartcher et al. 2003; Martie-Louise 2006). Islam et al. (2006) state:

It is noteworthy to mention that major accidents and emergencies rarely occur in SMEs although small losses, near misses, unsafe acts and unsafe conditions are common occurrences. But, problems, failures and mistakes as well as incorrect or ineffective actions, are very likely occurrences in the daily business of SMEs and for this reason, in practice, minor incidents and near misses are worth analyzing since in slightly different circumstances the consequences could have been quite serious. By monitoring even small problems and analyzing their underlying causes, it might be possible to discover causes for more serious problems and the existence of hazards. Therefore, no disturbance should be overlooked or should be allowed to happen again.

In the authors' knowledge, research works done on risk management have generally focused on particular industries such as nuclear, aviation, space exploration, chemical processing, and other areas where the consequence of a system breakdown is considered severe or catastrophic for human beings or the environment, and/ or where the potential financial loss is significant (Islam et al. 2006; Andrews and Moss 2002; Khan and Abbasi 1998; Milan 2000; Seastroma et al. 2004; Strupczewski 2003). In addition, research works on risk management in other areas, including financial sectors, medical science, transportation, and construction engineering, have also significantly expanded with time (Islam et al. 2006). In contrast to this, lower priority has been noticed in the literature concerning risk management in the SME sector. Most of the studies relevant to risk management in this sector indeed concentrate solely on the risks associated with safety and occupational health (Islam et al. 2006; Islam 2008). Protective practices such as occupational safety and health and other safety-related programs should, if properly implemented and practiced, ensure better health and working environments inside organizations. They do not, however, ensure the smooth running of the organization or minimize its risks operationally, technically, and/or financially.

Hazard identification within a system is the starting point of any risk identification or assessment process that emphasizes the critical components or factors that produce or could produce failure or harmful consequences for humans, assets, or the environment (Islam et al. 2006; Islam 2008). In this context, different techniques such as Hazard and Operability Analysis studies, Failure Mode and Effect Analysis, Failure Mode and Effect Critical Analysis, Hazard Analysis with Critical Control Points, Fault Tree Analysis, Event Tree Analysis, 'What if' analysis, and Checklists are widely used in practice (Islam et al. 2006; Khan and Abbasi 1998; Mushtaq and Chung 2000; Pearson and Dutson 1995; Tixier et al. 2002). All these techniques focus on the main hazard sources systematically, but none of them can produce a thorough list of important system failures, causes, consequences, and controls and can lend themselves to rigorous risk acceptability analysis (Islam et al. 2006). Furthermore, none of the techniques are necessarily effective in identifying and prioritizing the risks associated with multifaceted criteria. None of the abovementioned methods alone can readily be applicable for dealing with risks associated with operational disturbances, because of their complex nature. 'For example, a disturbance such as 'tool shortage' could be rooted in; erroneous planning of stock, misuse by the operator, unexpected breakage, or incorrect selection of tool for the particular task. Thus, the origin of the disturbance could either be strategic, operational or technical. This means that a detailed analysis of a particular disturbance is required to establish a suitable risk handling procedure' (Islam et al. 2006). In this connection, we have developed a strategic risk management model for SMEs and have published the model elsewhere (Islam et al. 2006; Islam et al. 2008). However, we conducted further study on the identification of specific risk determinants of SMEs and have discussed the identified determinates in this paper.

Case description
Research methodology
We choose an empirical investigation as it puts special emphasis on the affiliated research leading to the development of a strategic risk management framework in terms of operational and organizational aspects (Islam 2008; Glaser and Strauss 1980; Luis et al. 1999; Mills et al. 1995; Pettigrew et al. 1989). The empirical investigation was carried out by applying a multi-method approach (combination of case study and survey methods), called triangulation, which provided a relatively potent means of assessing the degree of convergence, as well as identifying divergences, between the results obtained (Islam 2008; Brewer and Hunter 2006; Jick 1979). In the triangulation method, the survey results improved the authors' understanding of the particular phenomenon (relationship between potential disturbances and their associated risks in this case). On the other hand, the case studies added to a more holistic and richer contextual understanding of the survey results. Thus, the multi-method approach is believed to be enhancing the credibility of the research results while reducing the risk of observations reflecting some unique artifact (Brewer and Hunter 2006; Denzin 1989).

Data collection methods and sample
For the empirical investigation, standard questionnaires were developed and verified by a panel of academic experts and subsequently by an industry focus group in a pilot study. The questionnaires were designed to explore the risk determinants (potential disturbances) and risk indicators (detrimental parameters to business performance) relating to existing practices in the studied organizations. The focal points of the questionnaire were (1) production-related activities associated with risks; (2) quality, reliability, and health- and safety-related issues of both assets and personnel; (3) major activities in the supply chain networks; and finally, (4) strategic issues relating to the current practices in risk management.

There were two phases in the data collection process. In the first phase, questionnaires were sent to 55 manufacturing SMEs (to 165 individual management personnel, to three tiers of management of each organization), and in the second, to 157 SMEs (to 417 management personnel). The respondents were given 1 month to return the completed questionnaires while an additional 3 weeks were allocated for telephoning and personal interviewing to acquire missing data in incomplete questionnaires. Out of 212 SMEs, 11 SMEs declined to participate in the questionnaire survey due to their organizational restructuring, busy scheduling of the management, absorption in other business sectors, or some other undisclosed reasons, though they mentioned their keen interest (in the response letters) to the research subject. Four sets of questionnaires were sent back to the researchers not finding the addressee. Five participating organizations provided partially completed questionnaires, which have been excluded in the analysis. Altogether, 96 usable responses from management personnel (top, middle, and front-line management), from 32 responding SMEs, were returned and have been analyzed, and presented in this paper. It is noted that the organization which returned three sets of completed questionnaire is only considered as responding SME. In this connection, the useful response rate of 18.27% from companies was considered satisfactory and representative of SMEs in New Zealand. The overall response rate of 23.08% from the selected SMEs indicates the substantial importance of the research topic, while past experience suggests that mail survey response rates are often low and appear to be declining among small business populations (Dennis 2003). However, before making any conclusive remarks on the survey findings, further verification was carried out by subsequent in-depth case studies involving five SMEs from among the participants in the mail survey. We choose the follow-up case study approach as '...an empirical inquiry that investigates a contemporary phenomenon within some real-life context and a methodology involving multiple sources of data which provides the fullest understanding of the phenomenon and improves the validity of research implications through triangulation' (Scudder and Hill 1998; Yin 1994). The case studies were conducted longitudinally over an 8-month period. The findings from the mail survey enabled us to develop a deeper understanding of the existing strategies and underlying practices in typical SMEs in New Zealand. During the case studies, ten elements of the operation, namely *premises, product, purchasing, people, procedures, protection, processes, performance, planning,* and *policy,* that represent the main risk areas to the success of a business were considered (Jeynes 2002). An ethnographic approach, which involves the sustained participation in, and observation of, the practical business settings which cover the day-to-day incidents and practical phenomena occurring within the organization, was applied in the case studies (Yin 1994; Bowman and Ambrosini 1997; Charmaz 2006). Apart from direct observation, relevant documents and diagrams from the studied organizations were reviewed and verified. Supplementary data were collected through formal interviews (with key senior executives who shape the firms' operations strategy) and informal discussions (with frontline managers, production supervisors, and some key employees on the shop-floor).

Validation of questionnaire

The mailed survey was carried out by the developed questionnaires. One questionnaire was designed for top management (senior executives), and the other was designed for middle and front-line management of each organization. The purpose of two separate questionnaires was to collect disturbance information from different areas of concerns of each management level. In total, 26 questions were formulated for the questionnaire of the top management and 34 questions were in the questionnaire for middle management. However, in the context of this paper, the questions that were directly related to the disturbances are presented in Additional file 1 for the clarity of the investigation. Most of these questions were of the 'multiple choice' kind. The answers of the questions comprised four-point rating scales for response. The four-point rating scale was chosen to prevent the occurrence of central tendency error.

A typical example of the questions related to an internal disturbance is, Over the last 12 months, how often did you notice 'absenteeism' in your organization? (4 = often, 3 = sometimes, 2 = rarely, and 1 = never). A typical example of the questions related to an external disturbance is, To what extent does 'skilled labor shortage' impede your business performance (profit/growth)? (4 = to great extent, 3 = to some extent, 2 = a little amount, and 1 = not at all). A typical example of the questions related to a risk indicator is, Over the last 12 months, how often did you notice 'lower than expected productivity'? (4 = often, 3 = sometimes, 2 = rarely, and 1 = never).

The questionnaires were designed in such a way that they were easy to understand and answer. They were pretested and carried out in two sequential stages. The first stage consisted of a review by a panel of academic experts and survey specialists who ensured that all necessary questions were included and ambiguous questions eliminated, and the categorization of the questions was set up properly to ensure that subsequent data analysis would provide the desired information. The second stage was a pilot study with ten participating SMEs. The responses from the pilot study allowed the authors to verify whether respondents were biased towards certain categories of questions or leaving questions unanswered. The study found that all respondents answered all questions and the responses on the ordinal scales were reasonably dispersed. Finally, the measuring scales were tested to verify the reliability of instrument with the help of Cronbach's alpha (α) (Hinton 2004; Black 1999). The values of α were 0.701 and 0.716, and 0.721 for the questions of internal and external disturbances, and risk indicators (consequential effect), respectively, that ensured the reliability and internal consistency of the measuring scales.

Characteristics of studied SMEs

The significance of the SME sector in New Zealand has been increasing, with further opportunities presented by globalization and technological development (Ministry of Economic Development 2004). New Zealand is a small nation state of 4.3 million people, ethnically diverse, with a strong culture of self-help and independence underpinning business development (Ministry of Economic Development 2004). New Zealand's size means that by international standards, its small businesses are very small but are the dominant sector in terms of employment, organizational structure, and social and economic cohesion. A recent report on SMEs states that in the context of policy consideration, the characteristics of small-sized businesses should typically include personal ownership and management, few specialist managerial staff, and not being part of a larger business enterprise (Ministry of Economic Development 2003). SMEs in New Zealand typically exhibit these characteristics, and it is in this context that our research has been designed to deal with companies with employment in the range of 10 to 100 employees (Islam 2008).

The list of SMEs selected for the mail survey and case studies was compiled from a variety of business databases; these were randomly chosen to represent a range of manufacturing groups. These groups covered the four sectors of (1) metal-based product and equipment manufacturers, (2) wood and wood-based product manufacturers, (3) paper- and plastic-based product manufacturers, and (4) textile and garment manufacturers. These groups were selected because of their economic importance to New Zealand. The characteristics of the participating SMEs in the mail survey are presented in Table 1.

Key findings and analysis

The key findings are categorized and presented in the following sections:

Table 1 Characteristics of the selected SMEs

Classification	Criteria	Number of organizations	Percentage of organizations (%)
Firm size	Small size (10 to 25 employees)	12	37.50
	Medium size (26 to 100 employees)	20	62.50
Annual turnover (New Zealand $)	Less than 5 million	4	12.50
	Between 5 and 25 million	20	62.50
	Between 25 and 50 million	6	18.75
	Over 50 million	2	6.25
Business category	Metal-based product and machinery manufacturing	16	50.00
	Textile and garment manufacturing	6	18.75
	Wood-based product and furniture manufacturing	6	18.75
	Plastic- and paper-based product manufacturing	4	12.50
Plant set-up	Single site	20	62.50
	Multi-domestic sites	9	28.13
	Multinational sites	3	9.38
Employment contracts	Nil	11	34.38
	Less than 5% of total employees	17	53.13
	Between 5% and 10% of total employees	2	6.25
	More than 10% of total employees	2	6.25
Total number of selected organizations		*32*	*100.00*

Risk indicators

Two principal measures of corporate performance are profit rate and growth rate (Freel 2000; Geroski and Machin 1992; Wynarczyk and Thwaites 1996). Needless to say, there are a number of ways to measure growth rate and profitability which are substantially linked to several variables of operational activities. Several studies have overwhelmingly indicated that effective employee management, along with other strategic measures, can lead to a competitive advantage in the form of a motivated workforce, improved operational and business performance, reduced employee turnover, and improved productivity, which in turn improve the net profit of a firm (Batt 2002; Macduffie 1995; Virdi 2005). Moreover, growth of a business would appear to play an important role in its sustainability in a dynamic business (Barbara et al. 2000). We could, therefore, interpret that dissatisfaction with net profit and in business growth (assuming that the business plan is realistic), as well as significant employee turnover rates, could be the results of inappropriate or inadequate strategic allocation and utilization of resources and that these should be treated as primary indicators of potential problems for an organization. Our research approach, however, was not to verify the measures of these categories. Rather, it tried to identify whether there is any correlation between business growth rate and net profit, and the potential disturbances. The research finds that approximately 32% of the SMEs are dissatisfied with their existing 'net profit'

(of which 10% are very dissatisfied) and about 40% are dissatisfied with 'business growth' (of which 10% are very dissatisfied). On the other hand, 9% of the organizations are very satisfied with both net profit and business growth. The study also finds that 30% of SMEs consider the existing 'employee turnover rate' as a substantial impediment to effective business operation, while 43% indicate the impediment from this factor to be small, and 26% indicate it to be negligible. These are apparently linked to operational risks of direct or indirect losses due to failures in systems, processes, and people or from external factors. Thus, dissatisfaction level with net profit and in business growth and employee turnover rate is considered as 'risk indicators' for our research. In addition to these three, 11 risk indicators which are linked to operational, occupational, and economic losses are identified from the study. Figure 1 shows the relative position of these risk indicators in terms of their emergence in the systems of the studied SMEs.

Operational disturbances

The risk indicators have potential linkages with day-to-day operational disturbances, which degrade business performance and the business environment. In consequence, the disturbances ultimately play a vital role in putting an organization at risk in terms of production, safety, and financial, resulting from both internal and external customer dissatisfaction (Islam 2008). These can lead to a loss of market share and eventually put the

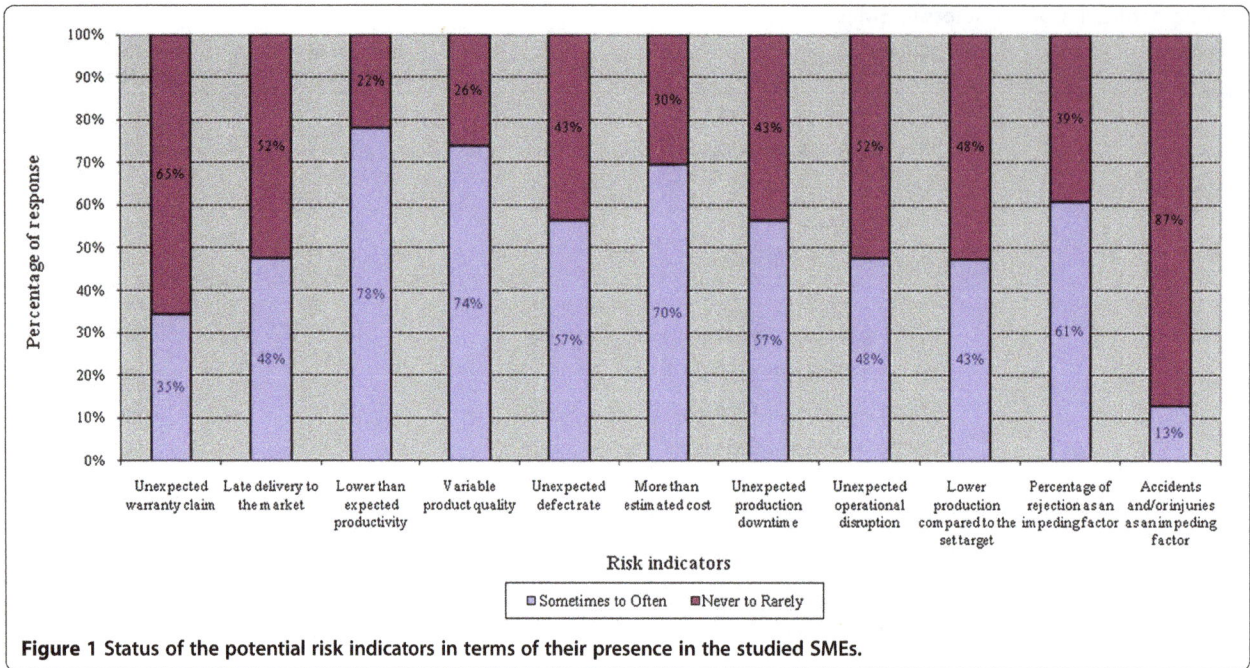

Figure 1 Status of the potential risk indicators in terms of their presence in the studied SMEs.

organization out of business, if they are not carefully treated. For this, a thorough investigation was conducted to identify key operational disturbances (in essence, driving risk factors) and their linkage to some risk indicators discussed in the previous section. We have identified a number of notable internal and external operational disturbances, which are summarized in Figures 2 and 3.

Among the internal disturbances, *absenteeism, machine malfunction, machine breakdown*, and *material handling disruption* were found to be the most significant disturbances, and *unexpected major hazards, unexpected*

accidents/injuries, and *tool shortage* were found to be the least significant ones, while the other disturbances were found to fall between these extremes. Among the external disturbances, *competition, delayed supply by the regular supplier*, and *skilled labor shortage* were found to be the most significant ones, while *financial obstacle* was found to be the least significant in terms of their influence on the operational system. However, despite the minimal influence of some disturbances, they were still considered for further analysis to find out their consequential effects.

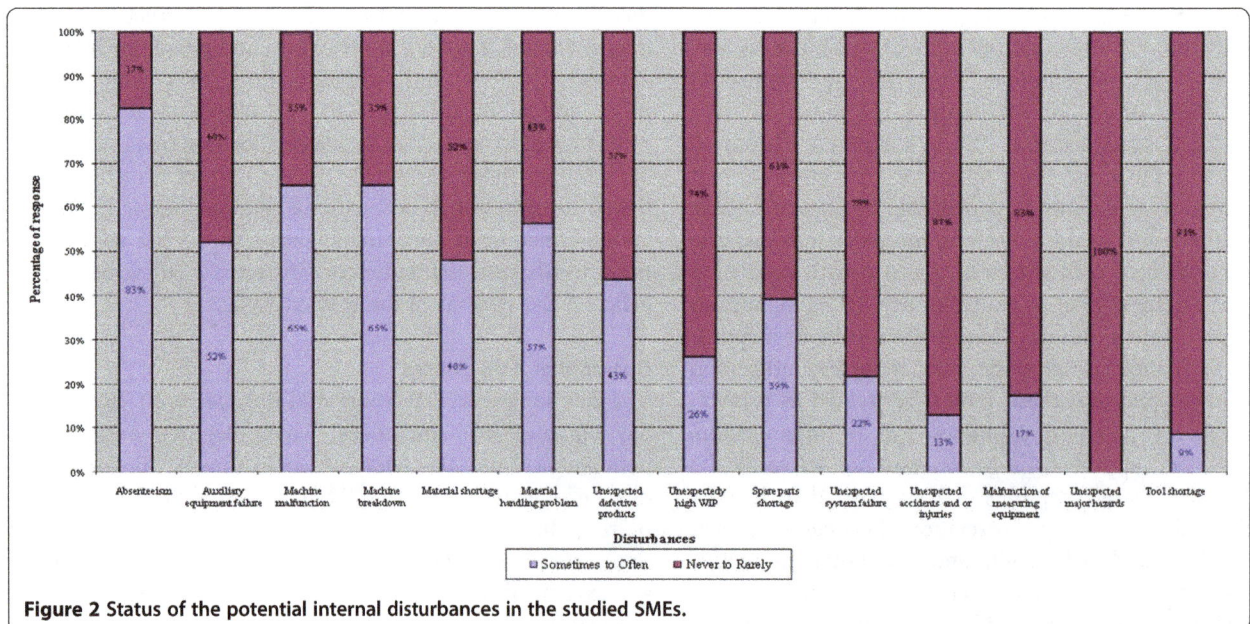

Figure 2 Status of the potential internal disturbances in the studied SMEs.

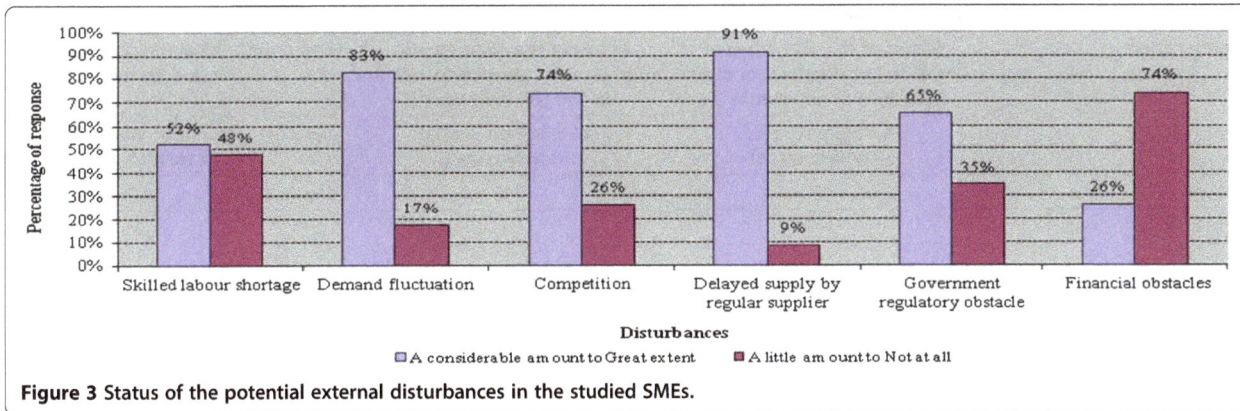

Figure 3 Status of the potential external disturbances in the studied SMEs.

Risk determinants

All disturbances presented in Figures 2 and 3 were considered for further analysis to determine whether they should be treated as risk determinants. The analysis included some statistical methods of parametric and non-parametric testing such as *t test*, *the Friedman test*, and the *Spearman correlation coefficient tests* (Hinton 2004) at two significant levels: $\alpha = 0.01$ (99% confidence level) and $\alpha = 0.05$ (95% confidence level). The results of the *t* test are presented in Table 2.

On the basis of their comparative occurrence in practice, the disturbances are assigned with relative scores. The disturbance which occurs most frequently is assigned with the highest score, while the disturbance which occurs least frequently is assigned with the lowest score. Thus, among the internal disturbances, 'absenteeism' scores the highest number of points, and 'tool shortage' and 'unexpected major hazard' jointly score the lowest.

The relative positions of the internal disturbances, based on their scores, are shown in the second column of Table 3. The final test results (based on Spearman's correlation coefficient, r_s) confirm the positive correlation between internal disturbances and risk indicators; the results are presented in Table 4. Based on the positive correlation of disturbances with a number of risk indicators, scoring is performed. The highest scorer is correlated with a maximum number of risk indicators, while the lowest one is correlated with a minimum number of risk indicators. Thus, all disturbances are assigned with scores and are presented in the third column of Table 3. Finally, on the basis of the product of two scores (one for appearance or occurrence and the other for correlation), final ranking is performed for the risk determinants. The determinant which scores the maximum value is assigned with the highest rank (1), and the determinant which scores the minimum value is assigned with the lowest rank (14). Accordingly, the

Table 2 Results of t tests with internal disturbances

	d2	d3	d4	d5	d6	d7	d8	d9	d10	d11	d12	d13	d14
Absenteeism (d1)	3.213**	3.026**	2.621**	3.346**	5.724**	2.421*	3.471**	4.592**	7.854**	5.738**	6.098**	11.754**	10.817**
Auxiliary equipment failure (d2)		−0.264	0.190	0.711	0.886	−0.549	1.283	1.899*	2.998**	3.696**	3.581**	6.789**	6.260**
Machine malfunctions (d3)			0.374	1.121	1.371	−0.264	1.513	2.197**	3.899**	4.243**	6.244**	7.414**	7.468**
Machine breakdown (d4)				0.514	0.753	−0.537	0.983	1.274	2.913**	2.731**	3.366**	6.249**	4.136**
Material shortage (d5)					0.309	−1.371	0.989	1.429	2.584**	3.931**	3.845**	6.969**	5.359**
Material handling problem (d6)						−1.429	0.789	1.045	2.989**	2.532**	3.581**	5.778**	6.054**
Unexpected defective product (d7)							3.308**	3.283**	3.329**	5.381**	4.333*	6.875**	8.125**
Unexpected work-in-progress (d8)								0.000	0.560	1.077	1.663	2.278**	3.696**
Spare parts shortage (d9)									0.783	1.899*	2.062*	2.954**	6.278**
Unexpected system failure (d10)										0.437	1.435	6.696**	3.280**
Unexpected accidents/injuries (d11)											1.208	2.249*	3.638**
Malfunctions of measuring equipment (d12)												0.711	1.986*
Unexpected major hazard (d13)													1.295
Tool shortage (d14)													

Numbers are the values of *t*'s; *$p < 0.05$; **$p < 0.01$.

Table 3 Potential risk determinants (internal)

Risk determinants	Scores of the disturbances based on the distribution of the frequency of occurrence (F)	Scores of the disturbances based on their positive correlation with the risk indicators (C)	Total score	Final ranks of the risk determinants
	(14 = highest score, 1 = lowest score)	(14 = highest score, 1 = lowest score)	(F × C)	(1 = most important, 14 = least important)
Absenteeism	14	14	196	1
Unexpected defective product	13	9	117	2
Machine malfunctions	12	5	60	5
Auxiliary equipment failure	11	4	44	7.5
Material shortage	9	12	108	3
Material handling problem	9	11	99	4
Machine breakdown	9	6.5	58.5	6
Spare parts shortage	7	3	21	10
Unexpected work-in-progress	5.5	2	11	12
Unexpected system failure	5.5	8	44	7.5
Unexpected accidents or injuries	4	10	40	9
Malfunctions of measuring equipment	3	1	3	14
Unexpected major hazard	1.5	6.5	9.75	13
Tool shortage	1.5	13	19.5	11

relative ranking for all risk determinants is established and is shown in the fifth column of Table 3. According to the final ranking, 'absenteeism' becomes the most important (number 1) risk determinant among the internal disturbances, while 'malfunctions of measuring equipment' becomes the least important one. Similar tests were conducted and relative measures were performed on the external disturbances, the results of

which are summarized in the second and third columns of Table 5.

Discussion and evaluation

The findings from the mail survey have been presented in the previous section. Most of the findings have strongly been supported by the findings from case studies. Both investigations confirm that there are some

Table 4 Correlation between internal disturbances and risk indicators

	D1	D2	D3	D4	D5	D6	D7	D8	D9	D10	D11	D12	D13	D14
Absenteeism	0.67**	IPC	0.39*	IPC	0.77**	IPC	0.30*	0.33*	0.40*	0.32*	IPC	0.48**	0.64**	0.60**
Auxiliary equipment failure	0.49*	IPC	IPC	0.47*	0.45**	0.50**	IPC	NC	0.31*	NC	NC	IPC	0.36*	IPC
Machine malfunctions	0.44*	IPC	IPC	0.33*	IPC	0.43*	NC	0.30*	IPC	NC	0.40*	IPC	0.41*	0.39*
Machine breakdown	0.42*	IPC	0.44*	IPC	0.40*	IPC	0.30*	IPC	0.55**	0.35*	0.51**	NC	IPC	IPC
Material shortage	0.54**	0.62**	0.33*	0.40*	0.38*	NC	0.53**	0.36*	0.51**	0.35*	IPC	IPC	NC	NC
Material handling problem	0.49**	0.53**	0.39*	0.44*	0.61**	IPC	IPC	0.33*	IPC	0.49**	0.43*	IPC	IPC	IPC
Unexpected defective product	0.36*	IPC	IPC	IPC	0.52**	NC	0.47**	NC	NC	0.56**	NC	0.33**	0.35*	0.45**
Unexpected work-in-progress	IPC	NC	IPC	NC	IPC	IPC	NC	0.41*	0.33*	IPC	NC	0.45**	IPC	NC
Spare parts shortage	0.35*	NC	0.38*	IPC	0.35*	IPC	IPC	NC	0.51**	NC	NC	0.34*	0.38*	IPC
Unexpected system failure	0.39*	IPC	0.43*	IPC	0.53**	IPC	0.36*	NC	NC	0.57**	0.40*	NC	IPC	0.45**
Unexpected accidents/injuries	0.04*	IPC	0.34*	0.37*	IPC	0.38*	IPC	0.43*	1.00**	NC	0.41*	0.44*	NC	NC
Malfunctions of measuring equipment	IPC	IPC	NC	NC	NC	NC	NC	0.50**	IPC	NC	0.52**	IPC	IPC	NC
Unexpected major hazard	0.39*	IPC	0.43*	IPC	0.49**	0.32*	0.32*	NC	0.31*	0.45**	0.41*	NC	IPC	NC
Tool shortage	0.48*	IPC	0.34*	0.37*	0.43*	0.50**	IPC	0.33*	0.37*	NC	IPC	0.49**	0.73**	0.45**

Numbers in the boxes are the values of r_s; IPC, insignificant positive correlation; NC, no correlation; D1, lower than expected productivity; D2, variable product quality; D3, unexpected defect rate; D4, more than estimated cost; D5, unexpected production downtime; D6, unexpected operational disruption; D7, lower production compared to the set target; D8, percentage of rejection at various levels; D9, accidents and/or injuries; D10, late delivery to the market; D11, unexpected warranty claim; D12, employee turnover rate; D13, dissatisfaction level with net profit; D14, dissatisfaction level with business growth; *$p < 0.05$; **$p < 0.01$.

Table 5 Potential risk determinants (external)

Risk determinants	Scores of the disturbances based on the distributions of the level of impediments on business (i)	Scores of the disturbances based on their positive correlation with the risk indicators (c)	Total score	Final ranking of the risk determinants
	(6 = highest score, 1 = lowest score)	(6 = highest score, 1 = lowest score)	(i × c)	(1 = most important, 6 = least important)
Delayed supply by the suppliers	6	2.5	15	2
Demand fluctuation	5	2.5	12.5	4
Competition	4	4.5	18	1
Skilled labor shortage	3	4.5	13.5	3
Government regulations	2	1	2	6
Financial obstacles	1	6	6	5

typical internal and external operational disturbances, which expose SMEs to operational risks. Comparative findings from the two investigations are depicted in Figures 4 and 5. The comparison for disturbances is made on an extended scale of 1 to 10 in terms of their frequency of occurrence (for internal disturbances) and of their detrimental effects on operational performance (for external disturbances). Figure 4 shows that both investigations identify 'absenteeism' as the most frequently occurring internal disturbance and 'tool shortage' as the least frequently occurring in the SMEs studied, while the others fall between these two extremes. Figure 5 shows that 'delayed supply by regular suppliers' (very closely followed by 'demand fluctuation' and 'competition') is the most detrimental external disturbance, and 'financial obstacles' is the least detrimental to the SMEs. Both investigations further confirmed a set of risk indicators, which can be used as the consequential effects resulting from the disturbances (Figure 6). These risk indicators are linked to operational, occupational, and economic losses. The findings of both investigations again converge on the same conclusions, in terms of

the overall ranking of the disturbances, even though there are slight, statistically insignificant variations in some cases.

The research study reveals that SMEs have, in general, inadequate measures and planned strategies in place to deal with such risk determinants. Thus, the identified set of internal and external risk determinants found from this study will play a vital role in ensuring that SMEs realize the strengths and weaknesses in their ability to cope with the identified internal factors, as well as the threats and opportunities arising from the identified external factors, while assisting them in formulating and implementing strategic measures to deal with the resulting operational risks. It is obvious that some disturbances are more detrimental than others. Moreover, the nature of the disturbance is found to be dynamic and idiosyncratic in nature. The dynamic behavior of a disturbance in different organizational settings and in different time frame leads us to a common understanding that the appearance of a particular disturbance varies from organization to organization and time to time. Moreover, the same disturbance produces different

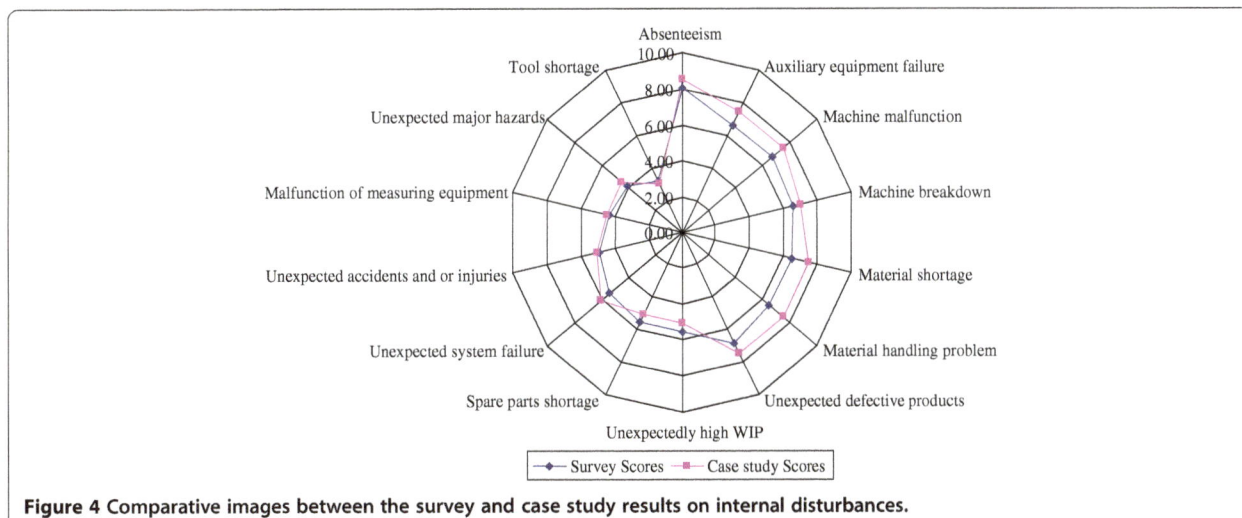

Figure 4 Comparative images between the survey and case study results on internal disturbances.

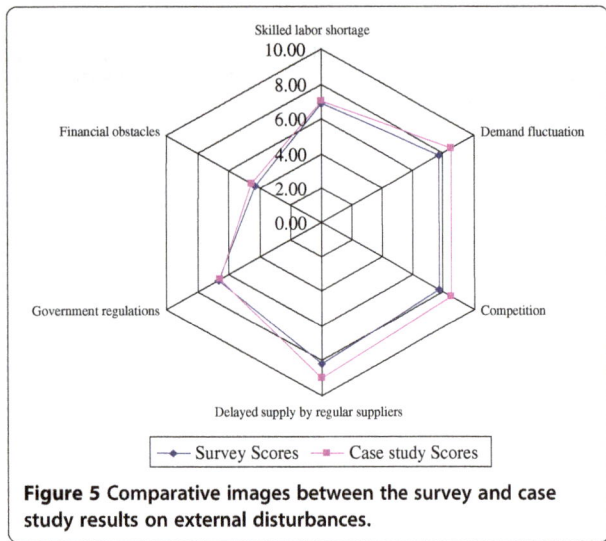

Figure 5 Comparative images between the survey and case study results on external disturbances.

optimization of response time to changes in the external environment becomes vital. At the same time, smooth and consistent operational performance in the internal environment is necessary to continue the business in this dynamic business world. Internal and external disturbances to its day-to-day operation put an SME at risk in terms of production, safety, and the business itself. The risks associated with disturbances can be detected by analyzing the negative or detrimental consequential effects, which are identified as risk indicators in the research. We have identified some typical internal and external operational disturbances that need to be considered as risk determinants for SMEs. It is found that some disturbances are positively correlated with a greater number of risk indicators and some with a lesser number of indicators. It is also found that every disturbance is significantly correlated with at least one of the risk indicators. This means that in terms of operational risks, an SME needs to consider all the identified disturbances (risk determinants) in its strategic decisions for managing operational risks successfully.

consequential effects to different organizations based on its time of occurrence and the duration of its existence in the system. An organization, therefore, needs to identify the characteristics of the various disturbances and their consequential effects over time, to develop a proactive strategy for managing operational risks.

Conclusions

An organization is basically a giant network of interconnected nodes. Changes in one part of an organization can affect other parts of the organization with surprising and often negative consequences. The minimization of delays in the system generally becomes an important issue in lean manufacturing. In this context, the

We find that the majority of the studied SMEs do not have systematic risk management strategies in place. It is discovered that the majority of SMEs used standard hazard identification forms, which comply with the requirements of the Health and Safety in Employment Act in New Zealand (Avery 1993). The current practices in SMEs regarding risk identification relies, almost exclusively, on the documented records of industrial injuries which these forms produce. Near misses are not generally recorded even though this is a requirement of the legislation. Moreover, the identification of root

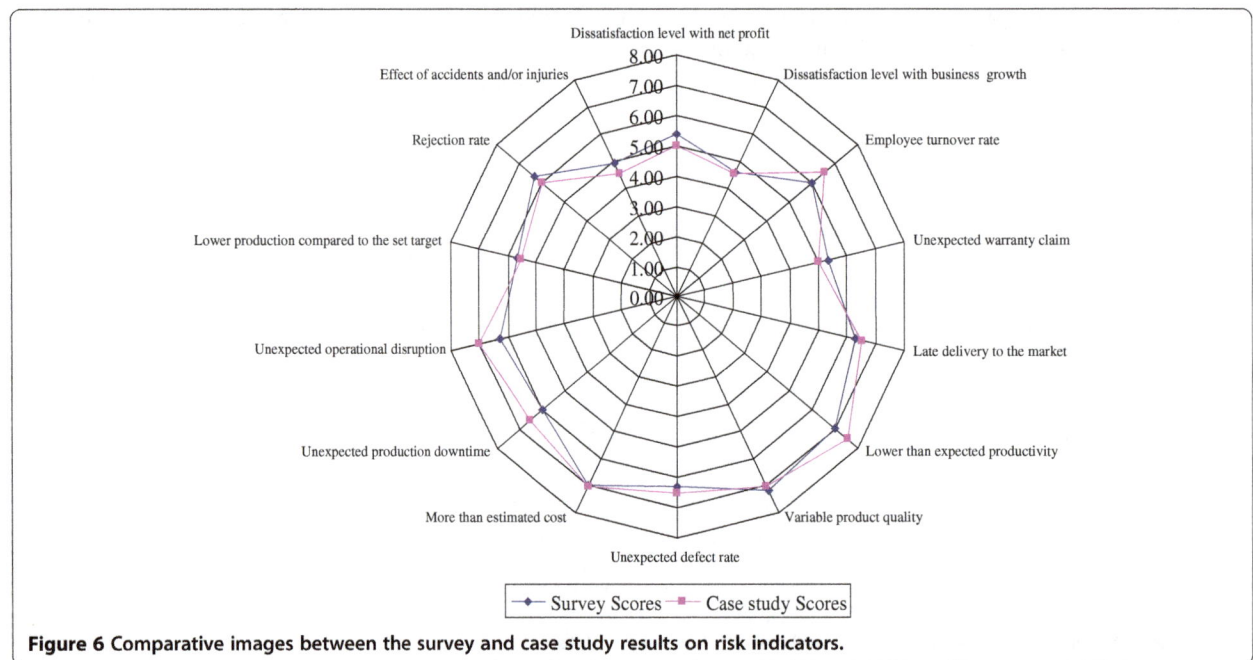

Figure 6 Comparative images between the survey and case study results on risk indicators.

causes of the risk determinants and their related origins is not practiced in the studied SMEs, and in SMEs where it is practiced to some extent, the flow of information tends to miss many of the relevant personnel. In addition, the disturbance handling systems in these organizations, in terms of data collection, information processing, information sharing, and decision making, are found to be relatively weak and very informal. With regard to the identification of external disturbances, most SMEs do not have assessment criteria in place to measure the consequences, nor have enough information available to help them determine their root causes.

The identified set of internal and external risk determinants should provide a quick reference or benchmark for SMEs. The struggle with the identification of operational risk determinants should be minimized by the identified set of determinants, obtained from a representative sample of SMEs in New Zealand. It is, however, relevant to note that the relative rankings of the identified risk determinants could vary from organization to organization based on their likelihood of occurrence and their impact on business performance. The individual business setting, including current strategic measures, practices, and vulnerability, would play a vital role in developing appropriate strategic plans and actions in each case. While it may be necessary for organizations to add or delete determinants to those identified in this research, depending on their particular situation, they should be able to apply the described methodology to assist them in identifying the risk determinants appropriate to them. In this way, they should be able to identify the extent of the risks associated with the determinants by incorporating the metrics of time, money, and asset loss due to these. In conclusion, the research findings presented in this paper will, hopefully, add to the body of knowledge on good practices in risk management resulting from operational disturbances which can affect SMEs and that may also be useful to both management professionals and researchers in the field of risk management.

Additional file

Additional file 1: Key questions in the questionnaire for top management and middle and front-line management.

Competing interests

The authors declare that they have no competing interests.

Authors' contributions

Dr. MAI designed the research, collected and analyzed the data, and drafted the manuscript. Dr. DT substantially contributed to the conception and design phase, modification of the questionnaires and analysis, and editing of the manuscript critically for its intellectual content. Both authors read and approved the final manuscript.

Authors' information

Dr. MAI is an associate professor in the Department of Industrial and Production Engineering, Shahjalal University of Science and Technology. He obtained his PhD degree from the University of Auckland, New Zealand, and has been actively doing research for more than 13 years. His principal research activities are in the areas of engineering management including risk management, quality management, and productivity improvement of manufacturing systems. Dr. DT is an associate professor in Engineering Management at the University of Auckland. He gained his PhD degree from the Queens University of Belfast and has been a practicing researcher for over 36 years. As a chartered engineer (C. Eng) and a member of the Institution of Engineering and Technology (IET), his principal research activities are in the areas of manufacturing systems and engineering management. He has been consulting widely with manufacturing and process industries in New Zealand, Australia, and the UK in areas directly related to productivity improvement and process optimization

Author details

[1]Department of Industrial and Production Engineering, Shahjalal University of Science and Technology, Sylhet 3114, Bangladesh. [2]Department of Mechanical Engineering, The University of Auckland, Auckland 1142, New Zealand.

References

Anderson RL, Dunkelberg JS (1990) Entrepreneurship: starting a new business. Harper and Row, New York

Andrews JD, Moss TR (2002) Reliability and risk assessment, 2nd edn. Professional Engineering, London

Avery M (1993) Health & safety laws at work: key issues. Teemay Consultants, New Zealand

Bahr NJ (1997) System safety engineering and risk assessment: a practical approach. Taylor & Francis, Washington, DC

Barbara JO, Sandy HS, Allan LR (2000) Performance, firm size, and management problem solving. Journal of Small Business Management 38(4):42–58

Batt R (2002) Managing customer services: human resource practices, quit rates, and sales growth. Academy of Management Journal 45:587–597

Black TR (1999) Doing quantitative research in the social sciences: an integrated approach to research design, measurement and statistics. Sage, Thousand Oaks

Bowman C, Ambrosini V (1997) Using single respondents in strategy research. British Journal of Management 8:119–131

Brewer J, Hunter A (2006) Foundation of multi-method research: synthesizing styles. Sage, Thousand Oaks

Charmaz K (2006) Constructing grounded theory - a practical guide through qualitative analysis. Sage, Thousand Oaks

Dennis WJ Jr (2003) Raising response rates in mail surveys of small business owners results of an experiment. Journal of Small Business Management 41(3):287–295

Denzin NK (1989) The research act: a theoretical introduction to sociological method, 3rd edn. Prentice Hall, Englewood Cliffs

Freel MS (2000) Do small innovating firms outperform non-innovators? Small Business Economics 14(3):195–210

Geroski P, Machin S (1992) Do innovating firms outperform non-innovators? Business Strategy Review Summer 3:79–90

Gibson B, Cassar G (2005) Longitudinal analysis of relationships between planning and performance in small firms. Small Business Economics 25(3):207–222

Glaser BG, Strauss AL (1980) The discovery of grounded theory - strategies for qualitative research, 11th printing. Aldine, New York

Hartcher J, Allan H, Scott H (2003) Perceptions of risks and risk management in small firms. Small Enterprise Research: The Journal of SEAANZ 11(2):71–92

Hinton PR (2004) Statistics explained, 2nd edn. Routledge, New York

Hoffman K, Milady P, Bessant J, Perren L (1998) Small firms' R&D, technology and innovation in the UK: a literature review. Technovation 18(1):39–55

Islam MA (2008) Risk management in small and medium-sized manufacturing organization in New Zealand, PhD Thesis, Department of Mechanical Engineering. The University of Auckland

Islam MA, Tedford JD, Haemmerle E (2006) Proceedings of the 2006 IEEE International Conference on Management and Innovation and Technology,

Singapore. In: Strategic risk management approach for small and medium-sized manufacturing enterprises (SMEs)—a theoretical framework, 2nd edn. Singapore, IEEE, pp 694–694

Islam MA, Tedford JD, Haemmerle E (2008) Managing operational risks in small- and medium-sized enterprises (SMEs) engaged in manufacturing–an integrated approach. International Journal of Technology, Policy and Management 8(4):420–441

Jeynes J (2002) Risk management: 10 principles. Butterworth-Heinemann, Oxford

Jick TD (1979) Mixing qualitative and quantitative methods: triangulation in action. Administrative Science Quarterly 24:602–610

Khan FI, Abbasi SA (1998) Techniques and methodologies for risk analysis in chemical process industries. Journal of Loss Prevention in the Process Industries 11:261–277

Luis EQ, Felisa MC, Serge W, Christopher OB (1999) A methodology for formulating a business strategy in manufacturing firms. International Journal of Production Economics 60–61:87–94

Macduffie JP (1995) Human resource bundles and manufacturing performance: organizational logic and flexible production systems in the world auto industry. Industrial and Labor Relations Review 48:197–221

Martie-Louise V (2006) Strategy-making process and firm performance in small firms. Journal of Management and Organization 12:209–222

Milan J (2000) An assessment of risk and safety in civil aviation. Journal of Air Transport Management 6:43–50

Mills J, Platts K, Gregory M (1995) A framework for the design of manufacturing strategy process: a contingency approach. International Journal of Operations and Production Management 15(4):17–49

Ministry of Economic Development (2003) SMEs in New Zealand: structure and dynamics. Ministry of Economic Development, Wellington

Ministry of Economic Development (2004) SMEs in New Zealand: structure and dynamics. Ministry of Economic Development and Statistics New Zealand, Wellington

Mitala A, Pennathurb A (2004) Advanced technologies and humans in manufacturing workplaces: an interdependent relationship. International Journal of Industrial Ergonomic 33:295–313

Monica PB, John RW (1999) HEDOMS—human errors and disturbance occurrence in manufacturing systems: toward the development of an analytical framework. Human Factors and Ergonomics in Manufacturing 9(1):87–104

Monostori L, Szelke E, Kadar B (1998) Management of changes and disturbances in manufacturing systems. Annual Reviews in Control 22:85–97

Mushtaq F, Chung PWH (2000) A systematic Hazop procedure for batch processes, and its application to pipeless plants. Journal of Loss Prevention in the Process Industries 13:41–48

Pearson AM, Dutson TR (1995) HACCP in meat, poultry and fish processing. Blackie Academic and Professional, New York

Pettigrew AM, Whipp R, Rosenfeld R (1989) Competitiveness and the management of strategic change process: a research agenda. In: Francis A, Tharakan M (eds) The competitiveness of European industry: country, policies and company strategies. Routledge, London, p 36

Scudder GD, Hill CA (1998) A review and classification of empirical research in operations management. Journal of Operations Management 16:91–101

Seastroma JW, Peercy RL Jr, Johnson GW, Sotnikov BJ, Brukhanov N (2004) Risk management in international manned space program operations. Acta Astronautica 54:273–279

Strupczewski A (2003) Accident risks in nuclear-power plants. Applied Energy 75:79–86

Tixier J, Dusserre G, Salvi O, Gaston D (2002) Review of 62 risk analysis methodologies of industrial plants. Journal of Loss Prevention in the Process Industries 15:291–303

Toulouse G (2002) Accident risks in disturbance recovery in an automated batch-production system. Human Factors and Ergonomics in Manufacturing 12 (4):383–406

Virdi AA (2005) Risk management among SMEs–executive report of discovery research. The Consultation and Research Centre of the Institute of Chartered Accountants in England and Wales, London

Waring A, Glendon AI (1998) Managing risk: critical issues for survival and success into the 21st century, 1st edn. International Thomson Business, London

Watson ML, Vuuren JJ (2002) Entrepreneurship training for emerging SMEs in South Africa. Journal of Small Business Management 40(2):154–161

Wynarczyk P, Thwaites A (1996) The financial performance of innovative small firms in the UK. In: Oakey R (ed) New technology based firms in the 1990s, 11th edn. Paul Chapman, London

Yin RK (1994) Case study research, 2nd edn. Sage, London

Zacharakis AL, Meyer GD, DeCastro J (1999) Differing perceptions of new venture failure: a matched exploratory study of venture capitalists and entrepreneurs. Journal of Small Business Management 37(3):1–14

5

Type-2 fuzzy set extension of DEMATEL method combined with perceptual computing for decision making

Mitra Bokaei Hosseini and Mohammad Jafar Tarokh[*]

Abstract

Most decision making methods used to evaluate a system or demonstrate the weak and strength points are based on fuzzy sets and evaluate the criteria with words that are modeled with fuzzy sets. The ambiguity and vagueness of the words and different perceptions of a word are not considered in these methods. For this reason, the decision making methods that consider the perceptions of decision makers are desirable. Perceptual computing is a subjective judgment method that considers that words mean different things to different people. This method models words with interval type-2 fuzzy sets that consider the uncertainty of the words. Also, there are interrelations and dependency between the decision making criteria in the real world; therefore, using decision making methods that cannot consider these relations is not feasible in some situations. The Decision-Making Trail and Evaluation Laboratory (DEMATEL) method considers the interrelations between decision making criteria. The current study used the combination of DEMATEL and perceptual computing in order to improve the decision making methods. For this reason, the fuzzy DEMATEL method was extended into type-2 fuzzy sets in order to obtain the weights of dependent criteria based on the words. The application of the proposed method is presented for knowledge management evaluation criteria.

Keywords: DEMATEL; Perceptual computing; Decision making; Interval type-2 fuzzy sets (IT2 FSs)

Introduction

Many decision making methods are being proposed to facilitate the decision making process. Decision making problems consist of several criteria, and each criterion is evaluated by some other subcriteria. The evaluation criteria are almost dependent based on the complexity and vagueness of the real world. Therefore, decision making methods that consider these interrelations between criteria are more desirable. The Decision-Making Trail and Evaluation Laboratory (DEMATEL) was proposed by the Battelle Memorial Institute through its Geneva Research Centre (Gabus and Fontela 1973). This method considers the causal relationships between criteria and illustrates the weights between criteria by diagraphs.

Lin and Wu (2004) proposed a fuzzy extension of the DEMATEL method. The judges are based on linguistic variables and triangular fuzzy numbers, and the final weights of criteria are crisp numbers. In their approach all decision makers used a specified linguistic variable that may have different meanings for them based on the vagueness of each word. Words mean different things to different people, so they are uncertain (Mendel and Wu 2010). After Zadeh (1965) introduced a fuzzy set theory to deal with vague problems, in which linguistic labels have been used within the framework of the fuzzy set theory. After he introduced the type-2 fuzzy sets (T2 FSs), the first concept of the fuzzy set was renamed to type-1 fuzzy sets (T1 FSs) (Zadeh 1965, 1975). The main difference between these two types is that the memberships of T1 FSs are crisp numbers, whereas the membership functions of T2 FSs are T1 FSs. The latter type has a sense of uncertainty. Zadeh (1999) proposed the paradigm of computing with words based on the T2 FSs that is a methodology in which the objects of computation are words and propositions drawn from a natural language. Mendel (2001, 2002, 2007) proposed the framework for perceptual computing based on computing with words. Words were the enabler of the

---
* Correspondence: mjtarokh@kntu.ac.ir
Department of IT, Faculty of Industrial Engineering, K. N. Toosi University of Technology, Tehran, 1631714191, Iran

perceptual computer; therefore, it could consider the uncertainty related to each word based on interval type-2 fuzzy sets (IT2 FSs). However, in perceptual computing, criteria were considered independent.

Therefore, the aim of this study was the IT2 FS extension of the DEMATEL method in order to obtain the criteria's weights based on the words. For this reason, perceptual computing was combined with the DEMATEL method to overcome the problem of interrelations between criteria in perceptual computing. The weights obtained from this study can be further used in perceptual computing judgments. The rest of this paper is organized as follows: In the 'Type-1 fuzzy DEMATEL method' section, we described the concepts of fuzzy DEMATEL. In the 'Interval type-2 fuzzy sets used in perceptual computing' section, a background about the IT2 FSs used in perceptual computing is represented. The IT2 FS extension of DEMATEL is proposed in the 'IT2 FSs DEMATEL method' section. In the 'Application of proposed method in defining weights for dependent criteria' section, an empirical study is illustrated to demonstrate that the proposed method is useful. Discussions are presented in the next section, and the conclusion is presented in the last section.

Type-1 fuzzy DEMATEL method

The DEMATEL method had been used successfully in many decision making problems. Also, many researchers used this method in combination with another multicriteria decision analysis (MCDM) method. For example, Jassbi et al. (2011) used the fuzzy DEMATEL method for modeling the cause and effect relationship of strategy map. Chang et al. (2011) used the fuzzy DEMATEL method for developing supplier selection criteria. Yang and Tzeng (2011) proposed a combined MCDM model based on DEMATEL and analytic network process (ANP). Also, Chen and Chen (2010) used DEMATEL, fuzzy ANP, and TOPSIS for evaluating innovation performance in Taiwanese higher education institutes.

Lin and Wu (2004) proposed their fuzzy DEMATEL method as a stepwise procedure:

1. Step 1: Identify the decision goal and form a committee.
2. Step 2: Develop evaluation criteria and design the fuzzy linguistic scale. Lin and Wu (2004) used fuzzy triangular numbers to propose the fuzzy DEMATEL method. They used five linguistic terms as {very high influence, high influence, low influence, very low influence, no influence}. These linguistic terms are shown in Table 1.
3. Step 3: Acquire and average the assessments of P decision makers. Every decision maker is asked to make pair-wise relationships between each pair of objects. Therefore, P fuzzy matrices $\tilde{Z}^1, \tilde{Z}^2, ..., \tilde{Z}^P$

Table 1 The correspondence of linguistic terms and linguistic values

Linguistic terms	Linguistic values (TFN)
Very high influence	(0.75,1.0,1.0)
High influence	(0.5,0.75,1.0)
Low influence	(0.25,0.5,0.75)
Very low influence	(0,0.25,0.5)
No influence	(0,0,0.25)

with triangular fuzzy numbers are obtained that show the pair-wise comparison of the objects based on the decision makers' perceptions. Equation 1 is then used to calculate the average matrix $\tilde{Z}$:

$$\tilde{Z} = \frac{\left(\tilde{Z}^1 \oplus \tilde{Z}^2 \oplus ... \tilde{Z}^P\right)}{P}. \tag{1}$$

The fuzzy matrix $\tilde{Z}$ is called the initial direct-relation fuzzy matrix as shown in Equation 2:

$$\tilde{Z} = \begin{bmatrix} 0 & \tilde{Z}_{12} ... & \tilde{Z}_{1n} \\ \tilde{Z}_{21} & 0 \quad 0 & \tilde{Z}_{2n} \\ \vdots & \vdots \quad \ddots & \vdots \\ \tilde{Z}_{n1} & \tilde{Z}_{n2} ... & 0 \end{bmatrix}. \tag{2}$$

In this matrix, $\tilde{Z}_{ij} = \left(l_{ij}, m_{ij}, u_{ij}\right)$ are triangular fuzzy numbers, and $\tilde{Z}_{ij}(i = 1, 2, ..., n)$ will be regarded as triangular fuzzy number (0, 0, 0) whenever is necessary (Jassbi et al. 2011).

4. Step 4: Normalizing initial direct-relation fuzzy matrix $\tilde{X}$ by Equation 4. The linear scale transformation is used as a normalization formula to transform the criteria scales into comparable scales (Lin and Wu 2004). Suppose $\tilde{a}_i$ shows each triangular fuzzy number in each cell of $\tilde{Z}_{ij}$ and suppose that r is the maximum summation of the third element of each triangular fuzzy number in each row in Equation 3. As in the crisp DEMATEL method, Lin and Wu (2004) assumed at least one i ($1 \le i \le n$) such that $\sum_{j=1}^{n} u_{ij} < r$. They claimed that this assumption is well satisfied in practical cases:

$$\tilde{a}_i = \sum_{j=1}^{n} \tilde{z}_{ij} = \left(\sum_{j=1}^{n} l_{ij}, \sum_{j=1}^{n} m_{ij}, \sum_{j=1}^{n} u_{ij}\right), r = \max_{1 \le i \le n}\left(\sum_{j=1}^{n} u_{ij}\right) \tag{3}$$

$$\tilde{X} = \begin{bmatrix} \tilde{X}_{11} & \tilde{X}_{12} ... & \tilde{X}_{1n} \\ \tilde{X}_{21} & \tilde{X}_{22} ... & \tilde{X}_{2n} \\ \vdots & \vdots \quad \ddots & \vdots \\ \tilde{X}_{n1} & \tilde{X}_{n2} ... & \tilde{X}_{nn} \end{bmatrix} \text{ where} \tag{4}$$

$$\tilde{x}_{ij} = \frac{\tilde{z}_{ij}}{r} = \left(\frac{l_{ij}}{r}, \frac{m_{ij}}{r}, \frac{u_{ij}}{r}\right).$$

Let $\tilde{x}_{ij} = \left(l'_{ij}, m'_{ij}, u'_{ij}\right)$ and define three crisp matrices, whose elements are extracted from $\tilde{X}$, as follows:

$$X_l = \begin{bmatrix} 0 & l'_{12} & \cdots & l'_{ln} \\ l'_{21} & 0 & 0 & l'_{2n} \\ \vdots & \vdots & \ddots & \vdots \\ l'_{n1} & l'_{n2} & \cdots & 0 \end{bmatrix};$$ (5)

$$X_m = \begin{bmatrix} 0 & m'_{12} & \cdots & m'_{ln} \\ m'_{21} & 0 & 0 & m'_{2n} \\ \vdots & \vdots & \ddots & \vdots \\ m'_{n1} & m'_{n2} & \cdots & 0 \end{bmatrix};$$

$$X_u = \begin{bmatrix} 0 & u'_{12} & \cdots & u'_{1n} \\ u'_{21} & 0 & 0 & u'_{2n} \\ \vdots & \vdots & \ddots & \vdots \\ u'_{n1} & u'_{n2} & \cdots & 0 \end{bmatrix}$$

5. Step 5: Compute total-relation fuzzy matrix $\tilde{T}$. Matrix $\tilde{X}$ was computed in the previous step. Based on the crisp DEMATEL method, total-relation fuzzy matrix $\tilde{T}$ can be computed through Equation 6:

$$\tilde{T} = \lim_{k \to \infty} \left(\tilde{X}^1 + \tilde{X}^2 + \cdots + \tilde{X}^k\right).$$ (6)

The elements of matrix $\tilde{T}$ contain triangular fuzzy numbers as shown in Equation 7:

$$\tilde{T} = \begin{bmatrix} \tilde{t}_{11} & \tilde{t}_{12} & \cdots & \tilde{t}_{1n} \\ \tilde{t}_{21} & \tilde{t}_{22} & \cdots & \tilde{t}_{2n} \\ \vdots & \vdots & \ddots & \vdots \\ \tilde{t}_{n1} & \tilde{t}_{n2} & \cdots & \tilde{t}_{nn} \end{bmatrix}, \text{ in which}$$ (7)

$\tilde{t}_{ij} = \left(l''_{ij}, m''_{ij}, u''_{ij}\right)$ and

$[l''_{ij}] = X_l \times \left(I - X_l^{-1}\right),$

$[m''_{ij}] = X_m \times \left(I - X_m^{-1}\right), [u''_{ij}] = X_u \times \left(I - X_u^{-1}\right).$

After acquiring matrix $\tilde{T}$, the next step is to calculate the $\tilde{D}_i + \tilde{R}_i$ and $\tilde{D}_i - \tilde{R}_i$, where $\tilde{D}_i$ and $\tilde{R}_i$ are the sum of the rows and the sum of the columns of $\tilde{T}$ (Lin and Wu 2004). To acquire the importance of the criteria and understand the causal relationship between criteria, $\tilde{D}_i + \tilde{R}_i$ and $\tilde{D}_i - \tilde{R}_i$ should be defuzzified. The $\left(\tilde{D}_i + \tilde{R}_i\right)^{\mathrm{def}}$ shows the relative importance of criterion i, and the $\left(\tilde{D}_i - \tilde{R}_i\right)^{\mathrm{def}}$ demonstrates the causal relationship. If the value of $\left(\tilde{D}_i - \tilde{R}_i\right)^{\mathrm{def}}$ is positive, the criterion belongs to the cause group, and if the value of $\left(\tilde{D}_i - \tilde{R}_i\right)^{\mathrm{def}}$ is negative, the criterion belongs to the effect group.

Interval type-2 fuzzy sets used in perceptual computing

The fuzzy extension of the DEMATEL method used linguistic terms for generating the initial direct-relation matrix. Therefore, decision makers are asked to compare the decision making criteria based on the codebook of words, e.g., Table 1. Zadeh (1999) proposed the paradigm of computing with words based on the T2 FSs that is a methodology in which the objects of computation are words and propositions drawn from a natural language. Computing with words is fundamentally different from the traditional expert systems which are simply tools to realize an intelligent system, but are not able to process natural language which is imprecise, uncertain, and partially true. As mentioned before, words mean different things to different people, so they are uncertain. Words in computing with words are modeled by T2 FSs that can model more uncertainties (Mendel and Wu 2007). Mendel and Wu (2010) used computing with words for making subjective judgments which was called perceptual computing. A perceptual computer consists of three parts: encoder (using interval approach (IA)), linguistic weighted average (LWA) engine, and decoder (Mendel 2001, 2002, 2007). Each part of the perceptual computer was utilized in the IT2 FS extension of the DEMATEL method.

In order to obtain an IT2 FS model for a word, IA was proposed by Mendel and Wu (2008). This approach had been referred to as T2 fuzzistics (Mendel 2007). In this approach, all decision makers ($i = 1, 2,..., n$) provide the end points of an interval associated with a word. The intervals need to be between 0 and 10. The mean and standard deviation are then computed for the end points. The intervals show the level of uncertainty associated to each word. This approach maps each evaluator's data interval into a prespecified T1 membership function (MF) and interprets the latter as an embedded T1 FS of an IT2 FS.

In this section mathematical definitions of IT2 FS are presented that is used in the rest of the paper.

An IT2 FS $\tilde{A}$ is characterized by the MF $\mu_{\tilde{A}}(x, u)$, where $x \in X$ and $u \in J_x \subseteq [0, 1]$, that is (Mendel and Wu 2010),

$$\tilde{A} = \left\{ \left((x, u), \mu_{\tilde{A}}(x, u) = 1\right) | \forall x \in X, \forall u \in J_x \subseteq [0, 1]\right\}.$$ (8)

Equation 8 can be expressed as (Mendel 2001; Mendel and John 2002)

$$\tilde{A} = \int_{x \in X} \int_{u \in J_x \subseteq [0,1]} 1/(x, u) = \int_{x \in X} \left[\int_{u \in J_x \subseteq [0,1]} 1/u\right]/x,$$ (9)

where x is called the primary variable with the domain of X. $J_x \subseteq [0, 1]$ is the primary membership of x, u is the secondary variable, and $\int_{u \in J_x} 1/u$ is the secondary MF at x. Note that Equation 9 means $\tilde{A} : X \rightarrow \{[a,b] : 0 \le a \le b \le 1\}$. Uncertainty about $\tilde{A}$ is conveyed by the union of all of the primary memberships, called the footprint of uncertainty of $\tilde{A}$ $(FOU(\tilde{A}))$, i.e., (Wu and Mendel 2007)

$$FOU(\tilde{A}) = \bigcup_{x \in X} J_x = \left\{ (x,y) : y \in J_x = \left[\underline{A}(x), \bar{A}(x) \right] \subseteq [0,1] \right\}.$$

(10)

An IT2 FS is shown in Figure 1. The FOU is shown as the shaded region. It is bounded by an upper MF (UMF), $\tilde{A}(x) \equiv \tilde{A}$, and a lower MF (LMF), $\underline{A}(x) \equiv \underline{A}$, both of which are type-1 fuzzy sets; consequently, the membership grade of each element of IT2 FS is an interval $[\underline{A}(x), \bar{A}(x)]$. It is also customary to use $\underline{\mu}_{\tilde{A}}(x)$ and $\bar{\mu}_{\tilde{A}}(x)$ for the LMF and UMF of $\tilde{A}$ (Mendel and Wu 2010):

$$FOU(\tilde{A}) = \bigcup_{\forall x \in X} \left[\underline{\mu_{\tilde{A}}}(x), \overline{\mu_{\tilde{A}}}(x) \right],$$

(11)

so $\tilde{A}$ can also be expressed in terms of its vertical slices as

$$\tilde{A} = {}^{1}/_{FOU(\tilde{A})}.$$

(12)

For discrete universe of discourse X and U, the embedded type-1 fuzzy set A_e has N elements, one each from $J_{x1}, J_{x2},..., J_{xN}$, namely $u_1, u_2,..., u_N$, i.e., (Wu and Mendel 2007)

$$A_e = \sum_{i=1}^{N} u_i/x_i \qquad u_i \in J_{xi} \subseteq U = [0,1].$$

(13)

The UMF and LMF of $\tilde{A}$ are two type-1 MFs that bound the FOU. UMF$(\tilde{A})$ is associated with the upper bound of FOU$(\tilde{A})$ and is denoted $\bar{\mu}_{\tilde{A}(x)}$, $\forall x \subseteq X$, and

LMF$(\tilde{A})$ is associated with the lower bound of FOU$(\tilde{A})$ and is denoted $\underline{\mu}_{\tilde{A}(x)}$, $\forall x \in X$, that is,

$$UMF(\tilde{A}) \equiv \bar{\mu}_{\tilde{A}}(x) = \overline{FOU(\tilde{A})} \quad \forall x \in X,$$

(14)

$$LMF(\tilde{A}) \equiv \underline{\mu}_{\tilde{A}}(x) = \underline{FOU(\tilde{A})} \quad \forall x \in X.$$

(15)

UMF contains four digits and LMF contains five digits, of which the fifth parameter is its height. Let $FOU(\tilde{A}) = \bigcup_{\forall \alpha} \left[\left[\bar{a}_1^\alpha, \underline{a}_1^\alpha \right], \left[\underline{a}_2^\alpha, \bar{a}_2^\alpha \right] \right]$ and $FOU(\tilde{B}) = \bigcup_{\forall \alpha} \left[\left[\underline{b}_1^\alpha, b_1^\alpha \right], \left[b_2^\alpha, \underline{b}_2^\alpha \right] \right]$ be the perfectly normal IT2 FN based on Equation 11 (Hamrawi and Coupland 2009; Kaufmann and Gupta 1985), and then according to Wu and Mendel (2008),

$$FOU(\tilde{A}) \circ FOU(\tilde{B}) = \begin{cases} \bigcup_{\forall \alpha} \alpha. \left(\left[[\bar{a}_1^\alpha \underline{a}_1^\alpha] \cdot [\bar{b}_1^\alpha \underline{b}_1^\alpha], [\underline{a}_2^\alpha, \bar{a}_2^\alpha] \cdot [\underline{b}_2^\alpha, \bar{b}_2^\alpha] \right] \right), \\ \qquad \text{if } 0 \le \alpha \le \min\left(h_{\tilde{A}}, h_{\tilde{B}} \right) \\ \bigcup_{\forall \alpha} \alpha. \left([\bar{a}_1^\alpha, \bar{a}_2^\alpha]^\circ [\bar{b}_1^\alpha \bar{b}_2^\alpha] \right), \\ \qquad \text{if } \min\left(h_{\tilde{A}}, h_{\tilde{B}} \right) < \alpha \le 1 \end{cases}$$

(16)

where $\circ = \{+, -, \times, \div\}$.

IT2 FS DEMATEL method

The procedure of developing the DEMATEL method by IT2 FSs is as follows:

1 Step 1: Identify the decision goals, criteria, and group of experts.
2 Step 2: Develop linguistic codebooks for decision making. In this step a codebook is designed, and decision makers are asked to define the interval end points for each word in the codebook. The codebook has the same words as in Table 1. Therefore, the codebook of words contains 'very high influence', 'high influence', 'low influence', 'very low influence', and 'no influence' . The IA is used to map these intervals into IT2 FSs (Mendel and Wu 2008). The DEMATEL method does not consider the difference between the levels of expertise for each expert, but in this paper we developed another codebook that considers the level of expertise for each expert. This codebook contains three words (low, moderate, and high) (Mendel and Wu 2007). Also, it is possible to put equal weights to the level of expertise for each expert.
3 Step 3: Acquire and compute the linguistic weighted average of the assessments. To measure the weights and causal relations between the criteria $C = \{C_i| \ i = 1, 2,..., n\}$, a group of p experts are asked to define

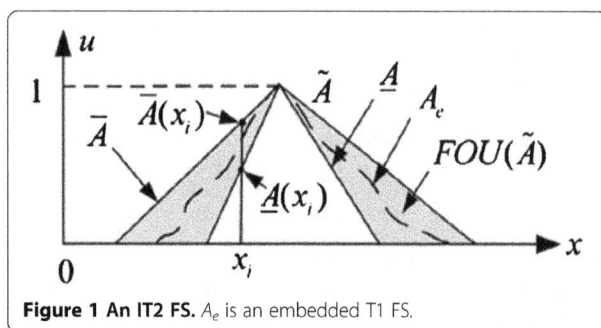

Figure 1 An IT2 FS. A_e is an embedded T1 FS.

Table 2 FOU data for all words in the influence codebook

Word	UMF	LMF
No influence	[0,0,0.137628,1.974745]	[0,0,0.091752,1.316497,1]
Very low influence	[0.37868,2,2.5,4.62132]	[0.585786,2.212445,2.212445,3.414214,0.849779]
Low influence	[2.37868,3.5,4.5,6.62132]	[2.792893,3.792893,3.792893,4.207107,0.585786]
High influence	[4.708759,7.770621,10,10]	[5.05051,8.724745,10,10,1]
Very high influence	[7.367007,9.816497,10,10]	[8.683503,9.908248,10,10,1]

The fifth parameter for the LMF is its height.

the influence relation between criteria based on the codebooks in step 2. Therefore, p pair-wise comparison IT2 FSs matrices $\tilde{Z}^1, \tilde{Z}^2, ..., \tilde{Z}^P$ are obtained. LWA that was proposed by Mendel and Wu (2007) was used to generate the IT2 FS average matrix that is called initial-direct-relation IT2 FS matrix.

In the previous section, we used the IA to encode each word from the codebook to an IT2 FS. The output of the previous section is used to activate the LWA. Each decision maker used a word from the codebook to transfer the influence of each criterion on another one. Each decision maker had a level of expertise that was assigned to him/her from a codebook of expertise weights that contained three words: 'low,' 'moderate,' and 'high.' Decision makers were asked to define the end points of an interval on the scale of 0 to 10 for each word in the codebook. Then the IA is used to encode the intervals into IT2 FSs. The LWA maps IT2 FSs into IT2 FSs. This method is based on the weighted average that is the most widely used form of aggregation. Suppose k is the number of decision makers ($k = 1, 2,..., p$) and $\tilde{z}_{ij}$ is ijth entry of initial-direct-relation IT2 FS matrix $\tilde{Z}$. The LWA matrix $\tilde{Z}$ can be obtained from Equation 17:

$$\tilde{z}_{ij} = \frac{\sum_{k=1}^{p} \tilde{Z}^k{}_{ij} \tilde{W}_k}{\sum_{k=1}^{p} \tilde{W}_k}. \quad (17)$$

4 Step 4: Establish the normalized initial-direct-relation matrix.

Let $\tilde{z}_{ij} = \left(\mathrm{UMF}(\tilde{Z}), \mathrm{LMF}(\tilde{Z})\right)$, and $\mathrm{UMF}(\tilde{Z}) = (a, b, c, d)$ and $\mathrm{LMF}(\tilde{Z}) = (e, f, g, i, h)$, of which the fifth element is its height. Therefore, $\tilde{z}_{ij}$ can be defined by nine matrices, whose elements are crisp numbers (Liu and Mendel 2008):

$$Z_a = \begin{bmatrix} 0 & a'_{12} & \cdots & a'_{1n} \\ a'_{21} & 0 & \cdots & a'_{2n} \\ \vdots & \vdots & \ddots & \vdots \\ a'_{n1} & a'_{n2} & \cdots & 0 \end{bmatrix},$$

$$Z_b = \begin{bmatrix} 0 & b'_{12} & \cdots & b'_{1n} \\ b'_{21} & 0 & \cdots & b'_{2n} \\ \vdots & \vdots & \ddots & \vdots \\ b'_{n1} & b'_{n2} & \cdots & 0 \end{bmatrix}, ...,$$

$$Z_h = \begin{bmatrix} 0 & h'_{12} & \cdots & h'_{1n} \\ h'_{21} & 0 & \cdots & h'_{2n} \\ \vdots & \vdots & \ddots & \vdots \\ h'_{n1} & h'_{n2} & \cdots & 0 \end{bmatrix}.$$

Z_d contains the forth element of $\mathrm{UMF}(\tilde{Z})$. All $\tilde{z}_{ij}$ are normal IT2 FSs; therefore, Z_d contains the greatest elements in the initial-direct-relation matrix. The normalized direct-relation matrix can be defined as

$$\tilde{X} = \begin{bmatrix} \tilde{x}_{11} & \tilde{x}_{12} & \cdots & \tilde{x}_{1n} \\ \tilde{x}_{21} & \tilde{x}_{22} & \cdots & \tilde{x}_{2n} \\ \vdots & \vdots & \ddots & \vdots \\ \tilde{x}_{n1} & \tilde{x}_{n2} & \cdots & \tilde{x}_{nn} \end{bmatrix} \text{ where}$$

$$\tilde{x}_{ij} = \frac{\tilde{z}_{ij}}{s} = \left(\frac{Z_a}{s}, \frac{Z_b}{s}, \frac{Z_c}{s}, \frac{Z_d}{s}, \frac{Z_e}{s}, \frac{Z_f}{s}, \frac{Z_g}{s}, \frac{Z_i}{s}, Z_h \right)$$

$$s = \max_{1 \le i \le n} \left(\sum_{j=1}^{n} X_{d_{ij}} \right). \quad (18)$$

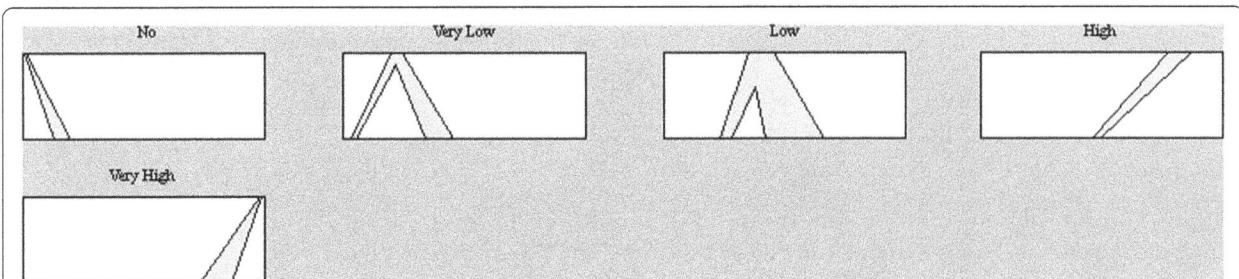

Figure 2 FOUs for the five words in the influence codebook.

Table 3 FOU data for all words in the expertise weight codebook

Word	UMF	LMF
Low	[0.085786,1.5,3,4.62132]	[1.792893,2.280847,2.280847,2.81066,0.404234]
Moderate	[3.585786,4.75,5.5,6.914214]	[4.858579,5.034231,5.034231,5.141421,0.273849]
High	[5.982233,7.75,8.6,9.517767]	[8.034315,8.357323,8.357323,9.165685,0.571004]

The fifth parameter for the LMF is its height.

Note that the fifth element of $\text{LMF}(\tilde{Z})$ (height) is normalized between 0 and 1; therefore, there is no need to normalize this element.

5 Step 5: Compute the total-relation IT2 FS matrix $\tilde{T}$. To compute the total-relation IT2 FS matrix $\tilde{T}$, we have to ensure the convergence of $\lim_{l \to \infty} \tilde{X}^l = 0$. The elements of $\tilde{X}^l$ are also IT2 FSs. $\tilde{X}$ can be defined by nine matrices, and the elements of these matrices are all crisp numbers.

Theorem 1. *Let*

$$\tilde{X}^l = \begin{bmatrix} \tilde{x}^l_{11} & \tilde{x}^l_{12} \cdots & \tilde{x}^l_{1n} \\ \tilde{x}^l_{21} & \tilde{x}^l_{22} \cdots & \tilde{x}^l_{2n} \\ \vdots & \vdots \ddots & \vdots \\ \tilde{x}^l_{n1} & \tilde{x}^l_{n2} \cdots & \tilde{x}^l_{nn} \end{bmatrix} \text{ where}$$

$$\tilde{x}^l_{ij} = \left(a^l_{ij}, b^l_{ij}, c^l_{ij}, d^l_{ij}, e^l_{ij}, f^l_{ij}, g^l_{ij}, i^l_{ij}, h^l_{ij} \right),$$

and further define eight matrices. There is no need to consider the ninth matrix that contains the heights of $LMF(\tilde{X})$.

$$\tilde{X}^l_a = \begin{bmatrix} \tilde{a}^l_{11} & \tilde{a}^l_{12} \cdots & \tilde{a}^l_{1n} \\ \tilde{a}^l_{21} & \tilde{a}^l_{22} & \tilde{a}^l_{2n} \\ \vdots & \vdots \ddots & \vdots \\ \tilde{a}^l_{n1} & \tilde{a}^l_{n2} & \tilde{a}^l_{nn} \end{bmatrix},$$

$$\tilde{X}^l_b = \begin{bmatrix} \tilde{b}^l_{11} & \tilde{b}^l_{12} & \tilde{b}^l_{1n} \\ \tilde{b}^l_{21} & \tilde{b}^l_{22} & \tilde{b}^l_{2n} \\ \vdots & \vdots \ddots & \vdots \\ \tilde{b}^l_{n1} & \tilde{b}^l_{n2} & \tilde{b}^l_{nn} \end{bmatrix}, ...,$$

$$\tilde{X}^l_i = \begin{bmatrix} \tilde{i}^l_{11} & \tilde{i}^l_{12} & \tilde{i}^l_{1n} \\ \tilde{i}^l_{21} & \tilde{i}^l_{22} & \tilde{i}^l_{2n} \\ \vdots & \vdots \ddots & \vdots \\ \tilde{i}^l_{n1} & \tilde{i}^l_{n2} & \tilde{i}^l_{nn} \end{bmatrix}.$$

Proof. The proof is straightforward; all the eight matrices contain crisp values, and the matrix multiplication is used to prove this theorem. Lin and Wu (2004) proved $\lim_{l \to \infty} \tilde{X}^l_u = O$ and $\lim_{l \to \infty} \left(I + X_u + X_u^2 + ... + X_u^l \right) = (1 - X_u)^{-1}$ based on $\sum_{j=1}^{n} X_{u_{ij}} < s$ for triangular fuzzy sets. We used this theorem for IT2 FS matrix $\tilde{T}$. Therefore, $\lim_{l \to \infty} \tilde{X}^l_d = O$ and $\lim_{l \to \infty} \left(I + X_d + X_d^2 + ... + X_d^l \right) = (1 - X_d)^{-1}$ based on $\sum_{j=1}^{n} X_{d_{ij}} < s$ and $\tilde{T} = \lim_{l \to \infty} \left(I + \tilde{X} + \tilde{X}^2 + ... + \tilde{X}^l \right)$. Then the total-relation matrix $\tilde{T}$ is acquired as follows:

$$\tilde{T} = \begin{bmatrix} \tilde{t}_{11} & \tilde{t}_{12} \cdots & \tilde{t}_{1n} \\ \tilde{t}_{21} & \tilde{t}_{22} \cdots & \tilde{t}_{2n} \\ \vdots & \vdots \ddots & \vdots \\ t_{n1} & \tilde{t}_{n2} \cdots & \tilde{t}_{nn} \end{bmatrix} \text{ where}$$

$$\tilde{t}_{ij} = \left(a^n_{ij}, b^n_{ij}, c^n_{ij}, d^n_{ij}, e^n_{ij}, f^n_{ij}, g^n_{ij}, i^n_{ij}, h^n_{ij} \right); \text{ then}$$

$$\left[a^n_{ij} \right] = X_a \times (1 - X_a)^{-1}, \quad (19)$$

$$\left[b^n_{ij} \right] = X_b \times (1 - X_b)^{-1}, ...,$$

$$\left[i^n_{ij} \right] = X_i \times (1 - X_i)^{-1}, \left[h^n_{ij} \right] = X_h.$$

To acquire the importance weight of each criterion, we calculated $\tilde{D}_i + \tilde{R}_i$, where $\tilde{D}_i$ shows the sum of the rows and $\tilde{R}_i$ shows the sum of the columns of the total-relation matrix $\tilde{T}$ and can be obtained through Equations 20 and 21:

$$D_i = \sum_{j=1}^{n} t_{ij} \quad (i = 1, 2, ..., n), \quad (20)$$

$$R_j = \sum_{i=1}^{n} t_{ij} \quad (j = 1, 2, ..., n). \quad (21)$$

Note that in Equations 20 and 21, t_{ij}, $i, j = 1, 2, ..., n$ are IT2 FS, and their addition must be based on Equation 16.

6 Step 6: Decode each IT2 FS into a word.

In the previous step, we calculated the weights for each criterion, but these weights are IT2 FSs and must be decoded into words. This process is called the decoder. The IT2 FSs obtained from the previous step were decoded into seven words: 'extremely low,' 'very low', 'low', 'fair', 'high', 'very high', and 'extremely high' . A decoding codebook is needed to store the FOUs for these seven words. Therefore, IA

Table 4 The pair-wise comparison matrix $\tilde{Z}^1$ for one of the decision makers

	C1	C2	C3	C4	C5	C6
C1	-	Low	Low	Low	Very high	Very high
C2	High	-	Low	Low influence	Very high	Very low
C3	High	High	-	High	High	High
C4	Low	Low	High	-	High	High
C5	High	Very high	Low	Low influence	-	Very high
C6	High	High	High	High	Low	-

Table 5 The initial direct-relation IT2 FS matrix $\tilde{Z}$

	C1	C6
C1	[0,0,0,0,0,0,0,0,0,0.2738]	[2.7652,5.6223,6.5244,8.1632,4.5201,5.0644,6.5135,6.8002,0.2738]
C2	[4.7088,7.7706,10,10,5.0505,6.0567,10,10,0.2738]	[0.3787,2,2.5,4.6213,0.5858,1.11,3.0269,3.4142,0.2738]
C3	[4.7088,7.7706,10,105.0505,6.0567,10,10,0.2738]	[4.7088,7.7706,10,10,5.0505,6.0567,10,10,0.2738]
C4	[1.0617,2.6951,3.5732,5.9383,1.6581,2.1852,3.5202,3.8219,0.2738]	[4.7088,7.7706,10,10,5.0505,6.0567,10,10,0.2738]
C5	[1.8574,4.6742,6.5244,8.1632,2.7550,3.5833,6.5135,6.8002,0.2738]	[7.3670,908165,10,10,8.6835,9.0189,10,10,0.2738]
C6	[3.1744,5.4791,7.4512,8.8462,3.88984.6585,7.0067,7.1855,0.2738]	[0,0,0,0,0,0,0,0,0.2738]

is used to map the intervals collected from the group of decision makers into IT2 FSs. In order to get the criteria weights based on the words in the codebook, the decoder must compare the similarity between two IT2 FSs so that the output of step 5 can be mapped into its most similar word in the codebook. These weights that are based on the words can be further used in the evaluation based on perceptual computing. Several similarity measures are introduced for IT2 (Bustince 2000; Gorzalczany 1987; Mitchell 2005; Wu and Mendel 2008, 2009). In this study we used the Jaccard similarity measure for IT2 FSs. This approach uses average cardinality. Equation 22 is used to calculate the Jaccard similarity measure for IT2 FSs.

To decode the IT2 FSs obtained from $\tilde{D}_i + \tilde{R}_i$, IT2 FSs must be mapped into [0,10]. For this reason, we used the min-max normalization method defined in Equation 23:

$$
\mathrm{sm}_j(\tilde{A}, \tilde{B}) = \frac{\sum_{i=1}^{N} \min(\bar{\mu}_{\tilde{A}}(x_i), \bar{\mu}_{\tilde{B}}(x_i)) + \sum_{i=1}^{N} \min\left(\underline{\mu}_{\tilde{A}}(x_i), \underline{\mu}_{\tilde{B}}(x_i)\right)}{\sum_{i=1}^{N} \max(\bar{\mu}_{\tilde{A}}(x_i), \bar{\mu}_{\tilde{B}}(x_i)) + \sum_{i=1}^{N} \max\left(\underline{\mu}_{\tilde{A}}(x_i), \underline{\mu}_{\tilde{B}}(x_i)\right)},
$$

(22)

$$
v' = \frac{v - \min A}{\max A - \min A} (new-\max A - new - \min A) + new - \min A.
$$

(23)

In this approach we acquired the criteria weights based on the interrelations between criteria. Further,

these weights can be used for evaluation based on the perceptual computing method. The weights used in perceptual computing were independent, but this study helped to extend perceptual computing using dependent criteria and defining weights for each of them.

Application of proposed method in defining weights for dependent criteria

We used the proposed method to define weights of criteria that were used to evaluate the knowledge management capability of organization.

1 Step 1: Identify the decision goals, criteria, and group of experts.
   For evaluating the knowledge management capability of organization based on perceptual computing, we had to define the weights for each criterion. Perceptual computing considers each criterion independent from the others. For this reason, the DEMATEL method was used to define the weights for criteria that were dependent and had interrelations. A group of three knowledge management experts were asked to compare the criteria. Six criteria were chosen for this reason including vision for change, culture, structure, infrastructure, support for change, and knowledge management processes.

2 Step 2: Develop linguistic codebooks for decision making.
   The codebook of words that was used for comparing the influence of criteria on each other contained 'very high influence,' 'high influence,' 'low influence,' 'very low influence,' and 'no influence.' The IA is

Table 6 The normalized direct-relation IT2 FS matrix $\tilde{X}$

	C1	C6
C1	[0,0,0,0,0,0,0,0,0,0.2738]	[0.0602,0.1224,0.1420,0.1777,0.0984,0.1102,0.1418,0.1480,0.2738]
C2	[0.1025,0.1692,0.2177,0.2177,0.1099,0.1318,0.2177,0.2177, 0.2738]	[0.0082,0.0435,0.0544,0.1006,0.0128,0.0242,0.0659,0.0743, 0.2738]
C3	[0.1025,0.1692,0.2177,0.2177,0.1099,0.1318,0.2177,0.2177, 0.2738]	[0.1025,0.1692,0.2177,0.2177,0.1099,0.1318,0.2177,0.2177,0.2738]
C4	[0.0231,0.0587,0.0778,0.1293,0.0961,0.0476,0.0766,0.0832, 0.2738]	[0.1025,0.1692,0.2177,0.21770.1099,0.1318,0.2177,0.2177,0.2738]
C5	[0.4004,0.1017,0.1420,0.1777,0.06,0.0780,0.1418,0.1480, 0.2738]	[0.1604,0.2137,0.2177,0.2177,0.1890,0.1963,0.2177,0.2177,0.2738]
C6	[0.0691,0.1193,0.1622,0.1926,0.0847,0.1014,0.1525,0.1564, 0.2738]	[0,0,0,0,0,0,0,0,0.2738]

Table 7 The total-relation IT2 FS matrix $\tilde{T}$

	C1	C6
C1	[0.0296,0.1573,0.4131,1.2882,0.0546,0.0882,0.3857,0.4404,0.2738]	[0.0996,0.2912,0.5439,1.4252,0.1646,0.2083,0.5262,0.5869, 0.2738]
C2	[0.1209,0.2971,0.5852,1.4177,0.1467,0.1960,0.5745,0.6302,0.2738]	[0.0501,0.2217,0.4716,1.3223,0.0837,0.1274,0.4796,0.5437, 0.2738]
C3	[0.1322,0.453,0.7340,1.6814,0.1603,0.2193,0.7058,0.7679, 0.2738]	[0.1476,0.3778,0.7504,1.6691,0.1814,0.2432,0.7345,0.7973, 0.2738]
C4	[0.0636,0.2460,0.5767,1.5072,0.0980,0.1473,0.5489,0.6103, 0.2738]	[0.1481,0.3678,0.7029,1.5614,0.1809,0.2403,0.6889,0.7447, 0.2738]
C5	[0.0759,0.2645,0.5504,1.4334,0.1218,0.1718,0.5325,0.5946, 0.2738]	[0.1865,0.3692,0.6065,1.4447,0.2416,0.2807,0.6007,0.6598, 0.2738]
C6	[0.0939,0.2672,0.5968,1.5117,0.1260,0.1742,0.5651,0.6256, 0.2738]	[0.0343,0.1803,0.4656,1.3369,0.0585,0.0952,0.4519,0.5110, 0.2738]

used to map these intervals into IT2 FSs. The FOUs for each word are presented in Table 2, and also, Figure 2 depicts the FOUs for the five words in the codebook.

The codebook used for the expertise weight is shown in Table 3. This codebook contains three words (low, moderate, high).

3　Step 3: Compute the linguistic weighted average of the assessments.

To measure the weights of each criterion based on the interrelationship between the six criteria, three knowledge management experts were asked to compare the criteria based on the codebook defined in Table 2. Therefore, three pair-wise comparison matrices $\tilde{Z}^1, \tilde{Z}^2, \tilde{Z}^3$ are obtained. Table 4 shows the relative comparison matrix for one of the decision makers based on the codebook defined in Table 2. The average of these three matrices is obtained from LWA using Equation 17. To compute the LWA mentioned in Equation 17, decision makers' expertise weights should be defined. In this study we assume the equal expertise weights ('moderate') for the decision makers. The weights for decision makers' expertise are shown in Table 3. The result of LWA is initial direct-relation matrix $\tilde{Z}$ that is shown in Table 5.

4　Step 4: Establish the normalized initial-direct-relation matrix.

We used Equation 18 to normalize the initial direct-relation IT2 FS matrix. The result is shown in Table 6.

5　Step 5: Compute the total-relation IT2 FS matrix $\tilde{T}$. Equation 19 was used to compute the total-relation

IT2 FS matrix $\tilde{T}$. The result is shown in Table 7. $\tilde{D}_i + \tilde{R}_i$ can be computed from Equations 21 and 22. Table 8 shows the result of $\tilde{D}_i + \tilde{R}_i$ for each criterion's nine numbers. Each set of nine numbers shows an interval type-2 fuzzy set that can be drawn and also can be decoded to a codebook of words.

6　Step 6: Decode each IT2 FS into a word.

The weights for each criterion were calculated in step 5 based on interval type-2 fuzzy sets. These weights can further be used in perceptual computing without decoding them to words. In addition, the weights can be decoded to words. The weight codebook was needed to decode the IT2 FSs obtained from step 5 into words. For this reason, a group of 30 people including the main decision makers in step 1 were asked to define end point intervals for the seven words in the codebook; then, the IA was used to map these intervals into IT2 FSs. The FOUs for the weight codebook are shown in Table 9. Also, Figure 3 depicts the FOUs for all seven words used for the weight codebook. As mentioned before, we used the Jaccard similarity measure to decode the FOUs obtained from the previous step into words from the weight codebook. In order to use the Jaccard similarity measure, $\tilde{D}_i + \tilde{R}_i$ should be normalized in [0,1]. For this reason, we used Equation 23. After normalizing $\tilde{D}_i + \tilde{R}_i$, values of the Jaccard similarity measure can be used to decode the IT2 FSs weights into words based on the Jaccard similarity measure that is shown in Equation 22. The result of the decoder for each criterion is shown in the last column of Table 8.

Table 8 FOU data for all words in the weight codebook

Word	UMF	LMF
Extremely low	[0,0,0.137628,1.974745]	[0,0,0.045876,0.658248,1]
Very low	[0.085786,1,2,3.414214]	[0.896447,1.353553,1.353553,1.603553,0.414214]
Low	[0.982233,2.75,3.75,4.81066]	[2.792893,3.353553,3.353553,4.207107,0.585786]
Fair	[2.87868,4.5,5.25,7.12132]	[4.292893,4.818667,5.207107,0.549337]
High	[4.585786,6,7.05,8.414214]	[5.792893,6.514348,6.514348,7.207107,0.573901]
Very high	[6.585786,8,9,9.789949]	[8.292893,8.630602,8.630603,9.207107,0.477592]
Extremely high	[7.367007,9.816497,10,10]	[9.473401,9.963299,10,10,1]

The fifth parameter for the LMF is its height.

Table 9 The values of $\tilde{D}_i+\tilde{R}_i$ and the decoded weights of criteria

	$\tilde{D}_i+\tilde{R}_i$	Decode
C1	[0.0371,1.776,3.1255,9.3084,0.2876,0.5925,2.9490,3.3510,0.2738]	Low
C2	[0,1.1046,2.9316,8.9268,0.2499,0.5425,2.8212,3.2277,0.2738]	Very low
C3	[0.1808,1.4706,3.6779,10,0.4287,0.7739,3.5119,3.9275,0.2738]	Low
C4	[0.0060,1.0729,2.9632,8.9265,0.2190,0.5105,2.8278,3.2186,0.2738]	Very low
C5	[0.1719,1.4384,3.4240,9.5869,0.4757,0.7974,3.3081,3.7270,0.2738]	Low
C6	[0.1177,1.3616,3.4275,9.5949,0.3695,0.6949,3.3075,3.7258,0.2738]	Low

Discussion

The purpose of this study was to extend DEMATEL and combine it with perceptual computing in order to consider the interrelations between weights in perceptual computing. According to the results, the IT2 FS extension of DEMATEL and the combination of perceptual computing and DEMATEL lead to the weights of evaluation criteria based on the codebook of words. Perceptual computing was used for decision making and subjective judgments. Words are the enabler of perceptual computing, and in this subjective judgment, IT2 FSs are used to model the words' uncertainty. In order to obtain the weights of dependent criteria, a codebook of words for evaluating the influence of criteria on each other was presented. We applied the proposed method to obtain the weights of criteria for knowledge management evaluation by perceptual computing. Decision makers were asked to associate the end points of intervals to each word. Then the intervals collected for each word were modeled into IT2 FSs with the use of the interval approach. The words used for defining the influences and their related IT2 FSs are shown in Table 2. Decision makers were asked to define the influence of criteria on each other through matrices. Three influence matrices are defined in this paper for six criteria. The difference between the fuzzy DEMATEL proposed by Lin and Wu (2004) and our approach is the effect of expertise weights on the aggregation of influence matrices. The linguistic weighted average was used to aggregate these matrices. The aggregated matrix was based on the level of expertise that contained IT2 FSs and is presented in Table 5. In order to decode the IT2 FSs into

words, we used the Jaccard similarity measure. The result of the decoder for the six criteria is shown in Table 8. However, the IT2 FSs of weights could be decoded into crisp numbers, but we mapped IT2 FSs to words to use them further in perceptual computing evaluations. Also, other methods can be used to decode the IT2 FSs into decision classes of words (Mendel and Wu 2010).

Conclusions

To improve the interrelations between decision making criteria in perceptual computing, we proposed an interval type-2 fuzzy set extension of the DEMATEL method. In this method, we combined the perceptual computing characteristics with the fuzzy DEMATEL in order to map the influence matrices defined by words into weights. In perceptual computing, words are mapped into IT2 FSs. IT2 FSs are able to show the uncertainty related to each word in the codebook; therefore, they are suitable to model the uncertainty associated to decision making in the real word. The DEMATEL method considers the interrelations between criteria and defines weights based on these relations. Therefore, the combination of these two methods leads to a decision making method that can consider the uncertainty related to decision making and also the interrelations between criteria. The weights obtained from the proposed method can further be used for evaluation based on words. In order to define the cause and relation between criteria, the IT2 FSs should be defuzzified into crisp numbers. However, other decoding methods can be used to map the IT2 FS into words.

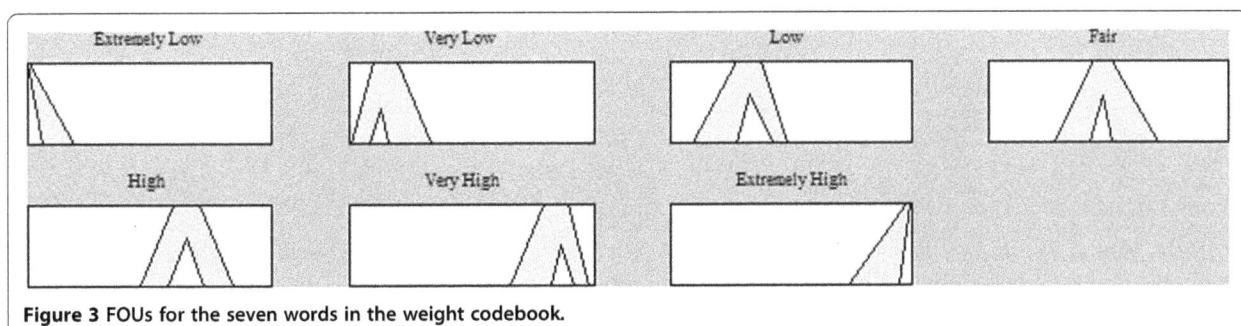

Figure 3 FOUs for the seven words in the weight codebook.

Competing interests
The authors declare that they have no competing interests.

Authors' contributions
MBH carried out the presentation of a solution to address the shortcoming of perceptual computing considering the relationship between decision making criteria. MJT carried out the whole research idea and the validation and verification of the research outcomes. Both authors read and approved the final manuscript.

Acknowledgments
The authors would like to acknowledge the industrial relation department of K. N. Toosi University of Technology to make relations between the MAPNA Co. and the university available; and special thanks to the reviewers who helped improve this manuscript.

References

Bustince H (2000) Indicator of inclusion grade for interval-valued fuzzy sets. Application to approximate reasoning based on interval-valued fuzzy sets. Int J Approx Reason 23(3):137–209

Chang B, Chang C-W, Wu C-H (2011) Fuzzy DEMATEL method for developing supplier selection criteria. Expert Syst Appl 38:1850–1858

Chen J-K, Chen I-S (2010) Using a novel conjunctive MCDM approach based on DEMATEL fuzzy ANP, and TOPSIS as an innovation support system for Taiwanese higher education. Expert Syst Appl 37:1981–1990

Gabus A, Fontela E (1973) Perceptions of the world problematique: communication procedure, communicating with those bearing collective responsibility (DEMATEL Report No. 1). Battelle Geneva Research Centre, Geneva

Gorzalczany MB (1987) A method of inference in approximate reasoning based on interval-valued fuzzy sets. Fuzzy Set Syst 21:1–17

Hamrawi H, Coupland S (2009) Type-2 fuzzy arithmetic using alpha-planes. In: Proceedings of the IFSA-EUSFLAT conference, Lisbon, 20–24 July 2009, 2009th edn

Jassbi J, Mohammadnejad F, Nasrokkahzadeh H (2011) A fuzzy DEMATEL framework for modeling cause and effect relationships of strategy map. Expert Syst Appl 38:5967–5973

Kaufmann A, Gupta M (1985) Introduction to fuzzy arithmetic: theory and applications. Van Nostran Reinhold, New York

Lin C-L, Wu W-W (2004) A fuzzy extension of the DEMATEL method for group decision making. Eur J Oper Res 156:445–455

Liu F, Mendel JM (2008) Encoding words into interval type-2 fuzzy sets using an interval approach. IEEE Trans Fuzzy Syst 16:1503–1521

Mendel JM (2001) Uncertain rule-based fuzzy logic systems: introduction and new directions. Prentice-Hall, Upper-Saddle River

Mendel JM (2002) An architecture for making judgments using computing with words. Int J Appl Math Comput Sci 12(3):325–335

Mendel JM (2007) Computing with words and its relationship with fuzzistics. Inform Sci 177:988–1006

Mendel JM, John RI (2002) Type-2 fuzzy sets made simple. IEEE Trans Fuzzy Syst 10:117–127

Mendel MJ, Wu D (2007) Perceptual reasoning: a new computing with words engine. In: Proceedings of the IEEE international conference on granular computing, Silicon Valley, 2–4 Nov., pp 446–451

Mendel MJ, Wu D (2008) Perceptual reasoning for perceptual computing. IEEE Trans Fuzzy Syst 16(6):1550–1564

Mendel MJ, Wu D (2010) Perceptual computing: aiding people in making subjective judgments. Wiley-IEEE, Hoboken

Mitchell HB (2005) Pattern recognition using type-II fuzzy sets. Inform Sci 170(2–4):409–418

Wu D, Mendel JM (2007) Aggregation using the linguistic weighted average and interval type-2 fuzzy sets. IEEE Trans Fuzzy Syst 15:6

Wu D, Mendel JM (2008) Corrections to "Aggregation using the linguistic weighted average and interval type-2 fuzzy sets". IEEE Trans Fuzzy Syst 16:6

Wu D, Mendel JM (2009) A comparative study of ranking methods, similarity measures and uncertainty measures for interval type-2 fuzzy sets. Inform Sci 179(8):1169–1192

Yang JL, Tzeng G-H (2011) An integrated MCDM technique combined with DEMATEL for a novel cluster-weighted with ANP method. Expert Syst Appl 38:1417–1424

Zadeh LA (1965) Fuzzy sets. Information and Control 8:338–353

Zadeh LA (1975) The concept of a linguistic variable and its application to approximate reasoning-I. Information Sciences 8:199–249

Zadeh LA (1999) From computing with numbers to computing with words—from manipulation of measurements to manipulation of perceptions. IEEE Trans Circ Syst Fund Theor Appl 4:105–119

6

The investigation of supply chain's reliability measure: a case study

Houshang Taghizadeh[1*] and Ehsan Hafezi[2]

Abstract

In this paper, using supply chain operational reference, the reliability evaluation of available relationships in supply chain is investigated. For this purpose, in the first step, the chain under investigation is divided into several stages including first and second suppliers, initial and final customers, and the producing company. Based on the formed relationships between these stages, the supply chain system is then broken down into different subsystem parts. The formed relationships between the stages are based on the transportation of the orders between stages. Paying attention to the system elements' location, which can be in one of the five forms of series namely parallel, series/parallel, parallel/series, or their combinations, we determine the structure of relationships in the divided subsystems. According to reliability evaluation scales on the three levels of supply chain, the reliability of each chain is then calculated. Finally, using the formulas of calculating the reliability in combined systems, the reliability of each system and ultimately the whole system is investigated.

Keywords: Supply chain, Reliability, Supply chain operational reference

Background

The supply chain includes all the activities related to the processing of materials and the conversion of goods from the stage of raw material to the stage of delivery to the final customer, as well as the informational and financial processes related to them, along with coordinated and integrated management (Shafia et al. 2008). In a broader sense, a supply chain consists of two or more organizations that could be companies which produce parts, constituents, and final products or they could even include the supply-and-distribute service providers or the final customer as well (Supply Chain Council 2008). The most important factor in the successful management of the supply chain is a reliable relationship among the partners in the chain in such a way that they can have mutual trust in each others' capabilities and activities. Therefore, in the development of any integrated supply chain, increasing the confidence and trust among the partners and devising the reliability for them are the crucial factors to achieve sustainable success (Ghazanfari and Fatholla 2006). In the current industries, choosing business partners and

establishing a successful and sustainable communication with them regarding the previous standards and criteria is not feasible. Hence, determining the quantitative criteria and parameters through which the most suitable partner could be chosen seems to be useful. The reliability factor is also one of the most effective criteria which mean the probability of the intact and flawless performance of the system for a definite and pre-scheduled period of time (Haj Shirmohammadi 2002). On these grounds, the present paper aims to study the reliability rate in the supply chain model and to determine whether the relationships within the supply chain have a high reliability rate or not. In order to study this, the 'supply chain operational reference' (SCOR), which is a valuable tool to analyze supply chains, has been used. The SCOR model supports the operational evaluation metrics at three levels. The metrics of level one provide an approach to supply chain in order to assess management, and the metrics of levels two and three include more specific and detailed criteria regarding the categories and elements of the processes. The metrics of level one are systematically divided into five operational criteria, three of which, reliability, flexibility, and responsiveness, are customer-facing attributes, and the other two, costs and assets, are internal-facing ones. Each of these metrics

* Correspondence: taghizadeh@iaut.ac.ir
[1]Department of Management, Tabriz Branch, Islamic Azad University, Tabriz, Iran
Full list of author information is available at the end of the article

is further divided into minor metrics at the lower levels (Supply Chain Council 2006). The rate of reliability, which is the operational criterion discussed in this article, is also assessed and measured at level one of the supply chain based on SCOR model through the metrics of perfect order fulfillment; at level two through the metrics of perfect order fulfillment, delivery performance to customer commit date, accurate documentation, and perfect condition of order (Stephan and Badr 2007). Level three of the supply chain under study also has minor and more detailed metrics for the assessment of the above-mentioned metrics. It is possible to calculate the reliability rate of each loop of the broad chain under study during different pre-scheduled time periods. In order to measure the reliability rate of the whole supply chain under study at a certain period of time, first, it is necessary to identify the type of the supply chain formed in one of the five positions-series, parallel, series/parallel, parallel/series, or composite, and then, based on the reliability formula of the related system, it is possible to calculate the reliability of the whole system at that period (Haj Shirmohammadi 2002).

In comparison with previous studies, by calculating reliability measurement metrics in different levels of the supply chain and identifying the impact value of each of these metrics on variances of reliability criterion in different periods, this research has been able to offer a new method for prioritizing decreasing reliability factors in a supply chain in order to reduce their effects. In addition, this research is a case study in Iran which is suitable for computing reliability of supply chain for Iranian organizations.

Case description

The case study provided in this paper is that of Tabriz Iran Khodro Factory (TIKF) and its suppliers and sale delegates. TIKF is a car-producing factory.

Review of literature

The scope and boundaries of cooperation among the constituents of the supply chain include various activities of which predicting the amount of the material needed, ordering the raw material, processing and carrying out orders, supervising the transport services, distributing the final product controlling the bill, and reviewing payment mode can be cited as examples (Bozarth and Handfield 2007). Figure 1 shows the main activities of the supply chain.

The existence of fault in meeting the needs and expectations in every part of a chain causes the progressive increase in problems, and defect in one part of the system creates problems in other parts. This chain-like state prevails and creates even more problems. One of the key indices in increasing the competitive and qualitative power of the products and production services of organizations

and institutes is creating and establishing reliable relationship with the chain of providers and suppliers of raw material and primary parts, and the careful assessment of the reliability rate of these relationships. The success or failure of each supply chain in the market is eventually determined by the final customer or the consumer. Thus, in order to establish a successful new relationship in the supply chain, assessing the reliability of the relationship is among the crucial factors in this field (Xujie 2009).

The SCOR model, which is a means of analyzing and configuring the supply chain, was devised by the Supply Chain Council. It was established by the Institute of Advanced Manufacturing Research, PRTM Counseling Company, and more than 65 major companies. It currently has over 850 members around the world.

The SCOR metrics are applied in relation with operation attributes. Operation attributes are the supply chain attributes through which it is possible to analyze and assess the company's supply chain strategy at each level separately and to compare it with other strategies. The metrics of level one are systematically divided into five classes, i.e., reliability, flexibility, and responsiveness, which are customer-facing attributes, and costs and assets, which are internal-facing attributes (Table 1; Supply Chain Council 2006).

Each of these metrics is further divided into minor metrics at the lower levels of the model, which are codified according to the format below (Table 2).

This is an easier way to eliminate errors during activities such as benchmarking of the supply chain and the like. The metrics' number format is $XX.y.z$, in which y is the metric level, z is the specific number, and XX is the operation attribute.

The possible values for XX include reliability (RL), responsiveness (RS), flexibility (AG), costs (CO), and assets (AS) (Supply Chain Council 2006). The operation attribute is the reliability of the attribute under discussion in this article. The codification and calculation manners of each metric at the three levels of reliability attribute have been shown in Tables 2 and 3, respectively.

As mentioned before, a system usually consists of a number of constituent elements, or a number of smaller systems, or subsystems; the juxtaposition of which as well as their dependence on each other will influence the reliability of the system (Han et al. 2007). The constituents of a system are linked to each other in one of the five positions: series, parallel, parallel/series, series/parallel, and composite. Table 4 shows the structure and calculation manner of the reliability of each position.

At present, the subject of supply chain is of great interest among the world researches and articles in this regard. These researches include introducing various types of mathematical models, different managerial techniques, methods of control, and other topics concerning industrial

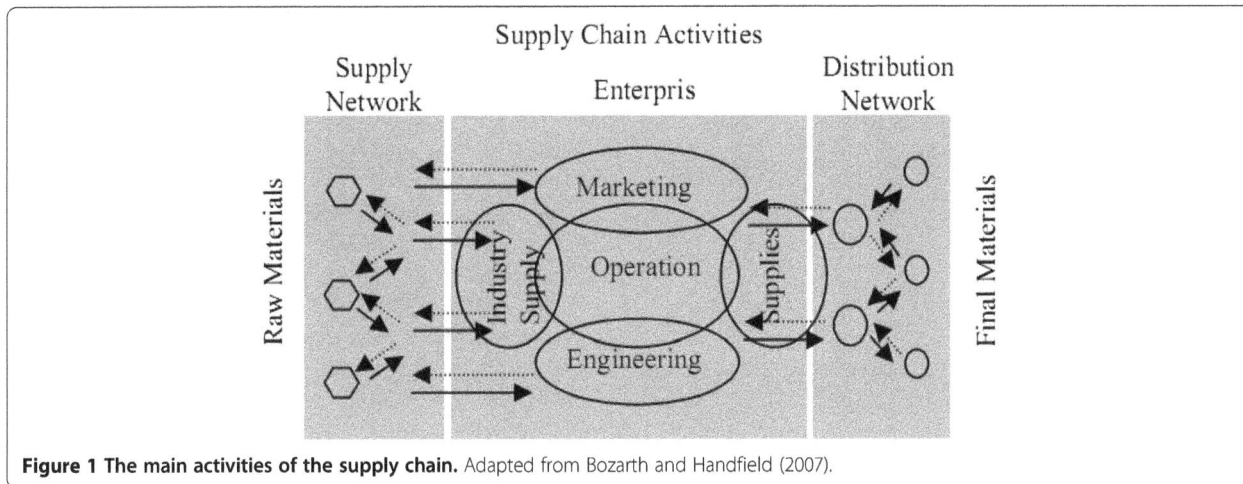

Figure 1 The main activities of the supply chain. Adapted from Bozarth and Handfield (2007).

engineering and management, especially supply chain management.

Some of the researches are as follows: Jabbour et al. (2011) have been studying to perform an empirical investigation about the constructs and indicators of the supply chain management practices framework. Banomyong and Supatn (2011) presented a supply chain performance assessment tool that measures the performance of key supply chain activities of a firm under different performance dimensions. Christopher et al. (2011) studied to understand how managers assess global sourcing risks across the entire supply chain and what actions they take to mitigate those risks. Seifbarghy et al. (2010) analyzed the supply chain using the SCOR model in a steel producing company. Tian (2009) researched on equilibrium of coordination reliability of supply chain and deepening in division of labor in the perspective of dilemma. Lin (2009) studied system reliability evaluation for a multistate supply chain

network with failure nodes using minimal paths. Xujie (2009) has done modeling and analyzing supply chain reliability by different effects of failure nodes. Jahandideh (2008) studied and assessed the process of managing car parts suppliers' chain in SAPCO Company. Klimov and Merkuryev (2008) presented a simulation model for supply chain reliability evaluation. Qing-kui (2008) studied the reliability analysis and evaluation on member enterprise of manufacturing supply chain based on BP neural network. Lirong Cui (2008) studied on reliability of supply chain based on higher order Markov chain. Hwang et al. (2008) evaluated the sourcing process in the SCOR model in the manufacturing industries of Taiwan. Stephan and Badr (2007) presented quantitative and qualitative approaches to manage risks in the supply chain operations reference. Han et al. (2007) reviewed and analyzed supply chain operations reference. Similarly, various studies have been done on supply chain and SCOR model in Iran,

Table 1 Level one metrics of an SCOR model

Level one metrics	Performance attributes				
	Customer facing			Internal facing	
	Reliability	Responsiveness	Flexibility	Costs	Assets
Perfect order fulfillment	*				
Order fulfillment cycle time		*			
Upside supply chain flexibility			*		
Upside supply chain adaptability			*		
Downside supply chain adaptability			*	*	
Supply chain management cost				*	
Cost of goods sold				*	
Cash-to-cash cycle time					*
Return on supply chain fixed assets					*
Return on working capital					*

Each asterisk shows the relationship between metric levels and performance attributes. Adapted from Supply Chain Council (2006).

Table 2 Codification of metrics at three levels of operation attribute of reliability in the SCOR model

Level	Code	Metric
One	RL.1.1	Perfect order fulfillment
Two	RL.2.1	Percentage of orders delivered in full
	RL.2.2	Delivery performance to customer commit date
	RL.2.3	Accurate documentation
	RL.2.4	Perfect condition
Three	RL.3.1	Delivery quantity accuracy
	RL.3.2	Delivery item accuracy
	RL.3.3	Customer commit date achievement time customer receiving
	RL.3.4	Delivery location accuracy
	RL.3.5	Shipping documentation accuracy
	RL.3.6	Compliance documentation accuracy
	RL.3.7	Payment documentation accuracy
	RL.3.8	Orders delivered damage free conformance
	RL.3.9	Orders delivered defect free conformance
	RL.3.10	Percentage of faultless installations
	RL.3.11	Warranty and returns

Adapted from Stephan and Badr (2007).

Table 3 Manner of calculating the metrics at three levels of the SCOR model

Code	Calculation
RL.1.1	[Total perfect orders]/[Total number of orders]
RL.2.1	[Total number of orders delivered in full]/[Total number of orders delivered]
RL.2.2	[Total number of orders delivered on the original commitment date]/[Total number of orders delivered]
RL.2.3	[Total number of orders delivered with accurate documentation]/[Total number of orders delivered]
RL.2.4	[Number of orders delivered at perfect condition]/[Total number of orders delivered]
RL.3.1	[Total perfect orders without item defect]/[Total number of orders]
RL.3.2	[Number of orders delivered without item defect quantity]/[Total number of orders]
RL.3.3	[Number of orders delivered without time defect]/[Total number of orders]
RL.3.4	[Number of orders delivered without location defect]/[Total number of orders]
RL.3.5	[Number of orders delivered without shipping documentation defect]/[Total number of orders]
RL.3.6	[Number of orders delivered without compliance documentation defect]/[Total number of orders]
RL.3.7	[Number of orders delivered without payment documentation defect]/[Total number of orders]
RL.3.8	[Number of orders delivered without damage in order]/[Total number of orders]
RL.3.9	[Number of orders delivered without defect in order]/[Total number of orders]
RL.3.10	[Total perfect orders delivered without installation problems in order]/[Total number of orders]
RL.3.11	[Total perfect orders delivered without warranty defect in order]/[Total number of orders]

Adapted from Stephan and Badr (2007).

some of which can be cited here. Liu et al. (2007) studied the performance of the supply chain in relation to assessing its reliability. Huan et al. (2004) developed in a case study in China a collaborative supply chain reference model. Hezarkhani (2006) focused on the promotion of supply chain performance, using SCOR model. Ren et al. (2006) suggested a framework based on SCOR model to manage supply chain performance. Satitsatian and Kapur (2005) devised an algorithm for reliability bound computation to assess supply chain networks. Shepherd and Gunter (2006) developed methods of determining supply chain reliability for a probable computation system based on the theory of reliability. Manavizadeh (2005) presented a system of measuring the performance in the supply chain in order to establish genuine production. Lockamy and McCormack (2004) examined the link between planning methods in the SCOR model for supply chain performance. Zarei Yaraki (2004) studied sharing information in the supply chain of the country's automobile industry. Riazy (1997) devised a decision-making method for evaluation, selection, and development of suppliers in supply chain management. Azimi (2001) focused on measuring supply chain performance. Teimouri (1999) expanded the model for suppliers' selection and distribution from the standpoint of supply chain management.

Research scope and data collection method

The supply chain structure shown in Figure 2 is a six-stage supply chain whose function is providing raw materials from the supplier, producing the product, and delivering it to the final customer. The first through third stages include the suppliers; the fourth stage includes the producing company, and the fifth and sixth stages include the primary and final customers. The research population and the research scope involve the second through the fifth stages. Since the main focus of the research is to study the reliability rate of the supply chain, the research method is descriptive, and in order to obtain the desired result, a combination of library studies including review of the available documents and evidence has been done.

Research model

If the chain under study is divided, based on the inter-stage relations, into three parts A, B, and C and each part is considered as a subsystem, then each of the created relations in each subsystems A, B, and C

Table 4 Manner of calculating the reliability of different positions in a systems

Position	Figure	Computation method
Series		$RL_s = P_1 \times P_2 \times \cdots \times P_k = \prod\limits_{i=1}^{k} P_i$
Parallel		$RL_s = 1 - \prod\limits_{j=1}^{n} \left(1 - P_j\right)$
Series/ parallel		$RL_s = \prod\limits_{i=1}^{k} \left[1 - \prod\limits_{j=1}^{n} \left(1 - P_{i,j}\right) \right]$
Parallel/ series		$RL_s = 1 - \prod\limits_{j=1}^{n} \left[\left(1 - \prod\limits_{i=1}^{k} P_{i,j}\right) \right]$
Composite	Composite complex	To calculate reliability in composite complex, first, a system must be divided to more subsystems, and then with the reliability computation of lesser subsystems, reliability of main systems is computable.

Adapted from Haj Shirmohammadi (2002).

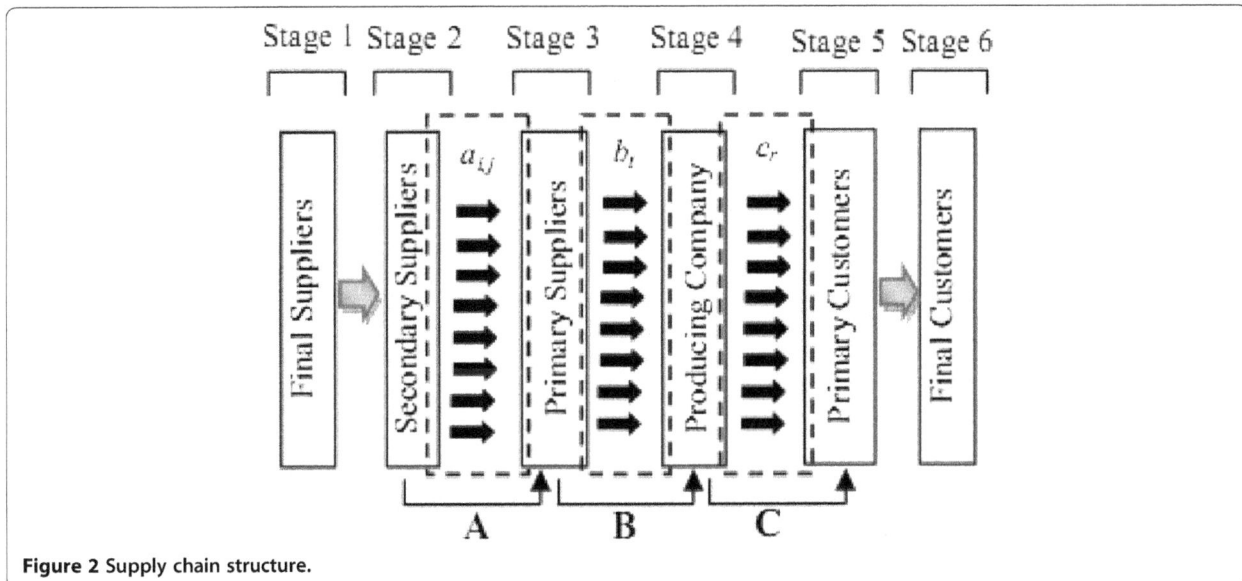

Figure 2 Supply chain structure.

will be shown by the symbols $a_{i,j}$, b_t, and c_r, respectively. Symbol $a_{i,j}$ serves to transfer order from the secondary supplier to the primary supplier; b_t functions to transfer the order from the primary supplier to the producing company, and c_r serves to transfer the order from the producing company to the primary customers.

In subsystem A, in case of failure in providing one kind of order from a secondary supplier for any reason, it is possible to obtain the order from another secondary supplier; thus, because of the inability to provide one kind of order, subsystem A will be inefficient. Therefore, in subsystem A, different types of orders are reciprocally dependent, while the suppliers of the same kind of order are independent of each other. Thus, $a_{i,j}$ forms a series/parallel structure. Likewise, in subsystem B, in case of inability to provide one kind of order, subsystem B will be inefficient. Therefore, in subsystem B, different kinds of orders are dependent on each other. As a result, b_t forms a series structure. However, in subsystem C, in case of inability to transfer the order from the producing company to a certain customer, subsystem C will not be inefficient; rather, it will be inefficient only if the transfer of order is not done to any of the primary customers. Thus, in subsystem C, the primary customers are independent of each other. As a result, c_r forms a parallel structure (Figure 3).

In this paper, in order to analyze the supply chain reliability based on the SCOR model, the reliability evaluation metrics are calculated at 12 different periods of time, with each period considered to last one month. The three-level metrics based on RL.y.z format have been shown in Table 2. Therefore, in order to obtain the values of the three-level metrics and to analyze the reliability of the whole chain, the RL.y.z values should be calculated for each subsystem (A, B, and C). Thus, the reliability of subsystems A, B, and C will be calculated through the relations (Equations 1, 2, and 3), respectively:

$$RL.y.z_A = \prod_{i=1}^{k} \left[1 - \prod_{j=1}^{n} \left(1 - RL.y.z_{a_{i,j}} \right) \right], \tag{1}$$

$$RL.y.z_B = \prod_{t=1}^{l} RL.y.z_{b_t}, \tag{2}$$

$$RL.y.z_C = 1 - \prod_{r=1}^{m} \left(1 - RL.y.z_{c_r} \right). \tag{3}$$

In those relations (Equations 1, 2, and 3), the following have been defined:

- RL.y.z
  is the reliability of the metric number z from level y.
- $a_{i,j}$
  is the subsystem transferring the order type i from the secondary supplier j.
- b_t
  is the subsystem transferring the order type t from the primary supplier.
- c_r
  is the subsystem transferring the order to the primary customer r.
- y
  is the number of the level ($y = 1, 2, 3, \ldots, p$).
- z
  is the number of the metric ($z = 1, 2, 3, \ldots, q$).
- i
  is the number of the order type transferred from the secondary supplier to the primary supplier ($i = 1, 2, 3, \ldots, k$).
- j
  is the number of the secondary supplier ($j = 1, 2, 3, \ldots, n$).
- t
  is the number of the order type transferred from the

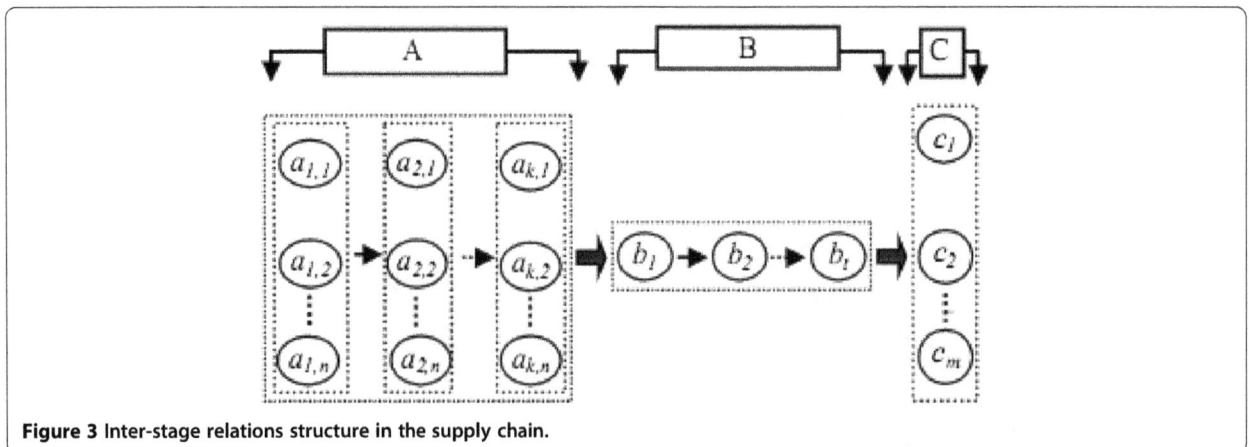

Figure 3 Inter-stage relations structure in the supply chain.

Table 5 RL.1.1 metric values in subsystems A, B, and C

Month	A	B	C
Farvardin	0.779	0.856	0.849
Ordibehesht	0.782	0.863	0.848
Khordad	0.794	0.83	0.844
Tir	0.824	0.901	0.844
Mordad	0.837	0.895	0.843
Shahrivar	0.828	0.893	0.84
Mehr	0.84	0.907	0.843
Aban	0.833	0.898	0.841
Azar	0.839	0.905	0.838
Day	0.839	0.901	0.843
Bahman	0.848	0.908	0.843
Esfand	0.836	0.903	0.848

primary supplier to the producing company ($t = 1, 2, 3, \ldots, l$)

- r

  is the number of the primary customer ($r = 1, 2, 3, \ldots, m$)

Now, with regard to the fact that subsystems A, B, and C are the independent and serial subsystems of the supply chain under study in this paper, in order to calculate each of the three-level metrics of the whole system, which is shown by the symbol RL.$y.z_T$ values, the formula for calculating the reliability of series systems is used:

$$RL.y.z_T = RL.y.z_A \times RL.y.z_B \times RL.y.z_C \qquad (4)$$

Discussion and evaluation

In order to obtain the values of the three-level reliability metrics in different months, at first, the values of RL.$y.z$ were separately studied in each subsystem (A, B, and C).

The RL.1.1 metric values in subsystems A, B, and C for 12 months are shown as an example in Table 5.

Next, by putting the metric values of each subsystem in the relation (Equation 4), the required value of that metric in the whole supply chain was obtained. The values of these metrics are shown in Table 6. In the last column of Table 6, the annual average of the reliability metric values of the whole supply chain has been given.

As is seen in the diagram of level one in Figure 4, in Shahrivar, Aban, and Esfand, the reliability rate regarding perfect order fulfillment shows a relatively low decrease as compared with the related value in the previous month. The reason can be understood by having a glance at the diagrams of levels two and three.

It is seen that in Shahrivar, the reliability is concerning metric (RL.2.4) at level two; at level three, the reliability is concerning metrics (RL.3.6), (RL.3.8), (RL.3.9), and (RL.3.11) which show a considerably sharp decrease, which causes the decrease in the whole supply chain reliability in Shahrivar. In addition, the decline in the rate of metrics (RL.2.3) and (RL.2.4) at level two and the metrics (RL.3.5), (RL.3.7), (RL.3.9), and (RL.3.11) at level three have caused the decrease in the total reliability metrics in Aban. Furthermore, in Esfand the decline in metrics (RL.2.1), (RL.2.2), and (RL.2.3) at level two and the metrics (RL.3.1), (RL.3.2), (RL.3.3), (RL.3.4), (RL.3.5),

Table 6 Reliability evaluation metric values in the whole supply chain under study

Code	Farvardin	Ordibehesht	Khordad	Tir	Mordad	Shahrivar	Mehr	Aban	Azar	Day	Bahman	Esfand	Average
RL.1.1$_T$	0.556	0.572	0.582	0.626	0.631	0.621	0.642	0.629	0.636	0.637	0.649	0.64	0.619
RL.2.1$_T$	0.916	0.922	0.919	0.936	0.934	0.935	0.922	0.934	0.927	0.926	0.936	0.929	0.928
RL.2.2$_T$	0.885	0.905	0.884	0.898	0.907	0.92	0.919	0.923	0.918	0.926	0.936	0.929	0.928
RL.2.3$_T$	0.831	0.832	0.845	0.873	0.864	0.869	0.884	0.864	0.887	0.881	0.89	0.885	0.867
RL.2.4$_T$	0.847	0.832	0.856	0.86	0.87	0.838	0.865	0.852	0.85	0.851	0.846	0.85	0.851
RL.3.1$_T$	0.964	0.975	0.97	0.977	0.973	0.975	0.955	0.97	0.97	0.969	0.971	0.969	0.97
Rl.3.2$_T$	0.951	0.945	0.947	0.958	0.957	0.959	0.957	0.964	0.955	0.956	0.965	0.959	0.956
RL.3.3$_T$	0.943	0.948	0.937	0.95	0.948	0.956	0.959	0.959	0.961	0.955	0.959	0.957	0.953
RL.3.4$_T$	0.939	0.956	0.943	0.946	0.959	0.963	0.954	0.962	0.955	0.97	0.968	0.965	0957
RL.3.5$_T$	0.939	0.942	0.948	0.953	0.951	0.952	0.959	0.951	0.971	0.963	0.963	0.953	0.954
RL.3.6$_T$	0.94	0.941	0.943	9.56	0.956	0.954	0.955	0.961	0.956	0.954	0.965	0.967	0.954
RL.3.7$_T$	0.941	0.94	0.945	0.96	0.951	0.959	0.964	0.946	0.956	0.96	0.957	0.96	0.953
RL.3.8$_T$	0.952	0.942	0.956	0.957	0.959	0.949	0.958	0.959	0.957	0.957	0.964	0.953	0.955
RL.3.9$_T$	0.958	0.953	0.956	0.958	0.967	0.963	0.967	0.952	0.955	0.961	0.955	0.959	0.959
RL.3.10$_T$	0.949	0.947	0.957	0.955	0.965	0.966	0.966	0.968	0.959	0.953	0.949	0.953	0.957
RL.3.11$_T$	0.979	0.978	0.979	0.975	0.978	0.962	0.97	0.965	0.971	0.972	0.969	0.976	0.973

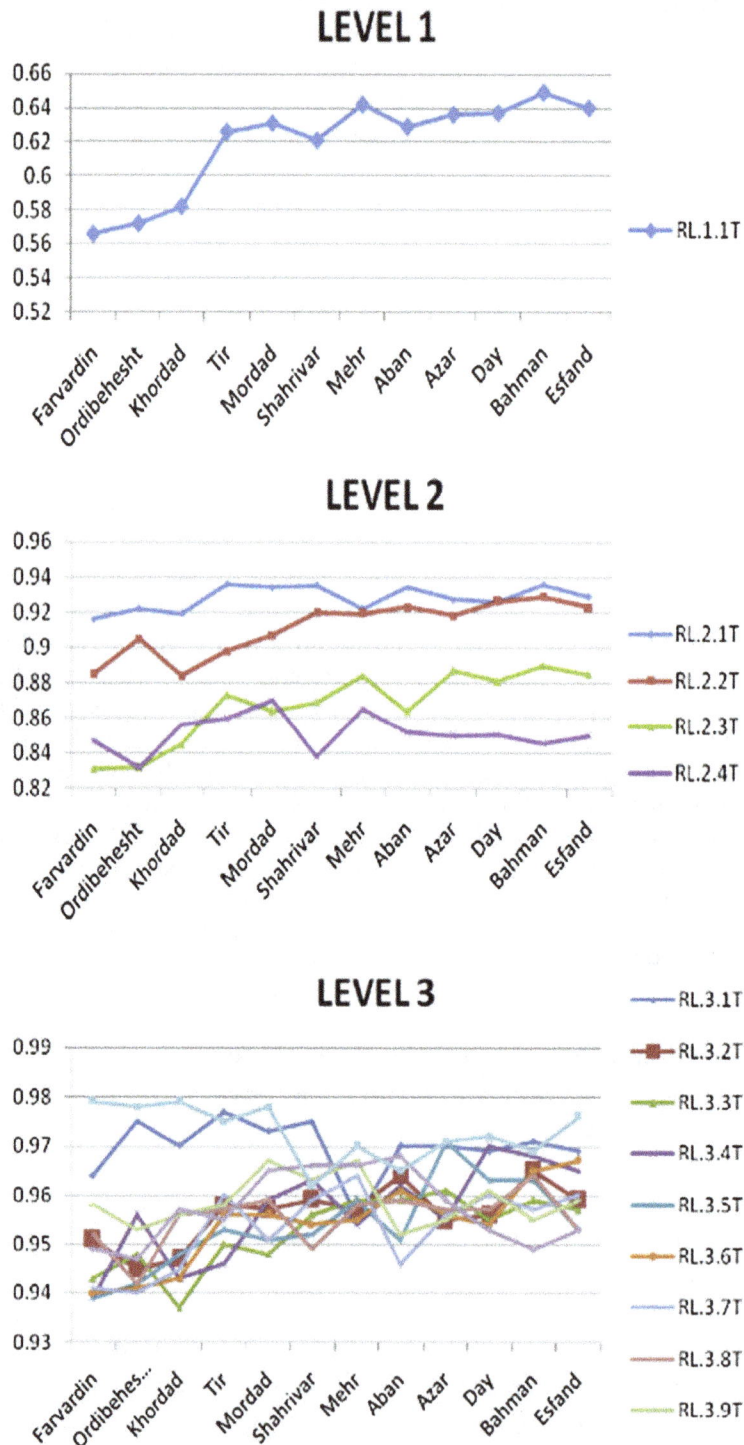

Figure 4 Process of changes in the three-level reliability evaluation metrics in 12 months.

and (RL.3.8) at level three has caused the decrease in the reliability metrics in the whole supply chain.

Conclusions

The reliability attribute is one of the most important means of measuring and assessing the performance in supply chains. The main purpose of this article is to investigate the reliability measure of the supply chain under study. The data applied in the research are the outcome of library studies including the review of evidence and documents. At first, the supply chain under study was divided into a number of stages, and the relations developed among these stages were studied based on the transfer of the order among the stages. Further, according to the relation developed among the different stages, the supply chain system under study was divided into the three subsystems: A, B, and C. Based on the configuration of the elements, the type of subsystems A, B, and C was then determined. Finally, based on the formula of calculating the reliability of compound systems, the reliability of each subsystem A, B, and C was calculated, according to which the reliability of the whole supply chain was assessed. The main conclusions of the research are as follows:

- The existence of fault in the unloading documentation, the existence of damage and defect in some orders, and the return of reliability warranty time caused the decrease in the reliability of the whole supply chain in Shahrivar by 0.01.
- The existence of fault in the loading or payment documentation, the existence of defect in the orders, and the return of some of the orders during warranty time caused the decrease in the reliability of the whole supply chain in Aban by 0.013.
- The existence of fault in material and quantity, inability to deliver to customer commit date, inability to deliver the orders at the accurate location, the existence of fault in the loading documentation, and the existence of defect in some of the orders caused the decrease in the reliability of the whole chain in Esfand by 0.021.

Competing interests
The authors declare that they have no competing interests.

Authors' contributions
Both HT and EH have contributed equally in the following sections: Background, Research scope and data collection method, and Research model. HT contributed most of the information in the 'Review of literature' and 'Conclusions' sections, with validating contributions from EH. Both authors read and approved the final manuscript.

Author details
[1]Department of Management, Tabriz Branch, Islamic Azad University, Tabriz, Iran. [2]Industrial Engineering - System Management and Productivity at the Non-Governmental and Private Higher Education Institution of ALGHADIR, Tabriz, Iran.

References
Azimi E (2001) Measuring the supply chain operation. M.Sc. Thesis. Faculty of Technology and Engineering, Teacher Training University, Tehran

Banomyong R, Supatn N (2011) Developing a supply chain performance tool for SMEs in Thailand. Supply Chain Management: An International Journal 16:20–31

Bozarth C, Handfield RB (2007) Introduction to operations and supply chain management, 2nd edn. Prentice Hall, New Jersey

Christopher M, Mena C, Khan O, Yurt O (2011) Approaches to managing global sourcing risk. Supply Chain Management: An International Journal 16:67–81

Ghazanfari M, Fatholla M (2006) A comprehensive look at supply chain management, 1st edn. Iran Science and Technology University Publications, Tehran

Han S, Chu CH, Yang S (2007) Developing a collaborative supply chain reference model: a case study in China. International Conference on Service Operations and Logistics and Informatics, Philadelphia, 27–29 August 2007

Haj Shirmohammadi A (2002) Programming maintenance and repair (Technical management in industry), 8th edn. Ghazal Publishers, Esfahan

Hezarkhani B (2006) Promoting the supply chain operation using SCOR model with Selecting and Expanding Supplier's in the Supply Chain Management. Doctoral Dissertations. Faculty of Technology, Tehran University, Tehran

Huan SH, Sheoran SK, Wang G (2004) A review and analysis of supply chain operations reference (SCOR) model. Supply Chain Management: An International Journal 9:23–29

Hwang YD, Lin YC, Lyu J Jr (2008) The performance evaluation of SCOR sourcing process—the case study of Taiwan's TFT-LCD industry. International Journal of Production Economics 115:411–423

Jabbour A, Filho A, Viana A, Jabbour C (2011) Measuring supply chain management practices. Measuring Business Excellence 15:18–31

Jahandideh D (2008) Evaluating the process of managing car parts supplier's chain in SAPCO. M.Sc. Thesis. Mazandaran University of Science and Technology, Mazandaran

Klimov R, Merkuryev Y (2008) Simulation model for supply chain reliability evaluation. Technological and Economic Development of Economy, Baltic Journal on Sustainability 14:300–311

Lin Y (2009) System reliability evaluation for a multistate supply chain network with failure nodes using minimal paths. IEEE Transactions on Reliability 58:34–40

Lirong Cui X (2008) A study on reliability of supply chain based on higher order Markov chain. IEEE International Conference on Service Operations and Logistics, and Informatics, IEEE/SOLI 2008 2:2014–2017, 40

Liu Y, Wu H, Luo M (2007) A reliability evaluation of supply chain: indicator system and fuzzy comprehensive evaluation. Springer, Boston

Lockamy A, McCormack K (2004) Linking SCOR planning practices to supply chain performance: An exploratory study. International Journal of Operations & Production Management 24:1192–1218

Manavizadeh N (2005) Presenting a measuring system in supply chain to establish genuine production. M.Sc. Thesis. Faculty of Technology, Tehran University, Tehran

Qing-kui C (2008) Reliability analysis and evaluation on member enterprise of manufacturing supply chain based on BP neural network. In: International Conference on Management Science and Engineering, 2008. ICMSE 2008. 15th Annual Conference Proceedings, Long Beach, 10–12 September 2008

Ren C, Dong J, Ding H, Wang W (2006) A SCOR-based framework for supply chain performance management. IEEE International Conference on Service Operations and Logistics and Informatics, Shanghai, 21–23 June 2006

Riazy A (1997) Designing a decision-making procedure for evaluation, selection and expansion of supplier in supply chain management. M.Sc. Thesis, Faculty of Industrial Management. Iran Science and Technology University, Tehran

Satitsatian S, Kapur KC (2005) An algorithm for reliability bounds computation to evaluate supply chain networks. Thesis, University of Washington, Seattle

Seifbarghy M, Akbari MR, Sajadieh MS (2010) Analyzing the supply chain using SCOR model in a steel producing company. 40th International Conference on Computers and Industrial Engineering (CIE), Awaji, 25–28 July 2010

Shafia MA, Jabal Ameli MS, Fathollah M (2008) A study of the effect of sharing costs in the supply chains based on SCOR. The International Journal of Industrial Management and Production Management, Iran Science and Technology University, 19:30

Shepherd C, Gunter H (2006) Measuring supply chain performance: current research and future directions. International Journal of Productivity and Performance Management 55:242–258

Supply Chain Council (2006) Supply chain operations reference (SCOR) model version 8.0. http://www.scribd.com/doc/2939600/ SCOR-Model-Version-8. Accessed 16 August 2012

Supply Chain Council (2008) Supply chain operations reference (SCOR) model version 10.0: overview. http://supply-chain.org/f/SCOR-Overview-Web.pdf. Accessed 16 August 2012

Stephan J, Badr Y (2007) A quantitative and qualitative approach to manage risks in the supply chain operations reference. 2nd International Conference on Digital Information Management 1:410–417

Teimouri E (1999) Developing a selection model for suppliers and distribution based on supply chain management approach. Ph.D. Dissertation. Faculty of Industrial Engineering, Iran Science and Technology University, Tehran

Tian G (2009) Research on equilibrium of coordination reliability of supply chain and deepening in division of labor in the perspective of dilemma. 2009 International Conference on Information Management, Innovation Management and Industrial Engineering 2:484–488

Xujie L (2009) Modeling and analyzing supply chain reliability by different effects of failure nodes. 2009 International Conference on Information Management, Innovation Management and Industrial Engineering 4:396–400

Zarei Yaraki A (2004) Information sharing in the supply chain of the country's automobile industry. M.Sc. Thesis. Faculty of Technology and Engineering, Teacher Training (Tarbiat Modarres) University, Tehran

7

Minimizing the total tardiness and makespan in an open shop scheduling problem with sequence-dependent setup times

Samaneh Noori-Darvish[1] and Reza Tavakkoli-Moghaddam[2*]

Abstract

We consider an open shop scheduling problem with setup and processing times separately such that not only the setup times are dependent on the machines, but also they are dependent on the sequence of jobs that should be processed on a machine. A novel bi-objective mathematical programming is designed in order to minimize the total tardiness and the makespan. Among several multi-objective decision making (MODM) methods, an interactive one, called the TH method is applied for solving small-sized instances optimally and obtaining Pareto-optimal solutions by the Lingo software. To achieve Pareto-optimal sets for medium to large-sized problems, an improved non-dominated sorting genetic algorithm II (NSGA-II) is presented that consists of a heuristic method for obtaining a good initial population. In addition, by using the design of experiments (DOE), the efficiency of the proposed improved NSGA-II is compared with the efficiency of a well-known multi-objective genetic algorithm, namely SPEA-II. Finally, the performance of the improved NSGA-II is examined in a comparison with the performance of the traditional NSGA-II.

Keywords: Open shop scheduling, Total tardiness, Makespan, Sequence-dependent setup times, NSGA-II, SPEA-II

Background

An open shop scheduling problem (OSSP) is a kind of shop scheduling such that the operations can be executed in any order. The open shop allows much flexibility in scheduling, but it is difficult to develop rules that give an optimum sequence for every problem (Sule 1997). This problem is a class of NP-hard ones (Gonzalez and Sahni 1976. In this paper, we consider a special feature in OSSPs, called sequence-dependent setup time. The process of preparing machines between jobs is considered as a setup. In fact, setup times affect on the completion time of each job. As a result, they also affect on tardiness, earliness and other important criteria. Allahverdi *et al.* (2008) surveyed the literature of setup times or costs in scheduling problems. They classified scheduling problems into those with batching and non-batching considerations as well as sequence-independent and sequence-dependent setup times. According to the

technology and the kind of machines used in a work environment and variety of products, setup times can be dependent on both machines and the sequence of jobs that should be processed on a machine.

In many practical production systems (e.g., chemical, printing, pharmaceutical and automobile manufacturing), the setup tasks (i.e., cleaning up and changing tools) are sequence-dependent (Zandieh *et al.*, 2006 and Roshanaei *et al.*, 2009). Low and Yeh (2008) addressed an open shop scheduling problem as a 0–1 integer programming model with the objective of minimizing the total job tardiness along with some assumptions, such as sequence-independent setup and sequence-dependent removal times. They proposed some hybrid genetic-based heuristics to solve the problem in an acceptable computing time. Mosheiov and Oron (2008) addressed batch scheduling problems on an open-shop with m machines and n jobs. Identical processing time jobs, machine-independent and sequence-independent setup times are the main assumptions of their problems. The objectives are to minimize the makespan and minimize

---
* Correspondence: tavakoli@ut.ac.ir
[2]Department of Industrial Engineering, College of Engineering, University of Tehran, PC: 14399-57131, Tehran, Iran
Full list of author information is available at the end of the article

flow time. They proposed an $O(n)$ time algorithm for the flow time minimization problem.

To achieve Pareto-optimal sets for medium to large-sized open shop problems using efficient meta-heuristic methods can be necessary and helpful. Naderi *et al.* (2011) considered an open shop with a set of parallel machines at each stage to minimize the total completion times. They proposed a mixed-integer linear programming (MILP) model for this problem. Moreover, they applied a memetic algorithm to solve the problem. Ahmadizar et al. (2010) addressed a stochastic group shop scheduling problem with known distributions for random release dates and random processing times. They formulated a stochastic programming problem and solved it by the use of an approach being a hybrid of an ant colony optimization (ACO) algorithm and a heuristic algorithm to minimize the expected makespan. Zhang and van de Velde (2010) considered an on-line two-machine open shop scheduling problem with time lags between the completion time and the start time of two consecutive operations of any job. They developed and analyzed the performance of a greedy algorithm to minimize the makespan. Mastrolilli et al. (2010) dealt with a concurrent open shop and proposed a primal–dual 2-approximation algorithm to minimize the total weighted completion times. They also considered several natural linear programming relaxations for the problem.

Fei et al. (2010) considered a weekly surgery schedule in an operating theatre. The objectives are to maximize the utilization of the operating rooms, minimize the overtime cost in the operating theatre, and minimize the unexpected idle time between surgical cases. They proposed a column-generation-based heuristic (CGBH) procedure and a hybrid genetic algorithm (HGA) for solving the planning problem as a set-partitioning integer-programming model and the daily scheduling problem as a 2-stage hybrid flow-shop problem, respectively. Matta (2009) developed two original mixed-integer programming (MIP) models (i.e., time-based model and sequence-based model) for the proportionate multiprocessor open shop scheduling problem and proposed a genetic algorithm (GA) to schedule the shop with the objective of the makespan minimization. Seraj and Tavakkoli-Moghaddam (2009) proposed a TS method to solve a new bi-objective mixed-integer mathematical programming model for an OSSP. This model seeks to minimize the mean tardiness and the mean completion time. Panahi and Tavakkoli-Moghaddam (2011) proposed a hybrid method based on multi-objective simulated annealing (SA) and ant colony optimization (ACO) to solve an open shop scheduling problem that minimizes bi-objectives, namely makespan and total tardiness. They compared their computational results with a well-known multi-objective genetic algorithm, namely NSGA II.

By considering the previous studies, we can conclude that there is only one paper considered the sequence-dependent setup time as an assumption for a single-objective OSSP without any mathematical model and written by Roshanaei *et al.* (2009). Therefore, in this paper, a bi-objective mixed-integer linear programming (MILP) model is designed with the sequence-dependent setup time as a constraint for the OSSP. To solve medium to large-sized problems, the well-known NSGA-II proposed by Deb *et al.* (2002) is improved by using a heuristic method in order to achieve good approximate Pareto-optimal frontiers.

The rest of this paper is organized as follows. The mathematical programming is presented in Section 2. The interactive multi-objective decision making approach is presented in Section 3. Section 4 elaborates the proposed improved NSGA-II. The numerical examples, computational results and performance analysis are given in Section 5. Finally, Section 6 includes the conclusion remark.

Mathematical Programming

The OSSP considered in this study includes n jobs to be processed on at most m machines. We extend the mathematical model proposed by Low and Yeh (2008) by changing it to a bi-objective model with sequence-dependent setup times.

Problem assumptions

The main assumptions of the presented model are as follows.

- Each job is to be processed on at most m machines.
- The processing sequence of each job is immaterial.
- A job is not processed by more than one machine simultaneously.
- Each machine processes at most one operation at a time.
- All jobs are available at time 0.
- No preemption is allowed.
- Machine-dependent and sequence-dependent setup times are considered for each operation.

Notations

The indices, parameters and decision variables used to formulate the mathematical model are introduced below.

Indices

N Set of jobs to be processed, $N=\{1, 2, \ldots, n\}; |N| = n$
L Set of machines, $L = \{1, 2, \ldots, m\}; |L| = m.$

i,k,g Job indices, $(i,k,g = 0,1,...,n)$, where job 0 is a dummy job.

j,h Machine indices, $(j,h = 1,2,...,m)$.

Parameters

M A large positive number.

O_{ij} Operation of job i on machine j ; $\forall i \in N$; $\forall j \in L$.

S_{kij} Setup time of job i on machine j immediately after job k.

p_{ij} Processing time of job i on machine j.

d_i Due date of job i.

H_j Dependent set up time matrix of machine j.

$$
H_j = \begin{array}{c} 0 \\ 1 \\ 2 \\ ... \\ n \end{array} \begin{bmatrix} - & S_{01j} & S_{02j} & ... & S_{0nj} \\ - & - & S_{12j} & ... & S_{1nj} \\ - & S_{21j} & - & ... & S_{2nj} \\ \vdots & \vdots & \vdots & \ddots & \vdots \\ - & S_{n1j} & S_{n2j} & ... & - \end{bmatrix} \begin{array}{c} 012...n \end{array}
$$

Decision variables

Ts_{ij} Starting time of a setup task for operation O_{ij}.

T_i Tardiness of job i.

C_{max} Makespan.

$$
Y_{ijh} = \begin{cases} 1 & if \ 0_{ij} \ precedes \ 0_{ih} \ for \ job \ i \\ 0 & Otherwise \end{cases}
$$
$$
\forall i \in N; \forall j,h \in L, j \neq h
$$

$$
X_{kij} = \begin{cases} 1 & if \ 0_{kj} \ precedes \ 0_{ij} \ on \ machine \ j \\ & Otherwise \end{cases}
$$
$$
\forall k \in N \cup \{0\}; \forall i \in N, i \neq k; \forall j \in L
$$

$$
Z_{kij} = \begin{cases} 1 & if \ 0_{kj} \ precedes \ 0_{ij} \ on \ machine \ j \\ 0 & Otherwise \end{cases}
$$
$$
\forall k \in N \cup \{0\}; \forall i \in N, i \neq k; \forall j \in L
$$

Mathematical model

As mentioned at the beginning of Section 2, the model designed in this paper is developed by extending and modifying the mathematical model proposed by Low and Yeh (2008). In this new model, the sequence-dependent setup time is used instead of the sequence-independent setup time considered in the traditional model. The makespan is added to the model as another criterion along with the total tardiness for the minimization purpose. According to these changes and for the validity of the extended model, all of the existing constraints are changed and four new constraints are added to the model. Thus, the bi-objective MILP (BOMILP) model is formulated. It should be noted that S_{0ij} is the setup time of job i on machine j when i is the first job on machine j. Moreover, if job i is the

first job on machine j, then $Z_{0ij} = 1$. The proposed model is as follows:

$$
Min \ Z_1 = \sum_{i=1} T_i \tag{1}
$$

$$
Min \ Z_2 = C_{max} \tag{2}
$$

s.t.

$$
Ts_{ij} + \sum_{k=0,k\neq i}^{n} (S_{kij} \times Z_{kij}) + p_{ij} \leq C_{max} \ ; \forall i \in N; \forall j \in L \tag{3}
$$

$$
Ts_{ij} + \sum_{k=0,k\neq 0}^{n} (S_{kij} \times Z_{kij}) + p_{ij} - M(1 - Y_{ijh} \leq Ts_{ih});
$$
$$
\forall i \in N; \forall j,h \in L, j \neq h \tag{4}
$$

$$
Ts_{ih} + \sum_{k=0,k\neq i}^{n} (S_{kih} \times Z_{kih}) + p_{ih} - M \times Y_{ijh} \leq Ts_{ij};
$$
$$
\forall i \in N; \forall j,h \in L, j \neq h \tag{5}
$$

$$
Ts_{ij} + \sum_{k=0,k\neq i \neq g}^{n} (S_{kij} \times Z_{kij}) + p_{ij} - M(1 - X_{igj}) \leq Ts_{gj};
$$
$$
\forall i,g \in N, i \neq g; \forall j \in L \tag{6}
$$

$$
Ts_{gj} + \sum_{k=0,k\neq i \neq g}^{n} (S_{kij} \times Z_{kij}) + p_{gj} - M \times X_{igj} \leq Ts_{ij};
$$
$$
\forall i,g \in N, i \neq g; \forall j \in L \tag{7}
$$

$$
Ts_{ij} + \sum_{k=0,k\neq i}^{n} (S_{kij} \times Z_{kij}) + p_{ij} - T_i; \forall i \in N; \forall j \in L \tag{8}
$$

$$
Y_{ijh} + Y_{ihj} = 1 \ ; \ \forall i \in N; \forall j,h \in L, j \neq h \tag{9}
$$

$$
X_{igj} + X_{gij} = 1; \forall i,g \in N, i \neq g; \forall j \in L \tag{10}
$$

$$
X_{kij} - Z_{kij} \geq 0 \ ; \forall k \in N \cup \{0\}; \forall i \in N, i \neq k; \forall j \in L \tag{11}
$$

$$
X_{kij} + Z_{ikj} \leq 1 \ ; \forall i,k \in N, i \neq k; \forall j \in L \tag{12}
$$

$$
\sum_{i=1,i\neq k}^{n} Z_{kij} \leq 1 \ ; \forall k \in N \cup \{0\}; \forall j \in L \tag{13}
$$

$$
\sum_{k=0,k\neq i}^{n} Z_{kij} = 1 \ ; \forall i \in N; \forall j \in L \tag{14}
$$

$$
Ts_{ij} \geq 0 \ ; \forall i \in N; \forall j \in L \tag{15}
$$

$$
T_i \geq 0; C_{max} \geq 0 \ ; \forall i \in N \tag{16}
$$

$$X_{kij}, Z_{kij} \in \{0,1\}; \forall k \in N \cup \{0\}; \forall i \in N, i \neq k; \forall j \in L \tag{17}$$

$$Y_{ijh} \in \{0,1\}; \forall i \in N; \forall j, h \in L, j \neq h \tag{18}$$

Two objective functions (i.e., total tardiness and makespan) are shown by Eqs. (1) and (2).Constraint (3) describes the makespan. Constraints (4) and (5) express the relationship between two operations of job i that do not require being consecutive. The starting time for setup task of operation O_{ih} is greater than or equal to the completion time of operation O_{ij}. Constraints (6) and (7) state the operational sequence of the operations which are processed on the same machine and do not require to be consecutive. The setup task of operation O_{gj} cannot be started until machine j has finished the processing task of operation O_{ij}. Constraint (8) describes the tardiness for job i. Constraint (9) expresses the order of any two operations of a job; if $Y_{ijh} = 1$ then $Y_{ihj} = 0$; otherwise, $Y_{ijh} = 0$ and $Y_{ihj} = 1$. Constraint (10)expresses the order of any operation pairs (O_{ij}, O_{gj}) on the same machine j; if $X_{igj} = 1$ then $X_{gij} = 0$, otherwise, $X_{igj} = 0$ and $X_{gij} = 1$. Constraints (11) and (12) describe the relationship between X_{kij} and Z_{kij}; if $X_{kij} = 1$, then $Z_{kij} = 1$ or 0; otherwise, $X_{kij} = 0$ and $Z_{kij} = 0$, by considering the dummy job 0. If $X_{kij} = 1$ then $Z_{ikj} = 0$; otherwise, $X_{kij} = 0$ and $Z_{ikj} = 1$ or 0; in this case, the dummy job 0 is not considered. Because the dummy job cannot be located after any job, it is only used for characterizing the first job on each machine to apply the corresponding relative setup time. Constraint (13) indicates that there is at most one job which can be processed immediately after job k on machine j. if job k is the last job on machine j, then $Z_{kij} = 0$. Constraint (14) indicates that there is only one job that can be processed immediately before job i on machine j by considering the dummy job 0. Constraint (15) expresses that all jobs should be available for scheduling at time 0.Constraints (16) to (18) define the continuous and binary decision variables, respectively.

Interactive MODM Approach

To solve the original bi-objective decision making (BODM) problem, an interactive fuzzy programming solution method, called the TH method proposed by Torabi and Hassini (2008), is used. This method is applied to achieve the Pareto-optimal solutions of the presented bi-objective crisp model. Torabi and Hassini (2008) proved that the TH method obtains efficient solutions for the original multi-objective model. According to the characteristics of the given problems, the steps of the TH method are as follows.

Algorithm1: TH method
Step 1
Calculate the positive ideal solution (PIS) and the negative ideal solution (NIS) for each objective function by solving the corresponding MILP model given below.

$$Z_1^{PIS} = \min \sum_{i=1}^{n} T_i \qquad \text{s.t. } X \in F_{(x)} \tag{19}$$

$$Z_1^{NIS} = \max \sum_{i=1}^{n} T_i \qquad \text{s.t. } X \in F_{(x)} \tag{20}$$

$$Z_2^{PIS} = \min C_{max} \qquad \text{s.t. } X \in F_{(x)} \tag{21}$$

$$Z_2^{NIS} = \max C_{max} \qquad \text{s.t. } X \in F_{(x)} \tag{22}$$

where X is a feasible solution vector consists of all of the continuous and binary variables in the original model and $F_{(x)}$ denotes the feasible area consists of Constraints (3) to (18).

Obtainment of the above ideal solutions requires solving four mixed-integer linear programs. In order to reduce the computational time and complexity, we can calculate the PIS for each objective function by solving the corresponding MILP models given in Eqs. (19) and (21). Then, the negative ideal solutions can be determined by the use of the following heuristic rule:

$$Z_h^{NIS} = Max(Z_h(x_k^*)); \quad h = 1, 2, k = 1, 2 \tag{23}$$

It is noted that x_h^* and $Z_h(x_h^*)$ denote the decision vector associated with the PIS of the h-th objective function and the corresponding value of the h-th objective function, respectively (Torabi and Hassini, 2008). The related results are shown in Table 1.

Step 2
For each objective function, determine a linear membership function by:

$$\mu_{Z_1}(X) = \begin{cases} 1, & Z_1 < Z_1^{PIS} \\ \dfrac{Z_1^{NIS} - Z_1}{Z_1^{Nis} - Z_1^{PIS}}, & Z_1^{PIS} \leq Z_1 \leq Z_1^{NIS} \\ 0, & Z_1 > Z_1^{NIS} \end{cases} \tag{24}$$

$$\mu_{Z_2}(X) = \begin{cases} 1, & Z_2 < Z_2^{PIS} \\ \dfrac{Z_2^{NIS} - Z_2}{Z_2^{Nis} - Z_2^{PIS}}, & Z_2^{PIS} \leq Z_2 \leq Z_2^{NIS} \\ 0, & Z_2 > Z_2^{NIS} \end{cases} \tag{25}$$

Figure 1 illustrates the graph of these membership functions.

Table 1 Payoff results

	Z_1	Z_2
X_1^*	Z_1^{PIS}	Z_2^{NIS}
X_2^*	Z_1^{NIS}	Z_2^{PIS}

Step 3

Transform the BOMILP model into an equivalent single-objective MILP using the following auxiliary formulation.

$$\text{Max } \lambda_{(X)} = \gamma\lambda_0 + (1-\gamma)\sum_h \theta_h \mu_{Z_h}(X) \qquad (26)$$

s.t.

$$\lambda_0 \le \mu_{Z_h}(X), \quad h = 1,2 \qquad (27)$$

$$X \in F_{(X)} \qquad (28)$$

$$\lambda_0, \gamma \in [0,1] \qquad (29)$$

According to the two objective functions considered in the problem, Constraint (27) can be written by:

$$Z_1^{NIS} - \sum_{i=1}^{n} T_i \ge \lambda_0 (Z_1^{NIS} - Z_1^{PIS}) \qquad (30)$$

$$Z_2^{NIS} - C_{max} \ge \lambda_0 (Z_2^{NIS} - Z_2^{PIS}) \qquad (31)$$

where $\mu_{Z_h}(X)$ is the satisfaction degree of the h-th objective function and $\lambda_0 = \min_h\{\mu_{Z_h}(X)\}$ is the minimum satisfaction degree of objectives. Also, θ_h denotes the importance level of the h-th objective function such that $\sum_h \theta_h = 1, \theta_h > 0$. The θ_h parameters are determined linguistically by the decision maker based on her preference. Moreover, γ is the coefficient of compensation. By changing the values of this parameter in the interval [0,1], the TH method can obtain both unbalanced and balanced compromised solutions. It means that for higher values of γ, the solution method results bigger

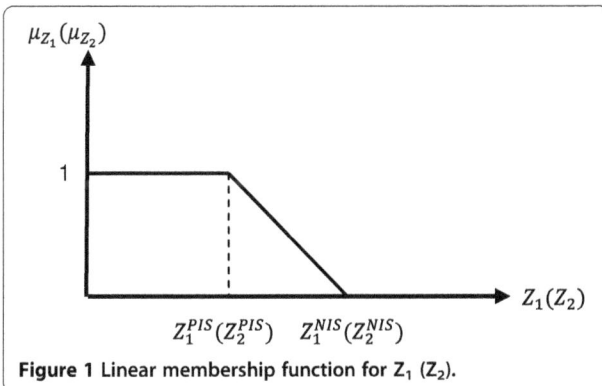

Figure 1 Linear membership function for Z_1 (Z_2).

lower bound for the satisfaction degrees of objectives (λ_0) for a given sample example. These solutions are balanced compromised solutions. These kinds of solution can be more appropriate when the importance levels of all objective functions are equal. On the other hand, for lower values of γ, the solution method resulting solutions with bigger satisfaction degrees for some objectives with higher importance levels than others. These solutions are unbalanced compromised solutions. These kinds of solutions can be more appropriate when the importance levels of the objective functions are different.

Step 4: Solve equivalent single-objective MILP model using the given coefficients (θ_h, γ). If the decision maker is satisfied with obtained efficient compromised solution, then stop; otherwise, change the value of some controllable parameters and then go to Step 2.

A set of non-dominated solutions as the Pareto-optimal solutions can be obtained for the BOMILP model by changing the values of controllable parameters of the MILP model given in Eqs. (26) to (31). As mentioned at the beginning of Section 3, it is proved that the TH method achieves efficient solutions for multi-objective optimization problems. It means that each optimal solution of the single-objective model given in Eqs. (26) to (31) is a Pareto-optimal solution of the original BOMILP model.

Proposed Method

The OSSP studied in this paper is the NP-hard problem. Thus, to solve medium to large-sized problems considered in this paper, an improved non-dominated sorting genetic algorithm II (NSGA-II) is proposed. Noori-Darvish and Tavakkoli-Moghaddam (2011) used the original NSGA-II for an OSSP. NSGA-II belongs to a class of multi-objective evolutionary algorithms (MOEAs). In most problems, it is able to find much better spread of solutions and better convergence near the true Pareto-optimal front compared to two other elitist MOEAs, namely Pareto-archived evolution strategy (PAES) and strength-Pareto evolutionary algorithm (SPEA), which pay special attention to creating a diverse Pareto-optimal front (Deb *et al.*, 2002).

Solution Representation

In our proposed method, a chromosome is an operation-based array or permutation list that has been a typical solution representation studied in OSSPs. As illustrated in Figure 2, the permutation list is a single-row array consisting of $n \times m$ operations. By using this encoding scheme, the basic assumptions of OSSPs are satisfied, in which the processing sequence of each job is immaterial and a job is not processed more than once by one machine. In this representation, operations are

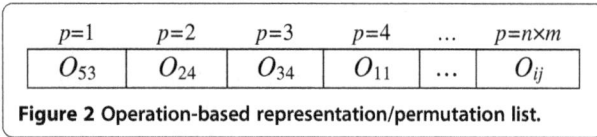

p=1	p=2	p=3	p=4	...	p=n×m
O_{53}	O_{24}	O_{34}	O_{11}	...	O_{ij}

Figure 2 Operation-based representation/permutation list.

listed in the relative order by which they are scheduled. p denotes the position of an operation in the list.

To decode the list presented in Figure 2 and obtain its equivalent solution in the solution space, we start from the beginning of the list and schedule the first operation. Then, according to the corresponding order of operations in the list, we schedule other operations one by one. It means, job 5 is first scheduled on machine 3. Then, job 2 is scheduled on machine 4 and so on. According to the rule that each operation is scheduled at the earliest time, there are two variables, namely 1) Max $\{C_{ij}\}$ for all i as the maximum completion time of jobs on machine j and 2) Max $\{C_{ij}\}$ for all j as the maximum completion time of job i. The starting time and completion time of each operation are computed by:

$$Ts_{ij} = \text{Max}\{0, \underset{\forall i}{\text{Max}}\{C_{ij}\}, \underset{\forall j}{\text{Max}}\{C_{ij}\}\} \quad (32)$$

$$C_{ij} = Ts_{ij} + \sum_{k=0,k\neq i}^{n}(S_{kij} \times Z_{kij}) + p_{ij} \quad (33)$$

Initialization

A good initial population can help an evolutionary algorithm to start multi-objective optimization from good solutions in the solution space and have a better performance that will be discussed in the next section. In our algorithm, a heuristic method is applied to generate the initial set of solutions or population. It first constructs a set of random sequences of operations as chromosomes (as many as the population size). By using the swapping of two adjacent or non-adjacent operations (see Figures 3 and 4), it generates all the feasible sequences in the neighborhood of each permutation list. For each chromosome, the best feasible solution obtained by these neighborhood searches is a member of the initial population. According to the dominance concept, the best solution is an efficient solution of the Pareto-optimal frontier. The set of Pareto-optimal solutions dominate any other solutions in the feasible area and improve all the objective functions simultaneously.

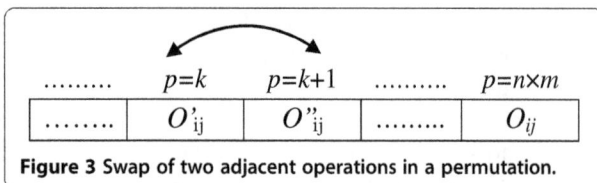

Figure 3 Swap of two adjacent operations in a permutation.

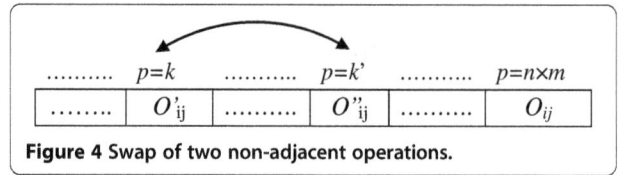

Figure 4 Swap of two non-adjacent operations.

It should be noted that $k,k' = 1,2,\ldots, n\times m$, as shown in Figures 3 and 4.

Crossover

A procedure proposed by Low and Yeh (2008) is used for the crossover operator.

Algorithm 2: Crossover operator

Step 1 Select two cut points randomly along the positions of the strings for each pair of parent strings (A and B). So, each permutation list is divided into three sections, called substring 1, substring 2 and substring 3.

Step 2 Exchange substring 1 of parent string A and substring 3 of parent string B. Similarly, exchange substring 3 of parent string A and substring 1 of parent string B. Do not change the elements of substring 2 of each parent string. Therefore, two proto-children are generated which are named A_0 and B_0.

Step 3 Legalize two generated offsprings A_0 and B_0. In order to do this, remove the elements of substring 1 and substring 3 of offspring string A_0, which are the same as substring 2, and then replace them with corresponding elements of substring 2 of offspring string B_0. Also, do the same procedure for offspring string B_0. Thus, two offspring strings (i.e., A_0 and B_0) are produced.

Mutation

As depicted in Figure 5, the insertion operator is applied for the mutation operator. In the permutation shown in this figure, $k,k' = 1,2,\ldots,n\times m$.

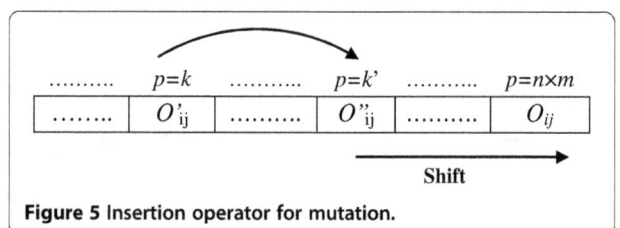

Figure 5 Insertion operator for mutation.

Table 2 Parameters of the NSGA-II

Population size	Maximum number of generations	Crossover rate	Mutation rate
100	300	0.95	0.02

Elitism and selection operator

Deb et al. (2002) designed an elitist NSGA including "Fast non-dominated sorting approach" and "Diversity Preservation" for solving multi-objective problems. In order to rank solutions (individuals) and sort them into different non-dominated frontiers, an approach called "Fast non-dominated sorting" is applied. By the use of this approach, a domination count (n_p) is calculated for each solution to specify its non-dominated frontier. The domination count of all solutions in the first non-dominated frontier ($i_{rank}=1$) is equal to zero. At the end of the multi-objective optimization process, solutions with $n_p=0$, dominate all other solutions in the solution space. It means they are the best global solutions. This approach results better convergence near the true Pareto-optimal frontier.

Along with the convergence to the Pareto-optimal set, it is also desired that an EA maintains a good spread of solutions in the obtained set of solutions (Deb et al., 2002). An operator called crowded-comparison operator is proposed for "Diversity Preservation". By considering the value of crowding distance of each solution ($i_{distance}$) in its frontier, this operator guides the selection process at the various stages of the algorithm toward a uniformly spread-out Pareto-optimal frontier. When the improved NSGA-II is iterated as many as the pre-specified iterations (i.e., maximum number of generations), the multi-objective optimization process is terminated (Noori-Darvish and Tavakkoli-Moghaddam, 2011).

Results and discussion

Several numerical examples in small to large sizes are generated randomly by the use of a classic approach of the literature in order to examine and analyze the validity and efficiency of the mathematical model presented in Section 2, the TH method presented in Section 3, and the performance of the improved NSGA-II presented in Section 4. The small-sized problems are solved exactly by the use of the Lingo 9 software and the results of the TH method are analyzed. If the number of jobs and the number of machines are more than 4, the sizes of the problems are medium to large. In

Table 3 Parameters of the SPEA-II

Population size	Maximum number of generations	Selection rate	Crossover rate	Mutation rate
100	400	0.8	0.8	0.2

Table 4 Values of parameters for numerical instances

Numerical instances	R	T
a	1	0.8
b	0.6	0.6
c	0.2	0.4

these conditions, even after several hours of running, Lingo cannot yield an optimal solution in a reasonable time. Thus, the improved NSGA-II is applied to solve these kinds of problems. Finally, the efficiency of this algorithm is first compared with the efficiency of a well-known multi-objective genetic algorithm, namely SPEA-II, by using the design of experiments (DOE). Then, the proposed algorithm is compared with the traditional NSGA-II. These algorithms are coded in Turbo C++4.5 on a PC (Main board P4, CPU E5200 2.5GHz, 6M Cache, RAM 2GB BUS 800).

SPEA-II

The strength Pareto evolutionary algorithm II (SPEA-II) proposed by Zitzler, et al. (2001) is specially designed for multi-objective optimization problems. This algorithm includes some special features (i.e., fitness assignment strategy) considering a number of solutions that dominate each individual and a number of solutions that each individual dominates the "nearest neighbor density estimation technique" that yields a density value for each solution, and the "archive truncation method" that preserves a specified number of solutions in the external non-dominated archive.

In our SPEAII, the solution representation is the permutation list and the initial population is generated randomly. Moreover, the binary tournament selection procedure is applied. When the SPEA-II is iterated as many as the pre-specified number of iterations (i.e., maximum number of generations), the multi-objective optimization process is terminated.

Generating numerical examples

In this section, we use a classical approach in the literature proposed by Loukil et al. (2005) to generate several numerical examples of small to large sizes

Table 5 Values of positive and negative ideal solutions

Z_2		Z_1		Sample
NIS	PIS	NIS	PIS	example
119	95	193	130	4×3-a
103	99	114	107	4×3-b
112	104	52	26	4×3-c
141	119	383	283	4×4-a
151	146	310	196	4×4-b
145	135	249	136	4×4-c

randomly. The processing times and due dates are uniformly distributed in the intervals $[0,100]$ and $\left[P\left(1 - T - \frac{R}{2}\right), P\left(1 - T + \frac{R}{2}\right)\right]$, respectively.

where $P = (m + n - 1)\bar{P}$ and $\bar{P} = \sum_{i=1}^{n}\sum_{j=1}^{m} p_{ij}/(n \times m)$ which is the mean values of the processing times. Parameters R and T take their values in the sets $\{0.2,0.6,1\}$, $\{0.4,0.6,0.8\}$,respectively. Also, the setup times are random variables between 10 and 50.

Parameter setting

Various sets of controllable parameters and different sizes of problems are considered. Then, many experiments are designed and run by using them. The effective values in terms of solution quality are determined. The parameter of the TH method are as follows. Parameter γ takes its values in the set $\{0,0.1,...,1\}$. Also, it is assumed that the preference information corresponding to the importance levels of the objective functions are specified linguistically by the decision maker as: $\theta_1 = \theta_2$ and $\theta_1 > \theta_2$. So, the values of these parameters in the first condition are $\theta_1 = \theta_2 = 0.5$, and in the second condition are $\theta_1 = 0.8$ and $\theta_2 = 0.2$. It should be mentioned that the values of controllable parameters (i.e., R and T) of the generating instances approach are given in the following subsections. Tables 2 and 3 depict the values are set for the parameters of the NSGA-II and SPEA-II, respectively. Moreover, each example is solved 10 times independently.

Solving small-sized problems

Two kinds of sample examples in small sizes with 4 jobs and 3 machines (i.e., 4×3), and 4 jobs and 4 machines (i.e., 4×4) are considered. For each type of these examples, three numerical instances are generated randomly. Tables 4 and 5 represent the values of controllable

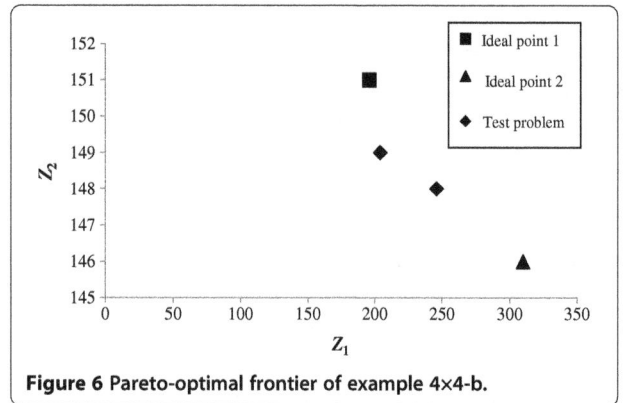

Figure 6 Pareto-optimal frontier of example 4×4-b.

parameters and the values of positive ideal solutions (Z_i^{PIS}) and negative ideal solutions (Z_i^{NIS}) for each numerical instances, respectively. Using the obtained values of the PIS and the NIS of each objective function, the final MILP model is exactly solved for each numerical instances by the Lingo 9 software. Table 6 illustrates the results of example 4×4-b.

The obtained resuts are analyzed as follows.

- In more cases with different importance levels of objective functions, the TH method performs correctly and efficiently. It means according to the importance levels , the solutions found by this method are unbalanced compromised solutions. However, in some cases, the TH method does not perform well and the satisfaction degree of the objective function with lower importance is more than that of the objective function with higher importance.

- When the importance levels are equal, the satisfaction degrees of objective functions are very close to each other, in some cases. Thus, the method approximately finds balanced compromised solutions.

- According to the discussion in Section 3, each optimal solution of the final MILP model is an efficient solution to the BOMILP model. Thus, in all cases, by changing the values of controllable parameters (i.e. γ and θ) two Pareto-optimal solutions are found by the TH method. Figure 6 indicates the Pareto-optimal frontier of example 4×4-b.

Table 6 Computational results of example 4×4-b

$\theta_1 = 0.8$, $\theta_2 = 0.2$				$\theta_1 = 0.5$, $\theta_2 = 0.5$				γ
μ_{Z_2}	μ_{Z_1}	Z_2	Z_1	μ_{Z_2}	μ_{Z_1}	Z_2	Z_1	
0.4000	0.9298	149	204	0.4000	0.9298	149	204	0
0.4000	0.9298	149	204	0.6000	0.5614	148	246	0.1
0.4000	0.9298	149	204	0.6000	0.5614	148	246	0.2
0.4000	0.9298	149	204	0.6000	0.5614	148	246	0.3
0.4000	0.9298	149	204	0.6000	0.5614	148	246	0.4
0.6000	0.5614	148	246	0.6000	0.5614	148	246	0.5
0.6000	0.5614	148	246	0.6000	0.5614	148	246	0.6
0.6000	0.5614	148	246	0.6000	0.5614	148	246	0.7
0.6000	0.5614	148	246	0.6000	0.5614	148	246	0.8
0.6000	0.5614	148	246	0.6000	0.5614	148	246	0.9
0.6000	0.5614	148	246	0.6000	0.5614	148	246	1
			$T = 0.6$		$R = 0.6$			

Table 7 Characteristic of medium to large-sized examples

Number of machines	Number of jobs	Representation	Sample example
5	10	10×5	1
7	14	14×7	2
9	20	20×9	3

Table 8 Computational results of the quality metric

Sample examples	Improved NSGAII										SPEAII									
	1	2	3	4	5	6	7	8	9	10	1	2	3	4	5	6	7	8	9	10
10×5-a	3	4	5	3	5	3	4	4	4	3	3	3	2	3	2	1	3	2	3	0
10×5-b	4	4	4	3	3	5	5	4	3	2	2	1	3	2	4	2	3	3	0	3
10×5-c	5	3	5	5	1	2	2	4	4	5	1	3	3	2	2	0	1	1	4	2
14×7-a	5	5	6	4	7	5	6	4	5	6	4	1	2	3	1	4	5	0	3	3
14×7-b	6	4	1	2	6	5	5	4	6	7	1	1	4	5	2	3	1	3	4	4
14×7-c	4	4	7	7	6	4	4	1	3	3	0	4	6	5	2	4	3	1	1	5
20×9-a	7	8	6	6	8	4	4	3	5	5	5	5	4	5	2	2	1	2	6	7
20×9-b	1	5	4	4	6	8	3	4	3	6	2	1	3	2	3	5	4	4	1	5
20×9-c	6	8	5	7	7	4	5	8	6	5	5	6	2	2	3	5	1	4	4	1

Solving medium to large-sized problems

In this section, three types of sample examples in medium to large sizes with different combinations of jobs and machines are designed. For each kind of these examples, three numerical instances are generated randomly. The values of controllable parameters for each instance are considered as shown in Table 4 and the characteristic of sample examples are presented in Table 7. All of these instances are solved independently by the proposed improved NSGA-II with the features described in Section 4 and SPEA-II. The performance of the improved NSGA-II is compared with the performance of the SPEA-II by using the design of experiments (DOE) based on three comparison metrics. Meanwhile, the instance 20×9-c is solved by traditional NSGA-II in order to compare the efficiency of the improved NSGA-II with the efficiency of the traditional NSGA-II.

Performance evaluation metrics

There are various metrics in the literature for evaluating the performance of multi-objective metaheuristics. Here we use three common and valid metrics to evaluate the performance of the improved NSGA-II, traditional NSGA-II and SPEA-II. These comparison metrics are defined as follows:

- *Quality Metric (QM)*: This kind of the metric was applied by Schaffer (1985). According to this metric, each algorithm that finds more Pareto solutions is considered to have a higher quality. However, some Pareto solutions of an algorithm may dominate those of another algorithm. Thus, a number of the final non-dominated solutions found by each algorithm are counted.
- *Diversity Metric (DM)*: This metric was applied by Zitzler (1999). To calculate the spread of the solutions in the final Pareto frontier found by each algorithm, the diversity metric is used and computed by:

$$D = \sqrt{\sum_{i=1}^{n} \max(x_i - y_i; \rightarrow x, \rightarrow y \in F)} \quad (34)$$

where, F is the set of obtained Pareto solutions, $\bar{x}$ and $\bar{y}$ are two solution vectors of Pareto frontier, and n denotes the dimension of the solution space that is equal to the number of objective functions. In this paper, n is equal to 2.

Table 9 Computational results of the diversity metric

Sample examples	Improved NSGA-II										SPEA-II									
	1	2	3	4	5	6	7	8	9	10	1	2	3	4	5	6	7	8	9	10
10×5-a	342.1	200.2	95.2	109.4	287.3	311.5	364	215.1	390.6	373.8	105.4	125.8	67.3	258.3	304.1	111.5	279.9	299.4	178.2	214.9
10×5-b	312.2	422.8	255	361.7	288.7	346.3	342.9	476.5	400.8	312.4	222.8	315.4	321.1	213.3	223.1	85.7	301	158.9	400.1	112
10×5-c	488.8	405.6	346.1	369	287	241.3	211.9	247.3	376.9	398.6	223	302.8	210.9	117.3	241.8	92.1	355	403.6	323.3	398.9
14×7-a	412.1	532.3	574	316.6	348.9	499	296.5	488.7	366.6	505.8	528.1	398.1	145.6	503.3	202.9	403.8	365.5	244.3	307.1	413.2
14×7-b	546.8	569	346.1	310.2	445.5	563.1	549.9	411	343.3	406.3	401.3	321.8	303.9	286	442.2	431.1	514.6	102.8	536.4	432.3
14×7-c	455.2	512.2	414.4	326	263.9	343.7	417.3	501.1	333.3	470.8	378.9	225.3	328.5	312.5	436.1	337.1	203.3	220	241.1	236.7
20×9-a	512.2	479.4	585.5	568.7	385.8	531.4	473.8	508.2	434.9	621.5	474.2	99.1	333.3	326.2	344.5	228.9	466.1	389	371.4	352.1
20×9-b	612.5	418.7	565.5	673.8	597.9	507.2	659.3	677.2	510.1	564	569.3	347.5	512.9	319	546.1	463.9	564.1	385.7	236.8	510.9
20×9-c	491.6	638.4	594.7	736.8	500.3	595.4	670.6	617.2	644.6	591.4	533.3	358.4	478.1	419.3	516.4	320.2	596.9	488.4	349.1	573.6

Table 10 Computational results of the spacing metric

Sample	Improved NSGA-II										SPEA-II									
examples	1	2	3	4	5	6	7	8	9	10	1	2	3	4	5	6	7	8	9	10
10×5-a	0.25	0.04	1.33	2.01	3	0.90	0.41	3	0.64	1.88	0.95	1.85	0.95	3	2	0.63	1.44	0.72	1.76	3
10×5-b	0.41	2	0.05	0.14	2	0.18	0.99	3	0.33	1.11	0.07	0.58	0.49	2.64	0.88	2	3	3	0.66	0.51
10×5-c	0.69	2	3	0.95	0.67	2.35	1.13	0.19	0.85	1.44	3	2.25	3	3	2	0.71	0.58	3	2.41	0.26
14×7-a	0.19	0.05	0.03	0.26	1.29	1.35	2	0.14	0.96	0.90	0.66	0.73	0.46	2.27	0.43	3	3	1.69	2	1.74
14×7-b	0.03	0.16	0.33	0.97	1.64	2.87	1.41	0.77	0.46	1.36	0.26	0.48	2.33	2.14	0.51	3	2	1.26	3	3
14×7-c	1.40	0.16	1.36	2	0.39	0.87	0.58	0.97	1.23	0.46	2.14	0.82	0.27	0.33	1.65	2	3	2	2.88	1.92
20×9-a	2.59	0.90	0.26	0.99	3	1.21	0.03	0.60	0.08	0.01	2.51	2.39	1.23	0.48	3	3	0.12	2	1.01	3
20×9-b	0.12	0.84	1.61	0.43	0.80	2.41	1.02	0.32	0.63	0.11	1.20	1.37	0.29	0.80	0.86	3	3	2	0.72	2.11
20×9-c	0.19	0.01	0.29	0.94	1.04	0	1.17	0.55	0.39	0.82	1.82	2	3	2	0.85	0.32	1.66	0.25	0.06	0.34

- *Spacing Metric (SM)*: This kind of the metric was applied by Srinivas and Deb (1994). This metric is used for estimating the uniformity of the spread of the points in the final Pareto solution frontier found by each algorithm and is calculated by:

$$S = \frac{\Sigma_{i=1}^{N-1} |d_i - \bar{d}|}{(N-1)\bar{d}} \quad (35)$$

where, d_i represents the Euclidean distance between the consecutive solutions of the obtained Pareto solution set, $\bar{d}$ is the mean value of all Euclidean distances, and N denotes the number of the final obtained Pareto solutions. For this metric, a higher quality value is a lower value and, of course, the best value is equal to 0.

Comparative results of the improved NSGAII and SPEAII by DOE

All of the medium to large-sized problems defined in Tables 4 and 7 are solved by the improved NSGA-II and the SPEA-II. In order to analyze the results, two-factor factorial experiments are designed to examine the effect of the two solution methods and nine test problems on each of three comparison metrics with 10 times executions of each algorithm for each test problem. This approach is similar to the approach given in, (Noori-Darvish et al. 2011). The computational results of three comparison metrics, which are calculated for the sets of Pareto solutions obtained by the improved NSGA-II and

Table 11 ANOVA result for the quality metric

P-Value	F_0	MS	SS	DF	Source of Variation
0.000	4.28	10.118	80.944	8	Test problem
0.000	62.45	147.606	147.606	1	Solution Method
0.714	0.67	1.593	12.744	8	Interaction
		2.364	382.900	162	Error
			624.194	179	Total

the SPEA-II in each time of running, are illustrated in Tables 8, 9 to 10. The following statistical linear model (Montgomery and Design and analysis of experiments, 2001) can represent the results of these tables.

$$y_{ijk} = \mu + \tau_i + \beta_j + (\tau\beta)_{ij}$$
$$+ \epsilon_{ijk} \begin{cases} i = 1, 2, \ldots, 9 \\ j = 1, 2 \\ k = 1, 2, \ldots, 10 \end{cases} \quad (36)$$

where μ is a common effect for the whole experiment, τ_i is the effect of the i-th test problem, β_j is the effect of the j-th solution method, $(\tau\beta)_{ij}$ is the interaction effect of the i-th test problem and the j-th solution method, and ϵ_{ijk} is the random error.

It should be noted that the adequacy of factorial designs is checked formerly. The hypothesis tests are considered as follows. The row treatment (i.e., test problems) effects are equal to 0. The column treatment (i.e., solution methods) effects are equal to 0. The interactions are equal to

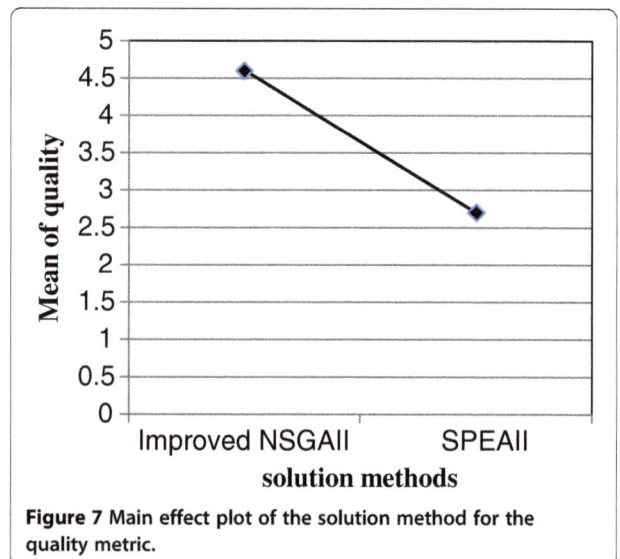

Figure 7 Main effect plot of the solution method for the quality metric.

Table 12 ANOVA result for the diversity metric

P-Value	F_0	MS	SS	DF	Source of Variation
0.000	19.86	200573	1604583	8	Test problem
0.000	56.35	568958	568958	1	Solution Method
0.455	0.99	10013	80101	8	Interaction
		10097	1635771	162	Error
			3889413	179	Total

Table 13 ANOVA result for the spacing metric

P-Value	F_0	MS	SS	DF	Source of Variation
0.310	1.19	1.0399	8.319	8	Test problem
0.000	23.44	20.5436	20.544	1	Solution Method
0.976	0.26	0.2318	1.854	8	Interaction
		0.8764	141.979	162	Error
			172.696	179	Total

0. In addition, the significance level (α) is set as 0.05.

$$H_0 : \tau_1 = \tau_2 = \cdots = \tau_9 = 0$$
$$H_1 : at\ least\ one\ \tau_i \neq 0 \tag{37}$$

$$H_0 : \beta_1 = \beta_2 = 0$$
$$H_1 : \beta_1 \neq \beta_2 \tag{38}$$

$$H_0 : (\tau\beta)_{ij} = 0$$
$$H_1 : at\ least\ one\ (\tau\beta)_{ij} \neq 0 \tag{39}$$

Table 11 depicts the ANOVA result for the QM. Considering the results, the p-value for the solution method and the test problem main effects are less than α=0.05. Thus, the effects of the solution method and the test problem are significant. It means that there is a significant difference between the mean values for the two solution methods and there is a significant difference between the mean values for the nine test problems. However, the interaction is not significant. Figure 7 indicates that more Pareto solutions are found by the improved NSGA-II, and this method performs better than the SPEA-II.

Table 12 shows the ANOVA result for the DM. Considering the results, the p-value for the main effects of the solution method and the test problem are less than α=0.05. Therefore, the method effects of the solution

method and the test problem are significant. However, the interaction is not significant. The diversity of solutions found by the improved NSGA-II is more than the SPEAII. Thus, the improved NSGA-II method performs better than the SPEA-II, as shown in Figure 8.

The ANOVA result for the SM, which are illustrated in Table 13, shows that the p-value for the main effect of the solution method is less than α=0.05. Therefore, the effects of the solution method is significant. Moreover, the main effect of the test problem and the interaction are not significant. The spacing metric value for solutions found by the improved NSGA-II is less than the SPEA-II, as shown in Figure 9. Thus, the performance of the improved NSGA-II is better than the SPEA-II.

Considering the above descriptions, all of the three hypothesis tests indicate the significant effect and better performance of the improved NSGA-II than the SPEA-II based on three performance metrics.

Comparative results of the improved and the traditional NSGAII

As discussed in Subsection 5.5, in order to examine if our improved NSGA-II performs better than the original NSGA-II, one of the large-sized instances (i.e., 20×9-c) is solved for 10 times independently by using the original NSGA-II. Then, three-mentioned comparison metrics

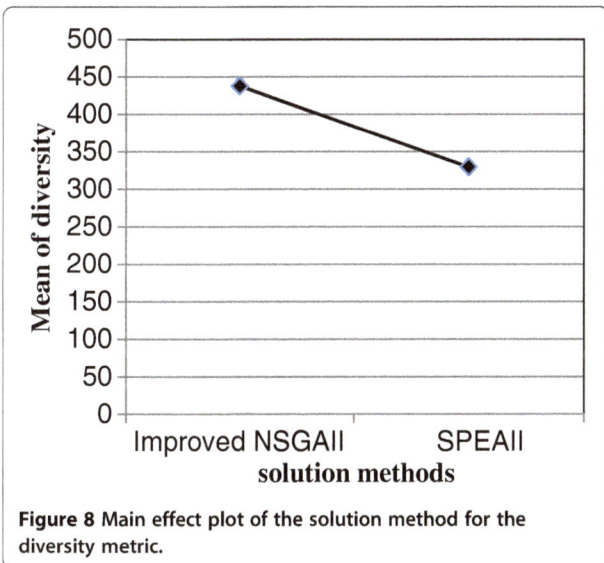

Figure 8 Main effect plot of the solution method for the diversity metric.

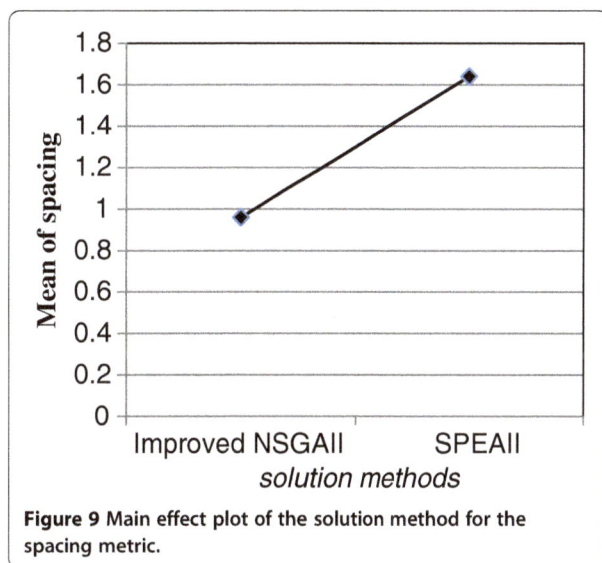

Figure 9 Main effect plot of the solution method for the spacing metric.

Table 14 Computational results of the QM, DM and SM of example 20×9-c solved by the improved NSGA-II

Comparison metrics	Improved NSGA-II										Mean value
	1	2	3	4	5	6	7	8	9	10	
QM	6	8	5	7	7	4	5	8	6	5	6.1
DM	491.6	638.4	594.7	736.8	500.3	595.4	670.6	617.2	644.6	591.4	608.1
SM	0.19	0.01	0.29	0.94	1.04	0	1.17	0.55	0.39	0.82	0.54

(i.e., QM, DM and SM) are calculated based on the set of Pareto solutions achieved by this algorithm in each time of running. These results are compared with the results of solving the instance 20×9-c by the improved NSGA-II, as shown in Tables 8 to 10. The computational results and the mean values of these three performance metrics are illustrated in Tables 14 and 15.

According to the obtained results, the mean values of the quality and diversity metrics calculated for the sets of Pareto solutions, which are found by the improved NSGA-II, are more than those of the original NSGA-II. In addition, the mean value of the spacing metric calculated for the sets of Pareto solutions obtained by improved NSGA-II are less than those of the traditional NSGA-II. Therefore, considering the above descriptions, all results indicate the better performance of the improved NSGA-II than the traditional NSGAII based on three performance metrics.

Conclusion

In this paper, an open shop scheduling problem with sequence-dependent setup times was examined. A novel bi-objective mathematical programming was designed in order to minimize the total tardiness and the makespan. An interactive multi-objective decision making (MODM) approach proposed by Torabi and Hassini (2008) was applied for solving small-sized instances optimally and obtaining Pareto-optimal solutions. Considering the results in more cases, the TH method has performed well according to the importance levels of the objective functions, and two Pareto-optimal solutions have been found. In order to achieve Pareto-optimal sets for medium to large-sized problems, an improved non-dominated sorting genetic algorithm II (NSGA-II) was presented by embedding a heuristic method for obtaining the good initial population. Finally, by using the design of experiments (DOE), the efficiency of the proposed improved NSGA-II

was compared with the efficiency of the SPEA-II. Moreover, the performance of the improved NSGA-II was compared with the performance of the traditional NSGA-II. The results have indicated the better performance of the improved NSGA-II based on three performance metrics.

Competing interests
The authors declare that they have no competing interests.

Authors' contributions
SND studied the literature review and proposed the problem solving methods. She analyzed the computational results and drafted the manuscript. RTM presented the mathematical model and was responsible for revising the whole manuscript. All authors read and approved the final manuscript.

Author details
[1]Department of Industrial Engineering, Allame Mohades Noori University, PC: 46415-451, Noor, Iran. [2]Department of Industrial Engineering, College of Engineering, University of Tehran, PC: 14399-57131, Tehran, Iran.

References
Ahmadizar F, Ghazanfari M, Fatemi Ghomi SMT (2010) Group shops scheduling with makespan criterion subject to random release dates and processing times. Comput Oper Res 37:152–162
Allahverdi A, Ng CT, Cheng TCE, Kovalyov MY (2008) A survey of scheduling problems with setup times or costs. European Journal Of Operational Research 187:985–1032
Deb K, Pratap A, Agarwal S, Meyarivan T (2002) A fast and elitist multi-objective genetic algorithm: NSGA-II. IEEE Trans Evol Comput 6:182–197
Fei H, Meskens N, Chu C (2010) A planning and scheduling problem for an operating theatre using an open scheduling strategy. Comput Ind Eng 58:221–230
Gonzalez T, Sahni S (1976) Open shop scheduling to minimize finish time. J ACM 23(4):665–679
Loukil T, Teghem J, Tuyttens D (2005) Solving multi-objective production scheduling problems using metaheuristics. Eur J Oper Res 161:42–61
Low C, Yeh Y (2008) Genetic algorithm-based heuristics for an open shop scheduling problem with setup, processing, and removal times separated. Robotics and Computer-Integrated Manufacturing 25:314–322
Mastrolilli M, Queyranne M, Schulz AS, Svensson O, Uhan NA (2010) Minimizing the sum of weighted completion times in a concurrent open shop. Oper Res Lett 38(5):390–395
Matta ME (2009) A genetic algorithm for the proportionate multiprocessor open shop. Comput Oper Res 36:2601–2618

Table 15 Computational results of the QM, DM and SM of example 20×9-c solved by the traditional NSGA-II

Comparison metrics	Traditional NSGA-II										Mean value
	1	2	3	4	5	6	7	8	9	10	
QM	6	5	7	7	6	4	6	4	7	6	5.8
DM	675.2	624.01	425.6	579.8	621.1	573.2	566.1	599.7	601.4	605.1	587.1
SM	0.28	0.39	0.26	0.47	0.31	1.01	0.7	1.23	0.15	0.92	0.57

Montgomery DC (2001) Design and analysis of experiments, 5th edn. John Wiley & Sons, INC, New York

Mosheiov G, Oron D (2008) Open-shop batch scheduling with identical jobs. Eur J Oper Res 187:1282–1292

Naderi B, FatemiGhomi SMT, Aminnayeri M, Zandieh M (2011) Scheduling open shops with parallel machines to minimize total completion time. J Comput Appl Math 235:1275–1287

Noori-Darvish S, Tavakkoli-Moghaddam R (2011) Solving a bi-objective open shop scheduling problem with fuzzy parameters. Journal of Applied Operational Research 3(2):59–74

Noori-Darvish S, Tavakkoli-Moghaddam R, Javadian N (2011) A multi-objective particle swarm optimization algorithm for possibilistic open shop problems with weighted mean tardiness and weighted mean completion times criterions. Iranian Journal of Operations Research (IJOR), Accepted for publication

Panahi H, Tavakkoli-Moghaddam R (2011) Solving a multi objective open shop scheduling problem by a novel hybrid ant colony optimization. Expert Syst Appl 38(3):2817–2822

Roshanaei V, SeyyedEsfehani MM, Zandieh M (2009) Integrating non-preemptive open shops scheduling with sequence-dependent setup times using advanced metaheuristic. Expert Syst Appl 37:259–266

Schaffer JD (1985) Multiple objective optimizations with vector evaluated genetic algorithms. In: Schaffer JD (ed) Genetic algorithms and their applications: Proceedings of the first international conference on genetic algorithms. Lawrence Erlbaum, Hillsdale, New Jersey, pp 93–100

Seraj O, Tavakkoli-Moghaddam R (2009) A tabu search method for a new bi-objective open shop scheduling problem by a fuzzy multi-objective decision making approach. Int J of Engineering, Transaction B: Application 22:269–282

Srinivas N, Deb K (1994) Multi-objective optimization using non-dominated sorting in genetic algorithms. Evol Comput 2:221–248

Sule DR (1997) Industrial scheduling. PWS Publishing Company, Boston, MA.

Torabi SA, Hassini E (2008) An interactive possibilistic programming approach for multiple objective supply chain master planning. Fuzzy Set Syst 159:193–214

Zandieh M, Fatemi Ghomi SMT, Moattar Husseini SM (2006) An immune algorithm approach to hybrid flowshops scheduling with sequence-dependent setup times. Appl Math Comput 180:111–127

Zhang X, van de Velde S (2010) On-line two machine open shop scheduling with time lags. European Journal Of Operational Research 204:14–19

Zitzler E (1999) Evolutionary algorithm for multi-objective optimization: methods and applications. Ph.D. Dissertation, Swiss Federal Institute of Technology (ETH), Zurich, Switzerland

Zitzler E, Laumanns M, Thiele L (2001) Improving the Strength Pareto Evolutionary Algorithm. In: Giannakoglou K, Tsahalis D, Periaux J, Papailou P, Fogarty T (eds) EUROGEN 2001, Evolutionary Methods for Design, Optimization and Control with Applications to Industrial Problems., Athens, Greece, 19-21 September 2011, Published by International Center for Numerical Methods in Engineering (Cmine), Barcelona, Spain, ISBN: 84-89925-97-6, pp 95–100

8

Supply chain network design problem for a new market opportunity in an agile manufacturing system

Reza Babazadeh, Jafar Razmi* and Reza Ghodsi

Abstract

The characteristics of today's competitive environment, such as the speed with which products are designed, manufactured, and distributed, and the need for higher responsiveness and lower operational cost, are forcing companies to search for innovative ways to do business. The concept of agile manufacturing has been proposed in response to these challenges for companies. This paper copes with the strategic and tactical level decisions in agile supply chain network design. An efficient mixed-integer linear programming model that is able to consider the key characteristics of agile supply chain such as direct shipments, outsourcing, different transportation modes, discount, alliance (process and information integration) between opened facilities, and maximum waiting time of customers for deliveries is developed. In addition, in the proposed model, the capacity of facilities is determined as decision variables, which are often assumed to be fixed. Computational results illustrate that the proposed model can be applied as a power tool in agile supply chain network design as well as in the integration of strategic decisions with tactical decisions.

Keywords: Supply chain management, Agile supply chain network design, Outsourcing, Responsiveness

Background

In recent years, the design and implementation of agile supply chain strategies have increasingly attracted interest and some companies such as Zara and Gina Tricot have achieved many advantages by employing agile strategy. Agile supply chain includes companies such as suppliers, production centers, and distribution centers, which are legally separate but, in terms of operations, are linked together by forward-flow materials and feedback information. Agile supply chain is focused on improving responsiveness, speed, and flexibility that is able to respond and react quickly and effectively to changing markets (Lin et al. 2006). In other words, agility is a term applied to an organization that has created the processes, tools, and training to enable it to respond quickly to customer needs and unforeseen market changes while still controlling costs and quality (Christopher et al. 2004). Traditional supply chains have long lead times and are forecast-driven, which makes these supply chains inventory-based, while the agile supply chain with quick response has shorter lead time and is demand-driven and information-based, in contrast with traditional supply chains. Additionally, while the market winning factor in agile supply chain is responsiveness and improved service level, in traditional supply chain, cost has been the market winning factor (Mason-Jones et al. 2000). Agile supply chain practices can be identified by four factors (Harrison et al. 1999):

1. Market sensitive. It is concerned with end customers in order to be able to specify customer needs and responds to them as soon as possible.
2. Virtual integration. It depends on information sharing along the supply chain.
3. Network-based. It provides flexibility by employing the strengths of specialism in each partner within the supply chain; therefore, it is critical to leverage the strengths and competencies of partners to realize quick responsiveness to market needs.

* Correspondence: jrazmi@ut.ac.ir
Department of Industrial Engineering, College of Engineering, University of Tehran, Tehran, Iran

4. Process integration. It is related to a high level of integration between partners within the supply chain and enables collaborative working methods such as joint product design. Therefore, the partners within the supply chain network will be able to improve a variety of products and deal with uncertainty.

Supply chain network design (SCND) decisions, as the most important strategic level decisions in supply chain management, are concerned with complex interrelationships between various tiers, such as suppliers, plants, distribution centers, and customer zones as well as determining the number, location, and capacity of facilities to meet customer needs effectively. Supply chain management integrates interrelationships between various entities through creating alliance, such as information system integration and process integration, between entities to improve response to customers in various aspects such as, higher product variety and quality, lower costs, and faster responses. One of the vital challenges for organizations in today's competitive markets is the need to respond to customer needs, which are very volatile and can occur in volume and in a variety of customer needs (Amir 2011). Agility with its various contexts is the most popular strategy that enables organizations to confront unstable and highly volatile customer demands. Since the SCND is the most important strategic level decision that affects the overall performance of supply chain, it is necessary to consider agility concepts such as response to customers in maximum allowable time, direct shipment, alliance (information and process integration) between entities in different echelons, discount to achieve competitive supply chain, outsourcing, using different transportation modes to achieve flexibility, and safety stock to improve responsiveness. It is evident that considering agility concepts in SCND plays an incredible role in agility of the overall supply chain. As yet, many researchers have tried to show the most important factors in agile supply chain management theoretically, and this context has been omitted in the mathematical modeling area especially in the supply chain network design area.

In this paper, to overcome literature gaps in agile supply chain network design, we present a mixed-integer linear programming (MILP) model that is able to consider agility concepts such as response to customers in maximum allowable time, direct shipments besides the traditional shipments, alliance between opened facilities in different tiers, safety stock, different transportation modes, discount, and outsourcing besides the traditional features of the SCND area. The reminder of this paper is organized as follows. The related works are reviewed. 'Description and formulation of the proposed model' includes the description and formulation of the proposed

model. The numerical results are reported in 'Results and discussion'. Finally, 'Conclusions' concludes this paper and offers some directions for further research.

Results and discussion
Computational results
In order to evaluate the performance of the presented model for designing agile supply chain network, two test problems are randomly generated according to the information specified in Table 1 inspired from literature. Interested readers can reach the detailed data set for the two test problems from the authors. Customer service level is considered to be 75% (i.e., $csl = 0.75$) in the two test problems. It should be mentioned that opening and closing a facility is a strategic decision, and a time-consuming and costly process. Therefore, changing facility location is impossible in the short run. On the other hand, determining the quantities of flow between the facilities of the network as a tactical decision is more flexible to change in the short run (Pishvaee et al. 2009). Therefore, strategic decisions (binary variables) should be determined independently from realizations, whereas tactical decisions can be changed and updated during the realizations.

The proposed model is solved by ILOG CPLEX 10.1 optimization software on a Pentium dual-core 2.60-GHz

Table 1 Data generation

Parameter	Value
d_k	~Unif (100, 200) unit
f_i	~Unif (1,700,000,4,000,000) $
g_j^r	~Unif (1,600,000,3,000,000)$
pr_i	~Unif (120,150)$
mh_j	~Unif (60,75) $
cs_i	~Unif (90,170)$
Ml_j	~Unif (80,100)$
MV_i	~Unif (90,110) unit
DP_i	~Unif (400,550) unit
CD_i	~Unif (10,500,14,500)$
a_k	~Unif (4,000,6,000)$
a_{ijn}	~Unif (40,60)$
b_{jkn}	~Unif (40,60)$
e_{ikn}	~Unif (90,130)$
ta_{ikn}	~Unif(10,16) day
tc_{jkn}	~Unif(6,12) day
te_k	~Unif(10,12) day
caw_j^r	~Unif (2,000,5,000) unit
cay_j^r	~Unif (2,000,3,500) unit
$CPA1_{im}$	~Unif (50,000,60,000)$
$CPA2_{jc}$	~Unif (30000,35,000)$
CPA_{ij}	~Unif (60,000,70,000)$

~Unif, uniform distribution.

Table 2 Computational results and complexity of presented models

Problem size $\|K\|*\|I\|*\|J\|*\|N\|*\|R\|$	O.F.V	Constraints	Variables	Run time(s)
10*5*8 *3*2	16,015,980	813	690	0.8
15*10*10 *3*2	22,917,570	1,578	1,861	29

O.F.V, objective function value.

computer with 4 GB RAM. To reduce the complexity of deterministic model, we relax the decision binary variables $PA1_{im}$, $PA2_{jc}$, and PA_{ij} as continuous variables. It should be noted that minimization of the objective function and binary variables W_i^r and Y_j^r in constraints (8) to (14) assures that the relaxed variables are set equal to 0 or 1.

Acquired results illustrated in Table 2 show that by increasing the size of the problem in question, objective function value (O.F.V) and complexity of the discussed problems increase. Table 3 demonstrates the share of different objective function components for the proposed model in details. As it is illustrated in Table 3, the fixed costs for opening facilities have the largest share in the objective function value for both models. Therefore, determining the number and location of facilities in each stage of the SCND process is the most important strategic decision which has effects on the overall performance of the SCN. Moreover, the total costs of production, outsourcing, inventory, and transportation are significant in the objective function. This is because the demand of customers must be satisfied based on predefined customer service level representing upstream of the supply chain management. Additionally, the share of outsourcing cost is lesser than production cost in the objective function value. It could be explained by the limitation on the majority of the products that could be outsourced (see constraint 8 in 'Description and formulation of the proposed model').

As described in previous sections, alliance between facilities in the supply chain network plays an important role in integration and success of the agile supply chain. The model pays alliance cost to improve information and process integration. In addition, although discount issue can improve competitiveness of the agile supply chain, its amount should be balanced with other costs.

As depicted in Figure 1, sensitivity analysis on demand of customers shows that the proposed model is sensitive to demand of customers. This observation can be

explained by the demand's effect on the overall supply chain network. In other words, the configuration of the supply chain network is constructed according to customer demands.

The considered agile supply chain network in this paper has a general structure that is able to support the agility concepts besides the traditional features of the SCND area and, therefore, could be applied to different kinds of industries with lead time restriction such as food industries and semiconductor industries. It can be concluded from the acquired results that the proposed model can be utilized as a power tool in practical cases with less shortage and high degree of responsiveness.

Conclusions

In this paper, we proposed a supply chain network design problem for a new market opportunity in an agile manufacturing system. The proposed agile SCND model is able to integrate production, outsourcing, discount, flexibility, and distribution activities by considering the most important factors of the agile supply chain. The experimental results show the efficiency of the proposed model in an agile supply chain network design. Also, it can be concluded from the obtained results that the locations of facilities in the supply chain network design are a strategic decision, and therefore, integration of facility location decisions with other decisions such as outsourcing, inventory control, production, etc. can improve supply chain performance and responsiveness.

Many possible future research avenues can be defined for future research directions. For example, addressing multi-product, multi-period agile supply chain network design under different kinds of operational and disruption risks is an attractive research direction with significant practical relevance. Moreover, time complexity is not addressed in this paper; however, this issue might be important in large-sized problems that the commercial solvers failed to solve; therefore, developing efficient exact, or heuristic solution methods can be interesting in this area.

Methods
Related works

In this section, we concisely review some SCND model developed recently. Melo et al. (2009) presented a general review on supply chain network design to identify basic features that such models must capture to support

Table 3 Share of different components of objective functions

Problem size $\|K\|*\|I\|*\|J\|*\|N\|*\|R\|$	Fixed costs	Production costs	Outsourcing costs	Inventory costs	Transportation costs	Alliance costs	Shortage costs	Discount amount
10*5*8 *3*2	13,403,300	1,220,666	30,400	12,130	801,739	583,819	0	36,076
15*10*10 *3*2	19,142,810	1,764,269	30,700	18,269	1,145,611	840,050	12,168	36,309

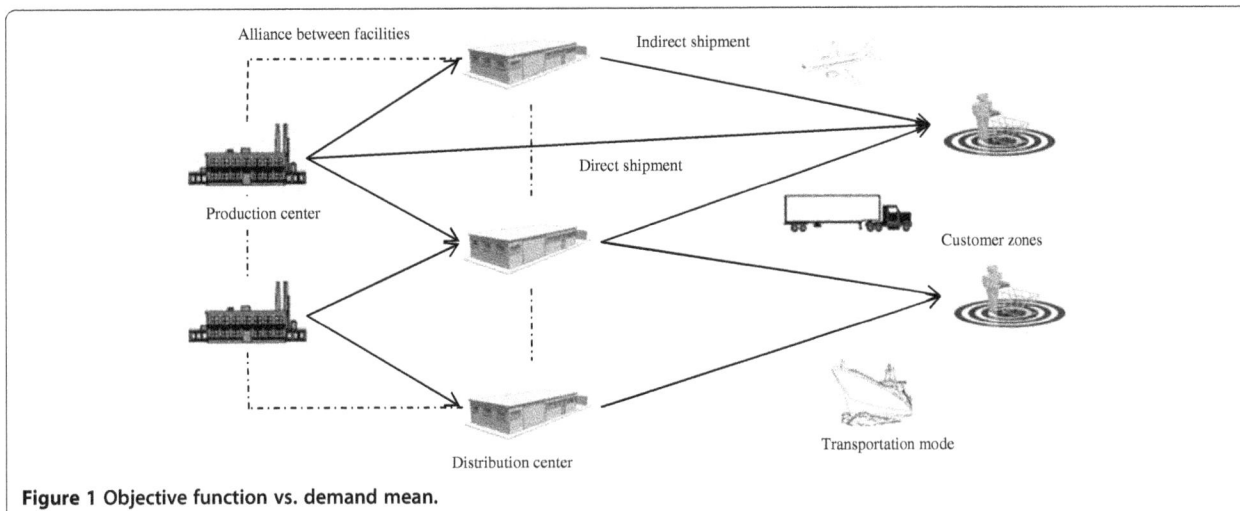

Figure 1 Objective function vs. demand mean.

decision making involved in strategic supply chain planning and to support a variety of future research directions. Other interesting reviews in this field can be found in Dullaert et al. (2007) and Snyder (2006). Most of the presented models in the literature focus on minimization of total costs and ignore other objectives such as responsiveness and flexibility, which are effective in the success of supply chain management. Zanjirani Farahani et al. (2010) gave a comprehensive review on multi-criteria facility location problems in the SCND. Their study shows that only approximately 8% of presented models have considered responsiveness as a determining factor in SCND. You and Grossmann (2008) developed a responsive bi-objective MINLP model by considering demand uncertainty and economic criterion. Their model was able to determine the safety stock levels to confront uncertainty. They used ε-constraint method to produce Pareto front. Rajabalipour Cheshmehgaz et al. (2011) presented a multi-objective, multi-stage, flexible model to design logistics network with the aim of minimizing response time and cost criteria. The efficient multi-objective evolutionary algorithm based on genetic algorithm (GA) was proposed. Ross and Jayaraman (2008) developed a MILP model to determine the location of the cross-docks in the SCND. Also, they utilized simulated annealing and Tabu search to solve the presented model. Amiri (2006) presented a MILP model to coordinate production and distribution activities. Additionally, the presented model was able to determine the optimum number, location, and capacity of facilities that should be opened. Altiparmak et al. (2006) developed a SCND model in a practical case and then proposed a GA by using priority-based encoding to escape from infeasible solutions. Thanh et al. (2008) proposed a dynamic MILP model for the facility location in the SCND. The proposed model includes strategic and tactical decisions.

As yet, among the numerous SCND models, which are presented to design SCN in an efficient way and improve supply chain management, agility as a winning factor in today's turbulent markets has been omitted. However, in recent years, some studies have tried to consider the agility concepts in SCND. Bachlaus et al. (2008) presented an integrated multi-objective MILP model to integrate production, distribution, and supply chain activities as strategic decisions considering agility as a key design criterion. They defined agility index in three levels including low, medium, and high for each facility, which should be satisfied in constraints. Additionally, they used flexibility as available capacity of each facility, which should be maximized in objective function. Pan and Nagi (2010) developed a robust optimization approach to deal with demand uncertainty in supply chain network in agile manufacturing. The presented model was able to consider alliance costs, which is one of the important factors of agile supply chain, between opened facilities. They proposed an efficient heuristic based on a k-shortest path algorithm to solve the presented model in large scales. However, their model doesn't cover distribution centers and delivery time to customer as well as one facility which should be determined in any echelon of supply chain network, and the robust optimization is based on scenarios. Hasani et al. (2011) presented a closed-loop supply chain network for perishable goods under uncertainty of demand and purchasing costs. Agility can be seen as handling products in their lifetime period. They used a bounded-box robust optimization approach to confront uncertainty in their model and solved it using LINGO.

Description and formulation of the proposed model

The concerned agile supply chain network in this paper is a multi-echelon and direct acyclic network, which integrates the production, outsourcing, flexibility, discount, safety stock, shortage, and distribution activities. In each echelon, there are some candidate facilities which should be determined to design the network in question. The discussed echelons are linked together with a forward shipment of products and backward flow of information in a pull system. As it is shown in Figure 2, the finished products manufactured in plants and semifinished products, which have been outsourced, are shipped to distribution centers or directly shipped to customer zones after processing in plants. The direct shipment of products includes shipping the products from production centers to the customer zones directly, while indirect shipment of products includes the shipping of goods from production centers to the distribution centers and then to the customer zones. Since direct shipment of products has higher costs with respect to indirect shipment, and there are budget limitations in this process, we assume that only a certain amount of products can be directly shipped. All direct and indirect shipments are performed in several transportation modes including land, sea, and/or air. Using different transportation modes improves the flexibility and responsiveness of the supply chain network in the sense that when utilizing a special transportation mode is not possible, the others could be efficiently used. Also, in critical conditions, the products could be delivered in maximum allowable time by using faster transportation modes to improve responsiveness. Some of the finished products in production centers shipped to distribution centers are stored, and the rest are shipped to customer zones according to customer needs. It should be noted that agile supply chains try to satisfy all customer demands; however, some customer needs may be not satisfied in the real world. Therefore, shortage can occur based on predefined customer service level.

To achieve flexibility and deal with disruption risks, the plants can perform outsourcing and use different transportation modes to deliver new products to customers. It is worthy to note that because of the

keeping independency, the maximum amount of outsourced products should be restricted so that the supply chain network will be able to respond to customer needs effectively when some of the suppliers are disrupted. Disruption of suppliers could occur because of natural disaster or other issues such as terrorist attacks and labors strike. As mentioned in previous sections, alliance between facilities is a key factor to improve agility of the supply chain. Creating alliance between opened facilities, selected to form the agile supply chain network, enables facilities to share their information, improve organizational educations, leverage organizational skills, and participate in new product development. Therefore, we assume that the supply chain network pays a certain cost to create alliance between opened facilities in different echelons. As regards delivery time to customers playing an incredible role in improving responsiveness to customers, we assume that all costumer demands should be met in their expected times. Another important issue in SCND is determining the capacity of facilities which should be opened. Most of the previously presented models consider fixed capacities for facilities, whereas determining the capacity of facilities is often difficult in practice (Wang et al. 2009). Therefore, it is assumed that the capacity level of facilities is determined as a decision variable to avoid additional and useless costs.

The following notation is used in the formulation of the proposed model.

Indices

I,M Index of candidate locations for plants ($i,m = 1, \ldots, I$)
J,C Index of candidate locations for distribution centers ($j,c = 1, \ldots, J$)
K Index of fixed locations of customers ($k = 1, \ldots, K$)
N Index of transportation modes ($n = 1, \ldots, N$)
R Index of capacity levels ($r = 1, \ldots, R$)

Parameters

d_k Demand of customer k
f_i^r Fixed cost of opening plant i with capacity level r
g_j^r Fixed cost of opening distribution center j with capacity level r
pr_i Production cost per unit of products at plant i
$CPA1_{im}$ Process and information (alliance) integration cost between plants i and m
$CPA2_{jc}$ Process and information integration cost between distribution centers j and c
CPA_{ij} Process and information integration cost between plants i and distribution center j
CD_i Discount amount by plant i to customers

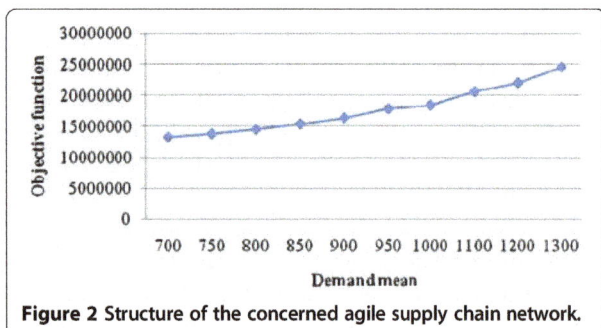

Figure 2 Structure of the concerned agile supply chain network.

α_k Penalty cost per unit of unsatisfied demand of customer k

mh_j Material handling and inventory cost per unit of products at distribution center j

oc_i Outsourcing cost per unit of products at plant i

a_{ijn} Unit transportation cost from plant i to distribution center j by mode n for product p

b_{jkn} Transportation cost per unit of products from distribution center j to customer k by mode n

e_{ikn} Transportation cost per unit of products from plant i to customer k by mode n

ta_{ikn} Delivery time from plant i to customer k by mode n

tc_{jkn} Delivery time from distribution center j to customer k

te_k Expected delivery time of customer k

MI_j Minimum inventory which should be held in distribution center j

MV_i Maximum amount of products which can be outsourced in plant i

DP_i Minimum amount required that plant i offers discount to customers

csl Percentage of predetermined customer service level

caw_i^r Capacity with level r for plant i

cay_j^r Capacity with level r for distribution center j

Decision variables

xu_i Quantity of products produced at plant i

In_j Quantity of products held at distribution center j

v_i Quantity of products outsourced at plant i

δ_k Quantity of unsatisfied demand of customer k

x_{ijn} Quantity of products shipped from plant i to distribution center j by mode n

Q_{jkn} Quantity of products shipped from distribution center j to customer k by mode n

L_{ikn} Quantity of products shipped directly from plant i to customer k by mode n

W_i^r 1 if plant i with capacity level r is opened; otherwise, 0

y_j^r 1 if distribution center j with capacity level r is opened; otherwise, 0

$PA1_{im}$ 1 if process and information integration is performed between plant i and plant m; otherwise, 0

$PA2_{jc}$ 1 if process and information integration is performed between distribution center j and distribution center c; otherwise, 0

PA_{ij} 1 if process and information integration is performed between plant i and distribution center j; otherwise, 0

λ_i 1 if plant i gives discount to customers, otherwise, 0

$$\min \sum_i \sum_r f_i^r w_i^r + \sum_j \sum_r g_j^r y_j^r + \sum_i (pr_i xu_i + oc_i v_i)$$
$$+ \sum_j mh_j In_j + \sum_i \sum_j \sum_n a_{ijn} x_{ijn}$$
$$+ \sum_j \sum_k \sum_n b_{jkn} Q_{jkn} + \sum_i \sum_k \sum_n e_{ikn} L_{ikn}$$
$$+ \sum_i \sum_{m \neq i} \frac{1}{2} CPA1_{im} PA1_{im} + \sum_j \sum_{c \neq j} \frac{1}{2} CPA2_{jc} PA2_{jc}$$
$$+ \sum_i \sum_j CPA_{ij} PA_{ij} + \sum_k \alpha_k \delta_k - \sum_i CD_i \lambda_i \quad \text{S.t.}$$
$$\tag{1}$$

$$\sum_i \sum_n L_{ikn} + \sum_j \sum_n Q_{jkn} + \delta_k \geq d_k \forall k \tag{2}$$

$$\delta_k \leq (1 - csl) d_k \forall k \tag{3}$$

$$\lambda_i \leq \frac{d_k}{DP_i} \forall i, k \tag{4}$$

$$\sum_i \sum_j \sum_n x_{ijn} + \sum_i \sum_k \sum_n L_{ikn} = \sum_i (xu_i + v_i) \tag{5}$$

$$\sum_i \sum_n x_{ijn} = \sum_k \sum_n Q_{jkn} + In_j \forall j \tag{6}$$

$$In_j \geq \sum_r y_j^r MI_j \forall j \tag{7}$$

$$v_i \leq \sum_r w_i^r MV_i \forall i \tag{8}$$

$$PA1_{im} \leq \sum_r w_i^r \forall i, m \neq i \tag{9}$$

$$PA1_{im} \geq \sum_r w_i^r + \sum_r w_m^r - 1 \forall i, m \neq i \tag{10}$$

$$PA2_{jc} \leq \sum_r y_j^r \forall j, c \neq j \tag{11}$$

$$PA2_{jc} \geq \sum_r y_j^r + \sum_r y_c^r - 1 \forall j, c \neq j \tag{12}$$

$$PA_{ij} \leq \sum_r w_i^r \forall i, j \tag{13}$$

$$PA_{ij} \leq \sum_r y_j^r \forall i, j \tag{14}$$

$$PA_{ij} \geq \sum_r w_i^r + \sum_r y_j^r - 1 \forall i, j \tag{15}$$

$$L_{ikn} ta_{ikn} \leq L_{ikn} te_k \forall i, k, n \tag{16}$$

$$Q_{jkn} tc_{jkn} \leq Q_{jkn} te_k \forall j, k, n \tag{17}$$

$$\lambda_i \leq \sum_r w_i^r \forall i \tag{18}$$

$$xu_i \leq \sum_r w_i^r caw_i^r \forall i \tag{19}$$

$$\sum_i \sum_n x_{ijn} \le \sum_r y_j^r \mathrm{cay}_j^r \forall j \tag{20}$$

$$\sum_j \sum_n x_{ijn} + \sum_k \sum_n L_{ikn} \le \sum_r w_i^r U \forall i \tag{21}$$

$$\mathrm{xu}_i \ge \mathrm{LB}_i \sum_r w_i^r \mathrm{caw}_i^r \forall i \tag{22}$$

$$\sum_i \sum_k \sum_n L_{ikn} \le \mathrm{UB} \tag{23}$$

$$\sum_r w_i^r \le 1 \forall i \tag{24}$$

$$\sum_r y_j^r \le 1 \forall j \tag{25}$$

$$\mathrm{xu}_i, \mathrm{In}_j, v_i, x_{ijn}, L_{ikn}, Q_{jkn}, \delta_k \ge 0 \ \forall i, j, k, n \tag{26}$$

$$w_i^r, y_j^r, \lambda_i, \mathrm{PA1}_{im}, \mathrm{PA2}_{jc}, \mathrm{PA}_{ij} \in \{0,1\} \forall i, j, r, m, c \tag{27}$$

Objective function (1), which minimizes the total costs, includes fixed opening costs, production cost, outsourcing cost, inventory holding cost, transportation and processing costs, alliance costs between opened facilities, and shortage costs as well as the amount of discount, is maximized to improve competitiveness of the agile supply chain network. Coefficient (1/2) is considered to avoid recalculation of alliance costs between opened facilities in the same echelon. Constraint (2) ensures that all demands of customers are not satisfied, and shortage is possible. Constraint (3) imposes the maximum allowable shortage based on predefined customer service level. Constraint (4) is inequality for decision making on giving discount to customers. It should be noted that since the discount amount is maximized in the objective function to improve the competency of supply chain network, the binary variable λ_i gives the value 1 when the customers purchase the certain amount of products specified to give discount (i.e., DP_i). Constraints (5) and (6) are the flow balances at the plants and warehouses, respectively. Constraint (7) imposes the minimum inventory which should be stored as safety stocks in opened warehouses. Constraint (8) assures that the amount of outsourced products doesn't exceed a certain amount to assure independency. Constraint (9) expresses that to create alliance between the production centers of the agile supply chain network, the production centers should be determined among the candidate locations with specified capacities in advance. Constraint (10) quarantines the creation of alliance between opened production centers in the first echelon. For example, assume a production center with capacity level 2 is established in location 3 (i.e., $w(3,2) = 1$), and another production center with capacity level 1 is established in location 4 (i.e., $w(4,1) = 1$), so the alliance between these production centers will be created

(i.e., $\mathrm{PA1}(3,4) \ge 1 + 1\ -1$). Note that the PA1 is a binary variable and so will be equal to 1. Constraints (11) and (12) are the same as constraints (9) and (10) to create alliance between opened distribution centers in the second echelon. Constraints (13) and (14) express that to create alliance between production and distribution centers in different echelons, they should be selected for establishing among the candidate locations in advance. Constraint (15) ensures that when the production center and distribution center are established in different echelons, the alliance will be created between them. Constraints (16) and (17) ensure that all the products are delivered to customers in the maximum allowable delivery time. Restriction (18) expresses that only opened plants are allowed to give discount to customers. Constraints (19) and (20) are capacity constraints in any facility. Constraint (21) ensures that only opened plants can send products to warehouses and customer centers. Constraint (22) expresses the utilization of lower-bound percentage of the capacity of opened plants. Constraint (23) expresses the limitation on the total number of products directly sent to customers. Constraints (24) and (25) ensure that any facility can be opened, at most, in one of the capacity levels. Finally, constraints (26) and (27) enforce the binary and non-negativity restrictions on corresponding decision variables.

Competing interests

The authors declare that they have no competing interests.

Authors' contributions

RB proposed the mathematical model and solved it. JR managed the study and was responsible for integrating and revising the manuscript. RG participated in the design of the study.

Acknowledgments

We would like to thank from anonymous referees for their constructive and valuable comments that improved the contents of the paper substantially. All authors read and approved the final manuscript.

References

Altiparmak F, Gen M, Lin L, Paksoy T (2006) A genetic algorithm approach for multi-objective optimization of supply chain networks. ComputInd Eng 51:197–216

Amiri A (2006) Designing a distribution network in a supply chain system: formulation and efficient solution procedure. Eur J Oper Res 171:567–576

Amir F (2011) Significance of lean, agile and leagile decoupling point in supply chain management. Journal of Economics and Behavioral Studies 3(5):287–295

Bachlaus M, Mayank KP, Chetan M, Ravi S, Tiwari MK (2008) Designing an integrated multi-echelon agile supply chain network: a hybrid taguchi-particle swarm optimization approach. J Intell Manuf 19:747–761

Christopher M, Lowson R, Peck H (2004) Creating agile supply chains in the fashion industry. International Journal of Retail & Distribution Management 32(8):367–376

Dullaert W, Braysy O, Goetschalckx M, Raa B (2007) Supply chain (re)design: support for managerial and policy decisions. Eur J Transp Infrastruct Res 7(2):73–91

Hasani A, Zegordi SH, Nikbakhsh H (2011) Robust closed-loop supply chain network design for perishable goods in agile manufacturing under uncertainty. Int J Prod Res. doi:10.1080/00207543.2011.625051

Harrison A, Hoek R, Christopher M (1999) Creating the Agile Supply Chain. Institute of Logistic &Transport, London

Lin CT, Chiu H, Chu PY (2006) Agility index in the supply chain. Int J Prod Econ 100(2):285–299

Mason-Jones R, Naylor JB, Towill DR (2000) Engineering the leagile supply chain. International Journal of Agile Manufacturing Systems 2(1):54–61

Melo MT, Nickel S, Saldanha-da-Gama F (2009) Facility location and supply chain management: a review. Eur J Oper Res 196:401–412

Pan F, Nagi R (2010) Robust supply chain design under uncertain demand in agile manufacturing. Comput Oper Res 37:668–683

Pishvaee MS, Jolai F, Razmi J (2009) A stochastic optimization model for integrated forward/reverse supply chain network design. J Manuf Syst 28:107–114

Rajabalipour Cheshmehgaz H, Ishak Desa M, Wibowo A (2011) A flexible three-level logistic network design considering cost and time criteria with a multi-objective evolutionary algorithm. J Intell Manuf. doi:10.1007/s10845-011-0584-7

Ross A, Jayaraman V (2008) An evaluation of new heuristics for the location of cross-docks distribution centers in supply chain network design. Comput Ind Eng 55:64–79. doi:10.1016/j.cie.2007.12.001

Snyder LV (2006) Facility location under uncertainty: a review. IIE Trans 38(7):547–564

Thanh PN, Bostel N, Peton O (2008) A dynamic model for facility location in the design of complex supply chains. Int J Prod Econ 113:678–693

Wang S, Watada J, Pedrycz W (2009) Value-at-risk-based multi-stage fuzzy facility location problems. IEEE transactions on industrial informatics 5:465–482

You F, Grossmann IE (2008) Design of responsive supply chains under demand uncertainty. Comput Chem Eng 32:3090–3111

Zanjirani Farahani R, Steadie Seifi M, Asgari N (2010) Multiple criteria facility location problems: a survey. Appl Math Model 34:1689–1709

9

Strategy-aligned fuzzy approach for market segment evaluation and selection: a modular decision support system by dynamic network process (DNP)

Ali Mohammadi Nasrabadi[1], Mohammad Hossein Hosseinpour[2] and Sadoullah Ebrahimnejad[3*]

Abstract

In competitive markets, market segmentation is a critical point of business, and it can be used as a generic strategy. In each segment, strategies lead companies to their targets; thus, segment selection and the application of the appropriate strategies over time are very important to achieve successful business. This paper aims to model a strategy-aligned fuzzy approach to market segment evaluation and selection. A modular decision support system (DSS) is developed to select an optimum segment with its appropriate strategies. The suggested DSS has two main modules. The first one is SPACE matrix which indicates the risk of each segment. Also, it determines the long-term strategies. The second module finds the most preferred segment-strategies over time. Dynamic network process is applied to prioritize segment-strategies according to five competitive force factors. There is vagueness in pairwise comparisons, and this vagueness has been modeled using fuzzy concepts. To clarify, an example is illustrated by a case study in Iran's coffee market. The results show that success possibility of segments could be different, and choosing the best ones could help companies to be sure in developing their business. Moreover, changing the priority of strategies over time indicates the importance of long-term planning. This fact has been supported by a case study on strategic priority difference in short- and long-term consideration.

Keywords: Market segmentation, Decision support system (DSS), Dynamic network process, Fuzzy logic, Risk

Background

Porter 1980) described a category scheme including three general types of strategies: Cost leadership, differentiation, and market segmentation which are commonly used by various businesses to achieve and maintain competitive advantages. These three generic strategies are defined along two dimensions: strategic scope and strategic strength. Strategic scope is a demand-side dimension and looks at the size and composition of the market you intend to target. Strategic strength is a supply-side dimension and looks at the strength or core competency of the firm. Market segmentation is narrow in scope when both cost leadership and differentiation are relatively broad in market

scope. Market segmentation divides the market into homogeneous groups of individual markets with similar purchasing response as a number of smaller markets have differences based on geography, demographics, firm graphics, behavior, decision-making processes, purchasing approaches, situation factors, personality, lifestyle, psychographics, and product usage (Aaker 1995; Bonoma and Shapiro 1983; Dickson 1993; Kotler 1997; Bock and Uncles 2002; Nakip 1999; File and Prince 1996). The results of segmentation could be improved considerably if information on competitors is considered in the process of market segmentation (Söllner and Rese 2001). Market segmentation allows the marketing program to focus on a special part of the market to increase its competitiveness by applying various strategies. These strategies can be new products development, differentiated marketing communications, advertisements creation, different customer services development, prospects

---
* Correspondence: ibrahimnejad@kiau.ac.ir
[3]Department of Industrial Engineering, Islamic Azad University, Karaj Branch, Alborz, Iran
Full list of author information is available at the end of the article

targeting with the greatest potential profits, and multi-channel distribution development. Many researchers developed the evaluation and selection of market segmentation methods to achieve more customer satisfaction by focusing on marketing programs designed to satisfy customer requirements efficiently. The vast majority of decision-making methods have focused on evaluating the different segmentation methods and techniques (Kuo et al. 2002; Lu 2003; Coughlan and Soberman 2005; Liu and Serfes 2007; Ou et al. 2009; Phillips et al. 2010; Tsai et al. 2011a, 2011b). In the market segment evaluation and selection, there are four stages or procedures that were introduced by Montoya-Weiss and Calentone (2001): problem structuring, segment formation, segment evaluation and selection, and description of segment strategy.

Distinction of segmentation at a strategic or at an operational level has been made by several authors such as Goller et al. (2002) and Sausen et al. (2005). The general assumption behind the dimension is that there is a fundamental difference in how the firm is affected by the segmentation (Clarke and Freytag 2008). At a strategic level, the consideration is on the top management level and concerns the creation of missions and strategic intent, and can become closely linked to the capabilities and nature of the organization (Jenkins and McDonald 1997). At the operational level, there is a concern for planning and operational schemes for reaching target segments with an effectively adjusted offering as well as monitoring the performance (Albert 2003). In a competitive market, strategies are critical points of business, which lead the companies towards their vision as their final destination. Strategy description and selection is an important part of strategic management process. Many approaches, techniques, and tools can be used to analyze strategic cases in this process (Dincer 2004). Ray (2000) applied strategic segmentation where, prior to price competition, each firm targets the information to specific consumers who are informed by a firm that they can buy from it.

Among the strategic tools, SPACE matrix (Rowe et al. 1982) is a common method. It is used as a strategy description and success evaluation technique that includes two dimensions: internal perspectives (financial strength (FS) and competitive advantage (CA)) and external perspectives (environmental stability (ES) and industry strength (IS)).

All marketing strategies include a search for competitive advantages (Bharadwaj and Varadarajan 1993; Day and Wensley 1988; Varadarajan and Cunningham 1995; Hunt and Arnett 2004). According to Söllner and Rese (2001), 'The consideration of competitive structure provides additional basic information on segment formation' and 'The consideration of competitive structure

facilitates the selection of promising segments'. SPACE matrix is a support tool for decision-making process, and it is very useful when the market competitiveness is a critical point of decision-making process. In this method, internal and external perspectives are evaluated according to the overall situation of the company in the market to build strategies basing on the factors in the four main groups (FS, CA, ES, and IS). These generic strategies are termed as 'defensive', 'aggressive', 'conservation', and 'competitive' which can be broken from the main strategies. Moreover, the SPACE matrix can indicate success possibility through the algebraic summation of the evaluated factor scores within its two dimensions (Figure 1).

In a dynamic and ever-changing world, the time frame is important for a segmentation process (Nakip 1999; Freytag and Clarke 2001). Market segments and strategies can be selected based on a set of factors and subfactors which vary over time. In competitive markets, the effects of time are more sensible on prioritization. It means that their priorities could be changed particularly when the factors are time-dependent. Saaty (2007) extended the analytic hierarchy process (AHP)/analytic network process (ANP) to deal with time-dependent priorities and referred them as dynamic hierarchy process (DHP)/dynamic network process (DNP). In his method, prioritization is done by considering the changes in the market over time, which affect the importance of factors. Moreover, the fuzzy concept has been applied to solve the problem due to the vagueness of the importance and the priorities of these factors. Below, Table 1 shows the recent works on this subject and summarizes their main considerable issues. As presented in the table, this work

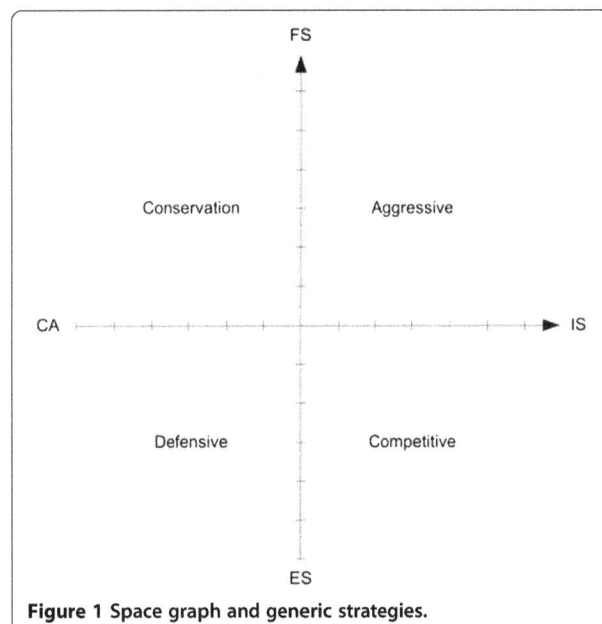

Figure 1 Space graph and generic strategies.

Table 1 Market segment evaluation and selection of literatures

Article	Segmentation by competitive factors	Strategy-aligned approach	Uncertainty issues	Interdependency	Risk analysis	Time-dependent decision making
Ou et al. (2009)	✓	✓	✓	-	-	-
Liu et al. (2010)	✓	-	-	-	-	-
Ren et al. (2010)	✓	✓	-	-	-	-
Tsai et al. (2011a)	✓	✓	-	-	-	-
Tsai et al. (2011b)	✓	-	-	✓	-	-
Xia (2011)	✓	✓	-	-	-	-
Aghdaie et al. (2011)	✓	-	✓	-	-	-
Shani et al. (2012)	✓	-	-	-	✓	-
Proposed model	✓	✓	✓	✓	✓	✓

'considers risk' and 'factors interdependency' in studying and in 'selecting the segment-strategies over time'.

As it is observed in the above table, recent researches have considered the important factors of this problem, but none of them provides a comprehensive model. Also, time-dependent decision making is an affective item that is provided in this paper, which was not considered in previous works.

In this paper, a strategy-aligned fuzzy approach is developed to select the best segment-strategy in market segment evaluation and selection problem. A modular decision-making process is implemented in two stages:

The first one selects the segments with more chance of success which has an acceptable risk in a competitive market according to their situation in the SPACE matrix. As the first contribution, by applying the SPACE matrix method, competition is taken into account by defining the distance of segments from the best situation of competitive advantage. On the other hand, this method can give an overall view of the competitive advantage of all segments with a risk evaluation of choosing the segments in a simple graph. The second stage is segment-strategy selection, considering that priorities change over time, by dynamic network process. As the second

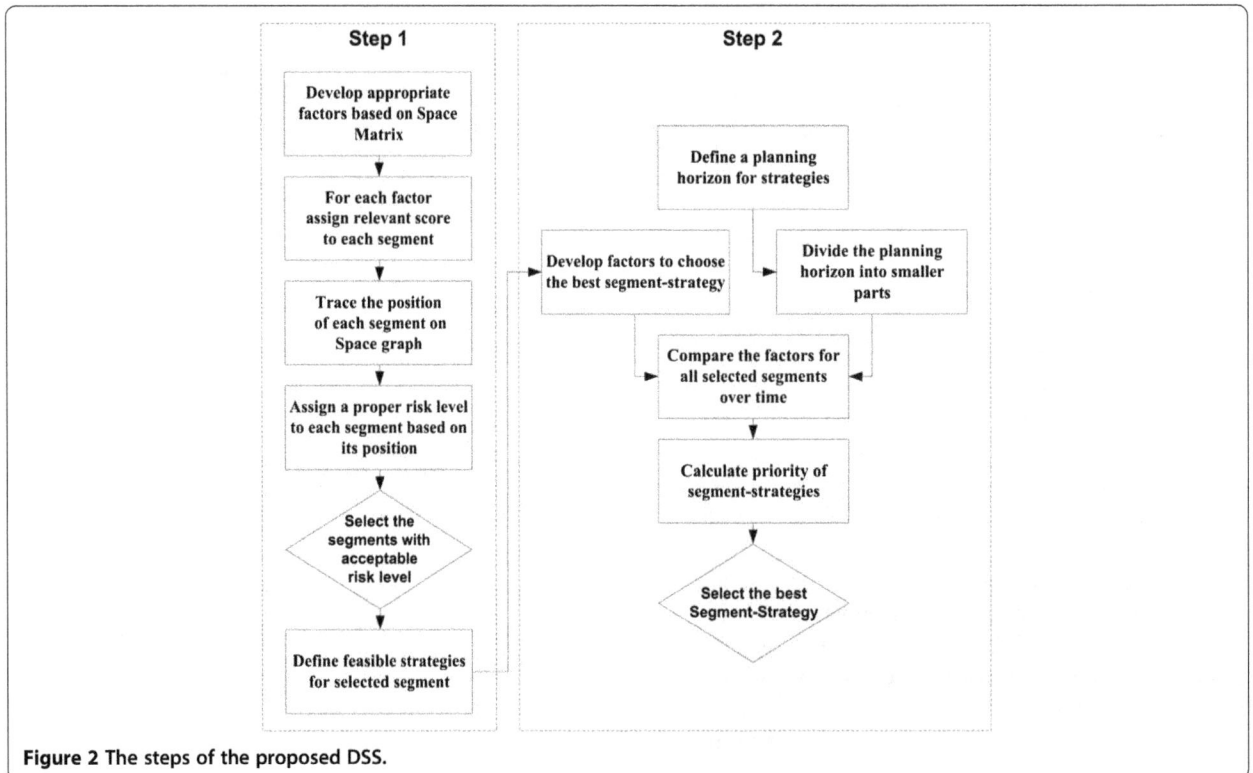

Figure 2 The steps of the proposed DSS.

Table 2 Mathematician's formulation of a dynamic judgment scale

Time dependent importance intensity	Description	Explanation
A	Constant for all t	No change in relative standing
$a_1t + a_2$	Linear relation in t, increasing or decreasing to a point, and then a constant value thereafter. Note that the reciprocal is a hyperbola	Steady increase in value of one activity over another
$b_1 \log(t + 1) + b_2$	Logarithmic growth up to a certain point and constant thereafter	Rapid increase (decrease) followed by slow increase (decrease)
$c_1 e^{c_2 t} + c_3$	Exponential growth (or decay if c_2 is negative) to a certain point and constant thereafter (not reciprocal of case c_2 is negative which results in a logistic S-curve)	Slow increase (decrease) followed by rapid increase (decrease)
$d_1 t^2 + d_2 t + d_3$	A parabola giving a maximum or minimum (depending on d_1 being negative or positive) with a constant value thereafter. May be modified for skewness to the right or left	Increase (decrease) to maximum (minimum) and then decrease (increase)
$e_1 t^n \sin(t + e_2) + e_3$	Oscillatory	Oscillates depending on n, $n \geq 0$ ($n \leq 0$) with decreasing (increasing) amplitude
Catastrophes	Discontinuities indicated	Sudden changes in intensity

contribution, segments are selected by considering the effect of time on the decision-making criteria. Moreover, the effects of strategies on changing the priorities are considered over time, and the trend of segment-strategy priorities can be determined in various time horizons. Porter's (1980) five force factors and sub-factors have been applied as well known decision-making criteria in dealing with competitive advantage. This approach defines the risk level of the segments; thus, decision makers (DMs) could select the appropriate segments according to their acceptable risk levels. Furthermore, they could select more exact strategies by focusing on selected segment. In addition, the proposed DNP method enables them to analyze segment-strategies over time, and this ability could affect their decision. The steps of the proposed DSS are shown in Figure 2.

The rest of this paper is organized as follows: In the 'Methods' section, the dynamic network process is shown including the explanation of its applications in the next section. The 'Fuzzy fundamental' section presents a brief overview of the fuzzy concepts. In section 'Fuzzy dynamic network process', the fuzzy DNP calculation method is presented. In the 'Results and discussion' section, a procedure for segment-strategy selection is introduced, including how to select an optimum solution. A case study with its computational results is also presented for the proposed model. The final section gives the conclusions and future works.

Methods

Time-dependent analytic network process

Market segment evaluation and selection can be classified as a multi-criteria decision-making (MCDM) problem. AHP is the well known and the most widely used method among several MCDM approaches such as SAW, MEW, TOPSIS, and ELECTRE. AHP was introduced by

Saaty (1980) for decision-making as a theory of relative measurement based on paired comparisons used to derive normalized absolute scales of numbers, the elements of which are then used as priorities. The ANP was developed and implemented by Saaty (1996) as an AHP with feedback. The ANP feedback approach replaces hierarchies with networks in which the relationships among the levels are not easily represented as higher or lower, dominant or subordinate, and direct or indirect (Meade and Sarkis 1999). In AHP and ANP, static and derived numbers are used to represent priorities. When the priorities vary across the time, AHP and ANP need to be dynamic through the use of numbers or functions and then derive either numbers that represent functions like expected values, or derive functions directly to represent priorities. Saaty (2007) extended the AHP/ANP to deal with time-dependent priorities and referred them as DHP/DNP. In this way, Saaty (2007) presented two methods: the (1) numerical solution of the principal eigenvalue problem by raising the matrix to powers and the (2) analytical solution of the principal eigenvalue problem by solving algebraic

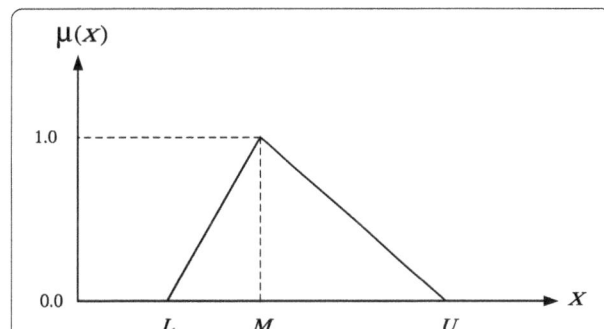

Figure 3 A triangular fuzzy number $\tilde{\xi} = (l, m, u)$. The broken line is a guide to present the position of the most promising value of a TFN (m), while the solid line denotes the membership values of a TFN.

Table 3 Linguistic scales for difficulty and importance

Linguistic scale for difficulty	Linguistic scale for importance	Triangular fuzzy scale	Triangular fuzzy reciprocal scale
Just equal	Just equal	(1, 1, 1)	(1, 1, 1)
Equally difficult	Equally important	(1/2, 1, 3/2)	(2/3, 1, 2)
Weakly more difficult	Weakly more important	(1, 3/2, 2)	(1/2, 2/3, 1)
Strongly more difficult	Strongly more important	(3/2, 2, 5/2)	(2/5, 1/2, 2/3)
Very strongly more difficult	Very strongly more important	(2, 5/2, 3)	(1/3, 2/5, 1/2)
Absolutely more difficult	Absolutely more important	(5/2, 3, 7/2)	(2/7, 1/3, 2/5)

equations of degree n. In ANP, the problem is to obtain the limiting result of powers of the super-matrix with dynamic priorities. Because its size will increase in the near future, the super-matrix would have to be solved numerically (Saaty 2007). In the numerical solution, the best fitting curves for the components of the eigenvector were obtained by plotting the principal eigenvector for the indicated values of t. In the analytical solution for the pairwise comparison judgments in dynamic conditions, Saaty (2007) purposed some functions for the dynamic judgments, which are given in Table 2.

To solve the problem and to obtain the time-dependent principal eigenvector, Saaty (2007) introduced the numerical approach by simulation, in which at first, the judgments express functionally but then derives the eigenvector from the judgments for a fixed instant of time, substitutes the numerical values of the eigenvectors obtained for that instant in a super-matrix, solves the super-matrix problem, and derives the priorities for the alternatives. This process is repeated for different values of time, which generates a curve for the priorities of the alternatives and then approximates these values by curves with a functional form for each component of

the eigenvector. This procedure is used in this paper to obtain the priorities of the alternatives in fuzzy dynamic network process (FDNP).

Why dynamic network process?

In a decision-making process of selecting market segments, priorities are calculated based on competitive factors with respect to some important criteria such as the effects of the interdependency among the factors and the trend of segment-strategy priorities in various time horizons. Dynamic network process as a powerful decision-making method can cover these important criteria by considering interdependency in networks in a dynamic decision-making process. Thus, DNP is a more useful method that can be applied to prioritize the alternatives in comparison with other decision-making processes.

Fuzzy fundamental

Fuzzy set theory was introduced by Zadeh (1965) to deal with the uncertainty caused by imprecision and vagueness in real world conditions. A fuzzy set is a class of objects with a continuum of grades of membership, which assigns to each object a grade of membership ranging between zero and one (Kahraman et al. 2003).

A triangular fuzzy number (TFN; $\tilde{\xi}$) with its membership function is shown in Figure 3. TFN can be denoted by (l, m, u), where the triplet (l, m, u) are crisp numbers and $l \leq m \leq u$. These parameters l, m and u denote

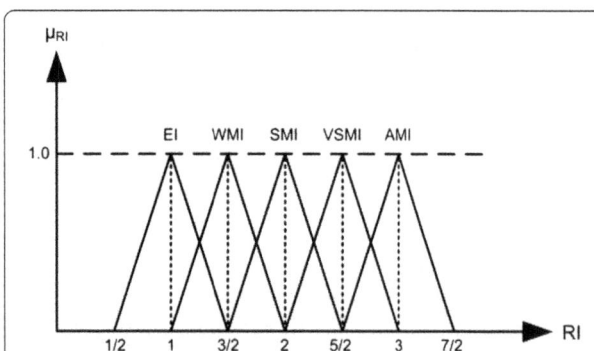

Figure 4 The membership functions of linguistic variables for importance weights. EI, equally important; WMI, weakly more important; SMI, strongly more important; VSMI, very strongly more important; AMI, absolutely more important. The broken line is a guide to present the position of the most promising value of a TFN (m), while the solid line denotes the membership values of a TFN.

Table 4 Linguistic values and mean of fuzzy numbers

Linguistic values for negative sub-factors	Linguistic values for positive sub-factors	The mean of fuzzy numbers
Very low	Very high	1
Low	High	0.75
Medium	Medium	0.5
High	Low	0.25
Very high	Very low	0

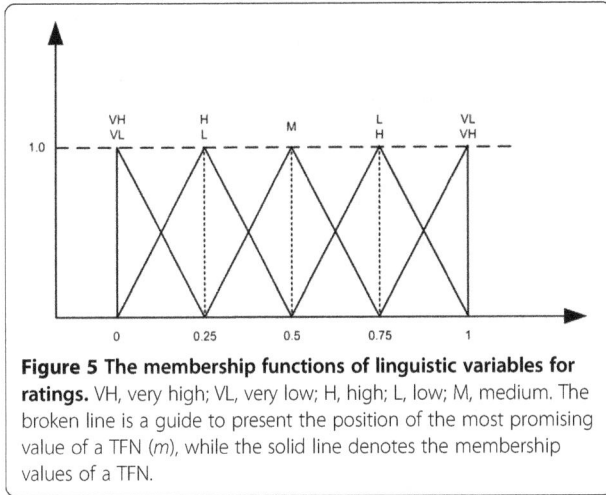

Figure 5 The membership functions of linguistic variables for ratings. VH, very high; VL, very low; H, high; L, low; M, medium. The broken line is a guide to present the position of the most promising value of a TFN (m), while the solid line denotes the membership values of a TFN.

the smallest possible value, the most promising value, and the largest possible value, respectively. The triplet (l, m, u) as a TFN has a membership function with following form:

$$\mu(x) = \begin{cases} \dfrac{x-l}{m-l} & \text{if } l \leq x \leq m \\ 1 & \text{if } x = m \\ \dfrac{u-x}{u-m} & \text{if } m \leq x \leq u \\ 0 & \text{Otherwise.} \end{cases}$$

Fuzzy operations for TFNs

Let $\tilde{A} = (l_A, m_A, u_A)$ and $\tilde{B} = (l_B, m_B, u_B)$ be two TFNs; there are some primary fuzzy operations as bellow (Keufmann and Gupta 1991; Kahraman et al. 2002):

1) Addition of two fuzzy numbers:

$$\tilde{A} \oplus \tilde{B} = (l_A + l_B, m_A + m_B, u_A + u_B)$$

2) Multiplication of two fuzzy numbers:

$$\tilde{A} \otimes \tilde{B} = (l_A l_B, m_A m_B, u_A u_B), \text{ where } l_A \text{ and } l_B \geq 0$$

Table 5 Segments in the coffee market

Segments	Remarks
S_1	Supplies Iran market with branded products
S_2	Exports branded products to the Middle East
S_3	Supplies Iran market with bulk products
S_4	Exports products in bulk to the Middle East
S_5	Produces branded products for other brands

3) Multiplication of a crisp number k and a fuzzy number:

$$k.\tilde{A} = (kl_A, km_A, ku_A), \text{ where } l_A \text{ and } l_B \geq 0$$

4) Division of two fuzzy numbers:

$$\tilde{A} \Delta \tilde{B} = (l_A/u_B, m_A/m_B, u_A/l_B), \text{ where } l_A \text{ and } l_B \geq 0$$

5) Addition of two fuzzy numbers:

$$\tilde{A} \oplus \tilde{B} = (l_A + l_B, m_A + m_B, u_A + u_B)$$

6) Multiplication of two fuzzy numbers:

$$\tilde{A} \otimes \tilde{B} = (l_A l_B, m_A m_B, u_A u_B), \text{ where } l_A \text{ and } l_B \geq 0$$

7) Multiplication of a crisp number k and a fuzzy number:

$$k.\tilde{A} = (kl_A, km_A, ku_A), \text{ where } l_A \text{ and } l_B \geq 0$$

8) Division of two fuzzy numbers:

$$\tilde{A} \Delta \tilde{B} = (l_A/u_B, m_A/m_B, u_A/l_B), \text{ where } l_A \text{ and } l_B \geq 0$$

Linguistic variables and fuzzy numbers

Linguistic variables represent an opinion independent of measuring system. While variables in mathematics usually take numerical values, in fuzzy logic applications, the non-numeric linguistic variables are often used to facilitate the expression of rules and facts (Zadeh et al. 1996). The fuzzy scale regarding the relative importance to measure the relative weights is proposed by Kahraman et al. (2006). This scale was used to solve fuzzy decision-making problems (Kahraman et al. 2006; Tolga et al. 2005; Dağdeviren and Yüksel 2010) in the literature of strategic management. This scale was later used by Mikhailov (2000, 2003) in fuzzy prioritization approach. Linguistic scales for difficulty and importance are shown in Table 3 and Figure 4, and the linguistic values and the mean of fuzzy numbers are shown in Table 4 and Figure 5.

Why fuzzy logic?

In most of cases, pairwise comparisons are vague because every decision has its special specifications. Using fuzzy numbers is a powerful tool to overcome the uncertainty and vagueness of data. On the other hand, pairwise comparisons with linguistic variables are easier for experts. Fuzzy set theory was introduced by Zadeh (1965) to deal with uncertainty due to imprecision and vagueness; since then, many applications have been developed in fuzzy decision-making processes. For computational efficiency, trapezoidal or triangular fuzzy numbers are usually used to represent fuzzy numbers (Klir and Yuan 1995). In this paper, TFNs are used to make the mathematics manageable and easy to understand, and to facilitate presentation of the case.

Fuzzy dynamic network process

Mikhailov (2000, 2003) developed a fuzzy prioritization approach with the advantage of the measurement of consistency indexes for the fuzzy pairwise comparison matrices. In other methods (Buckley 1985; Chang 1996;

Table 6 Factors and sub-factors of the risk definition model

Factors	Sub-factors	S1	S2	S3	S4	S5
Environmental stability (ES)	Demand variation (ES_1)	5	4	4	3	5
	Competitor prices (ES_2)	3	3	2	2	4
	Inflation rate (ES_3)	5	3	5	3	4
	Technology improvement rate (ES_4)	3	4	3	3	3
	Elasticity of demand (ES_5)	4	3	5	3	3
Industry strength (IS)	Supply chain management (IS_1)	2	3	4	3	5
	Potential growth ability (IS_2)	3	4	3	3	3
	Profitability (IS_3)	6	4	3	2	3
	Optimal resources consumption (IS_4)	4	4	4	4	4
	Optimal capacity usage (IS_5)	3	4	3	4	3
Financial strength (FS)	Liquidity power (FS_1)	2	3	3	4	3
	Investment returns (FS_2)	6	4	3	2	3
	Working capital (FS_3)	5	6	3	3	2
	Cash flow (FS_4)	5	4	4	3	4
	Ease of leaving the market (FS_5)	3	3	4	3	2
Competitive advantage (CA)	Market share (CA_1)	3	4	2	3	2
	Product quality (CA_2)	4	5	4	5	3
	Customer loyalty (CA_3)	3	4	3	3	4
	Technology (CA_4)	2	3	2	3	2
	Product distribution power (CA_5)	4	3	2	3	1

Cheng 1997; Deng 1999; Leung and Cao 2000), it is not possible to determine the consistency ratios of fuzzy pairwise comparison matrices without conducting an additional study. Mikhailov (2000, 2003) introduced three stages:

1) Statement of the problem
2) Assumptions of the fuzzy prioritization method
3) Solving the fuzzy prioritization problem that has survived as follows:

In a decision making problem with n elements, decision maker provides a set of $F = \{\tilde{a}_{ij}\}$ of $m \le n$ $(n - 1)/2$ pairwise comparison judgments, where $i = 1, 2, ..., n - 1, j = 2, 3, ..., n, j > i$, represented as triangular fuzzy numbers $\tilde{a}_{ij} = (l_{ij}, m_{ij}, u_{ij})$. A crisp priority vector $w = (w_1, w_2, ..., w_n)$ could reach from the problem with the fuzzy condition as follows:

$$l_{ij} \tilde{\le} \frac{w_i}{w_j} \tilde{\le} u_{ij}, \tag{1}$$

where the symbol $\tilde{\ }$ denotes 'fuzzy equal or less than' and with a membership function of inequality shown as follows:

$$\mu_{ij}\left(\frac{w_i}{w_j}\right) = \begin{cases} \dfrac{(w_i/w_j) - l_{ij}}{m_{ij} - l_{ij}} & \dfrac{w_i}{w_j} \le m_{ij} \\ \dfrac{u_{ij} - (w_i/w_j)}{u_{ij} - m_{ij}} & \dfrac{w_i}{w_j} \ge m_{ij} \end{cases}. \tag{2}$$

There are two main assumptions that the solution of prioritization is based on. The first one is the existence of non-empty fuzzy feasible area P on the $(n - 1)$ dimensional simplex Q^{n-1}:

$$Q^{n-1} = \left\{ (w_1, w_2, ..., w_n) | w_i > 0, \sum_{i=1}^{n} w_i = 1 \right\}, \tag{3}$$

where the membership function of the fuzzy feasible area is given by:

$$\mu_p(w) = \min_{ij} \left\{ \mu_{ij}(w) | \begin{array}{l} i = 1, 2, ..., n-1; \\ j = 2, 3, ..., n; j > i \end{array} \right\}. \tag{4}$$

Table 7 Risk levels based on Euclidean distance proportion

Linguistic variables	Euclidean distance proportion
Very low	Between 0 and 1/5
Low	Between 1/5 and 2/5
Medium	Between 2/5 and 3/5
High	Between 3/5 and 4/5
Very high	Between 4/5 and 1

Table 8 Proper risks for each segment with regard to Euclidean distance

Segments	Euclidean distance	Proper risks
S_1	0.377	Low
S_2	0.395	Low
S_3	0.486	Medium
S_4	0.507	Medium
S_5	0.648	High

The best situation = (6, 6), and the worse situation = (−6, −6).

The second one is a priority vector that is selected with having the highest degree of membership in the aggregated membership function (4).

$$\lambda^* = \mu_p(w^*) = \max_{w \in Q^{n-1}} \min_{ij} \left\{ \mu_{ij}(w) \right\} \qquad (5)$$

The maximum decision rule from the Game Theory is used to solve the fuzzy prioritization problem. The maximum prioritization problem (5) is extended as follows:

Maximize λ (6)

subject to:

$$\lambda \le \mu_{ij}(w), i = 1, 2, .., n-1, j = 2, 3, ..., n, \ i < j$$

$$\sum_{i=1}^{n} w_i = 1$$

$$w_i > 0, i = 1, 2, ..., n.$$

With regard to the membership function (2), problem (6) can be transferred into another form that is shown as follows:

Maximize λ (7)

$$\left(m_{ij} - l_{ij} \right) \lambda w_j - w_i + l_{ij} w_j \le 0$$
$$\left(u_{ij} - m_{ij} \right) \lambda w_j + w_i - u_{ij} w_j \le 0$$
$$\sum_{k=1}^{n} w_k = 1, w_k > 0, k = 1, 2, ..., n.$$
$$i = 1, 2, ..., n-1 \qquad j = 2, 3, ..., n \qquad j > i$$

The non-linear problem (7) will be optimized where $\lambda = \lambda^*$ and $W = W^*$, and the fuzzy judgment will be satisfied if the λ^* is positive. Also, it can be applied as the consistency measure of the initial set of fuzzy judgments. When the value of λ^* is negative, the solution ratios approximately satisfy all double-side inequalities (1), that means, the fuzzy judgments are inconsistent. To obtain

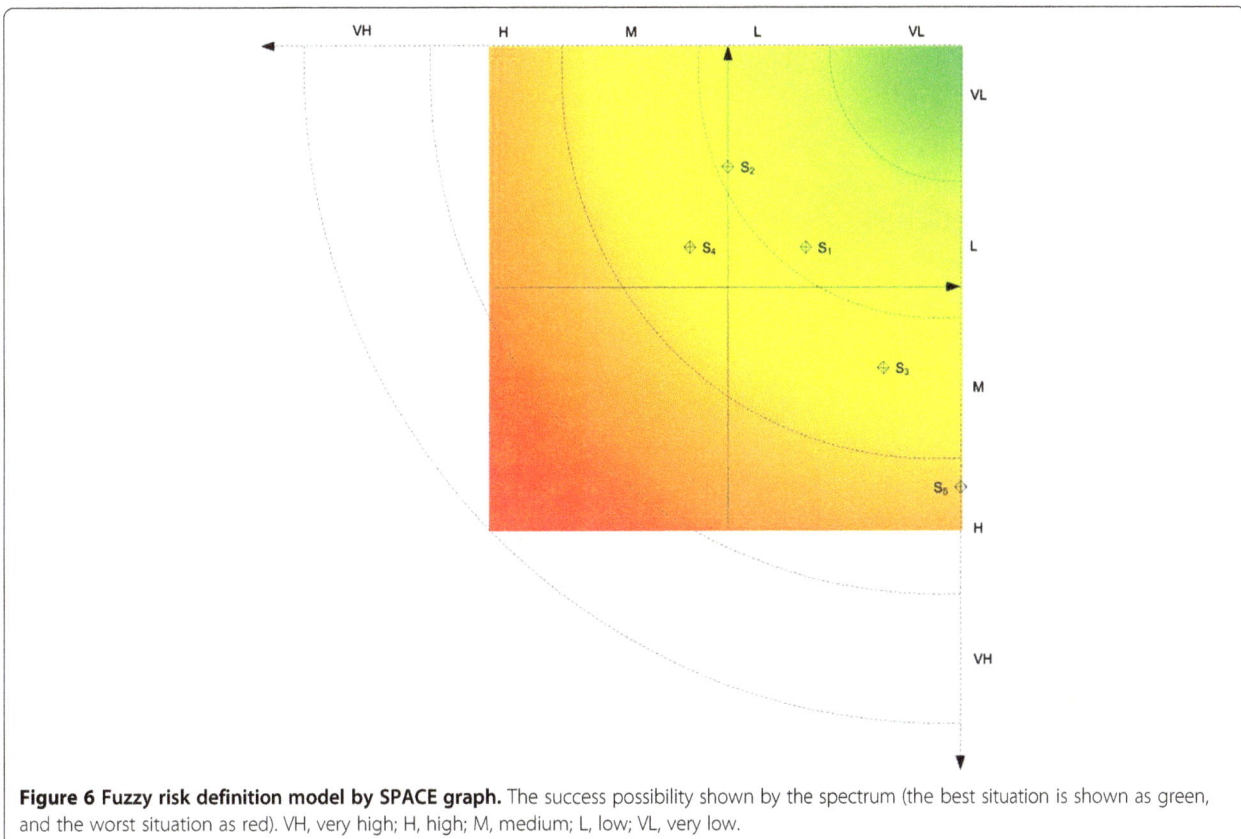

Figure 6 Fuzzy risk definition model by SPACE graph. The success possibility shown by the spectrum (the best situation is shown as green, and the worst situation as red). VH, very high; H, high; M, medium; L, low; VL, very low.

Table 9 Factors and sub-factors of the five forces model

Factors	Sub-factors
The bargaining power of supplier (F_1)	Supplier concentration (F_{11})
	Importance of order volume to supplier (F_{12})
	Presence of substitute inputs (F_{13})
	Switching cost of firms in the industry (F_{14})
	Differentiation of inputs (F_{15})
	Threat of perceived level of product (F_{16})
The bargaining power of customer (F_2)	Price sensitive (F_{21})
	Substitutes available (F_{22})
	Buyer concentration (F_{23})
	Product differentiation (F_{24})
	Brand identification (F_{25})
The threat of substitute products (F_3)	The quality of substitute products (F_{31})
	Buyer inclination to substitute (F_{32})
	Relative price performance of substitute (F_{33})
The threat of new entrants (F_4)	Brands (F_{41})
	Access to distribution (F_{42})
	Access to input (F_{43})
	Government policy (F_{44})
	Capital requirement (F_{45})
The intensity of competitive rivalry (F_5)	Exit barriers (F_{51})
	Fixed cost and value added (F_{52})
	Number of competitors (F_{53})
	Brand identification (F_{54})
	Product differences (F_{55})
	Switching cost (F_{56})

the time-dependent principal eigenvector, W^* should be calculated for different values of time N_t. These eigenvectors (W^*) are used to generate a curve that shows the alternative priority in each period. The alternative curves are gathered in a graph that could help DMs to select the best option.

Results and discussion
Procedure of segment-strategy selection
In this section, a procedure for segment-strategy selection is developed in ten steps to select the best potential segment with its strategies by considering an acceptable risk and in five competitive forces factors which have been developed by Porter (1980). According to this procedure, the market segments and strategies are selected in two main modules of a decision support system. In the first step, the risk amount is assigned by SPACE matrix method, and the segments are filtered based on special acceptable risk level which has been defined by DMs. In the second step, there are some segments which come from the first step. For every segment, some strategies are defined according to their position on SPACE matrix. DNP method in fuzzy environment has been applied to rank the segment-strategies.

Regarding this model, DM will be able to select the segments that have more chance of success according to their risk amount and to select proper strategies in each segment with competitive conditions. These steps are defined as follows:

Step 1. Segment filtering based on risk amount

1. Develop appropriate factors based on SPACE dimensions including internal perspectives (FS and CA) and external perspectives (ES and IS)
2. Assign relevant scores for each factor of segments and compute the total score in each dimension (internal and external); then, trace the position of each segment on SPACE graph
3. Assign a proper risk level for each segment and omit the segments which are out of the acceptable risk level (ARL)
4. Define feasible strategies for each segment and make a list of segment-strategy

Step 2. Select the best segment-strategy

1. Develop proper factors to choose the best segment-strategy, considering the vision statement
2. Compare factors for each alternative by considering the time variation and determine the effect of factors on each other
3. Calculate the score by FDNP for each segment-strategy according to the five competitive force factors
4. Make a discussion based on the score and choose the best segment-strategy

A case study is illustrated to select an optimum segment-strategy for a special coffee product in Iran market with regard to the procedure that was introduced before. While coffee is not technically a commodity, coffee is bought and sold by roasters, investors, and price speculators as a tradable commodity insofar as coffee has been described by many, including historian Pendergrast (1999), as the world's 'second most legally traded commodity'. Decaffeination is the act of removing caffeine from coffee beans. As of 2009, progress towards

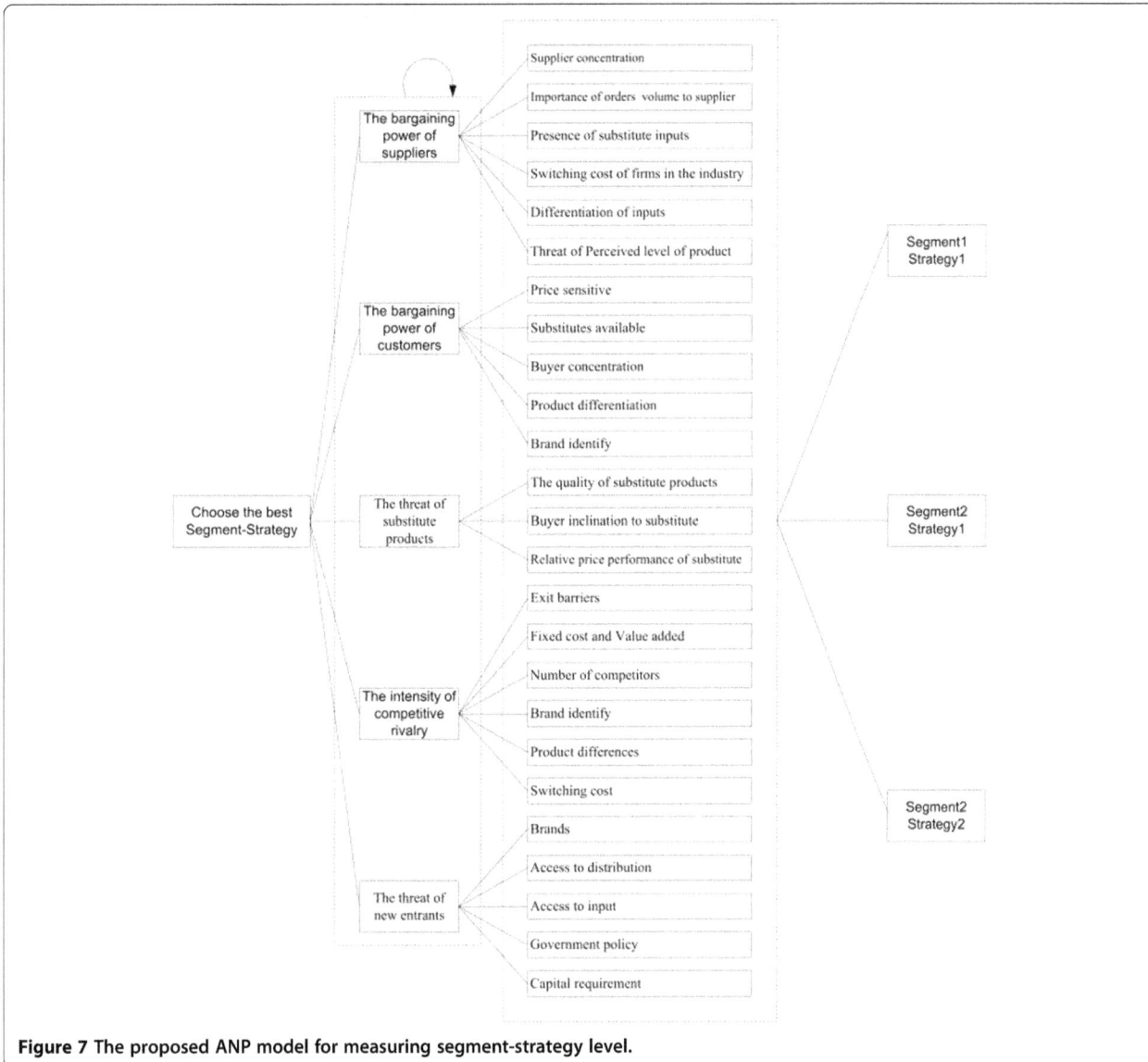

Figure 7 The proposed ANP model for measuring segment-strategy level.

growing coffee beans that do not contain caffeine is still continuing (Mazzafera et al. 2009). Consumption of decaffeinated coffee appears to be as beneficial as caffeine-containing coffee in regard to all-cause mortality, according to a large prospective cohort study (Brown et al. 1993). Decaffeinated products are produced in a coffee firm in Iran as a special product that can be put into the narrow markets from a demand perspective, particularly in the Middle East area. In Middle East, tea is a more popular beverage than coffee. This decreases the demand of coffee as a substitute product (especially decaffeinated coffee which has not existed before) in comparison with tea.

To focus on a special part of the market to increase competitiveness, a committee defines five segments

Table 10 Pair-wise comparison matrix of factors with local weights

Factor	F₁	F₂	F₃	F₄	F₅	Weight
The bargaining power of supplier (F_1)	(1, 1, 1)	(1/2, 2/3, 1)	(1/2, 2/3, 1)	(2/5, 1/2, 2/3)	(1, 3/2, 2)	0.1577
The bargaining power of customer (F_2)	(1, 3/2, 2)	(1, 1, 1)	(1/2, 1, 3/2)	(1/2, 2/3, 1)	(3/2, 2, 5/2)	0.2172
The threat of substitute products (F_3)	(1, 3/2, 2)	(2/3, 1, 2)	(1, 1, 1)	(1/2, 2/3, 1)	(3/2, 2, 5/2)	0.2172
The threat of new entrants (F_4)	(3/2, 2, 5/2)	(1, 3/2, 2)	(1, 3/2, 2)	(1, 1, 1)	(2, 5/2, 3)	0.2947
The intensity of competitive rivalry (F_5)	(1/2, 2/3, 1)	(2/5, 1/2 ,2/3)	(2/5, 1/2, 2/3)	(1/3, 2/5, 1/2)	(1, 1, 1)	0.1132

$\lambda = 0.7889$.

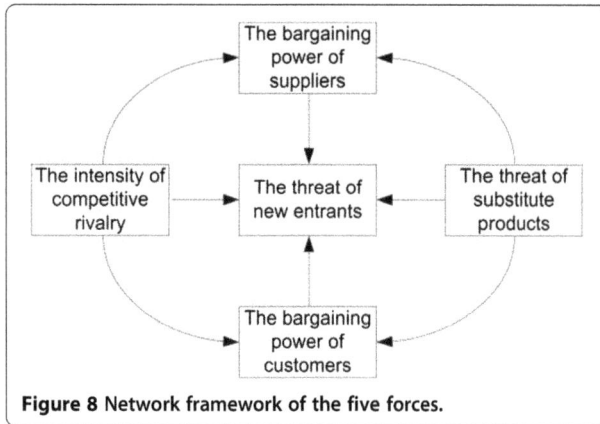

Figure 8 Network framework of the five forces.

(Table 5) to develop decaffeinated coffee around the Middle East. This committee includes business and market experts which have more than eight years of experience in sales or marketing in the Middle East. This committee consists of six managers within the company, who are professional in market with high experience in strategy development. All data have been collected by a team of market research experts to present to the committee to evaluate and segment the market, define and select factors and sub-factors, develop strategies, and execute pairwise comparisons in the decision-making process.

Segment filtering based on risk amount
Definition of segment positions
After developing the appropriate factors based on SPACE dimensions, DMs assign a relevant score to each sub-factor for each segment (Table 6).

The scores should be between 0 to 6, where 6 indicates the best condition and 0 indicates the worst for positive factors (financial strength and industry strength) and vice versa for negative factors (environmental stability and competitive advantage). For example, the amount of *product distribution power* (CA_5) which is a sub-factor of *competitive advantage* as a negative factor is 1 in S_5, which means there are suitable conditions to distribute the products in S_5 in comparison with the

competitors. On the other hand, the amount of *Profitability* (IS_3) as a positive factor of *industry strength* is 6 in S_1, which means there are suitable conditions to produce the product with high profitability in S_1 in comparison with other products in the other segments. According to these scores, total scores are calculated in each dimension of SPACE matrix using (8) and (9). The position of each segment is traced on SPACE graph according to the obtained pairs. It could assign a proper risk amount to each segment. Segment filtering will be done according to the assigned risk amounts and by a certain acceptable risk level.

(x^j, y^j) shows the position of segment j, where x and y are horizontal and vertical dimensions of the SPACE matrix, respectively. These pairs are calculated based on the sub-factor scores in two dimensions, where x is calculated by (8) and y by (9).

$$x^j = IS^j - CA^j = \sum_{i=1}^{N} \left(IS_i^j - CA_i^j \right)$$
$$i = 1, 2, ..., N \text{ (number of sub} - \text{factors)} \qquad (8)$$
$$j = 1, 2, ..., 5$$

$$y^j = FS^j - ES^j = \sum_{i=1}^{N} \left(FS_i^j - ES_i^j \right) \qquad (9)$$
$$i = 1, 2, ..., N \text{ } (N = \text{number of sub} - \text{factors)}$$
$$j = 1, 2, ..., 5.$$

The position of each segment has been calculated based on the sub-factor scores:

$$x^1 = IS^1 - CA^1 = \sum_{i=1}^{5} \left(IS_i^1 - CA_i^1 \right)$$
$$= (2 + 3 + 6 + 4 + 3) - (3 + 4 + 3 + 2 + 4) = 2$$

$$y^1 = FS^1 - ES^1 = \sum_{i=1}^{5} \left(FS_i^1 - ES_i^1 \right)$$
$$= (2 + 6 + 5 + 5 + 3) - (5 + 3 + 5 + 3 + 4) = 1.$$

Table 11 The inner dependence matrix of the factors with respect to 'F_4'

F_4	F_1	F_2	F_3	F_5	Weight
F_1	(1, 1, 1)	(1/2, 2/3, 1)	(2/5, 1/2, 2/3)	(2/3, 1, 2)	0.1875
F_2	(1, 3/2, 2)	(1, 1, 1)	(1/2, 2/3, 1)	(1, 3/2, 2)	0.2724
F_3	(3/2, 2, 5/2)	(1, 3/2, 2)	(1, 1, 1)	(1, 3/2, 2)	0.3081
F_5	(1/2, 1, 3/2)	(1/2, 2/3, 1)	(1/2, 2/3, 1)	(1, 1, 1)	0.2320

$\lambda = 0.3478.$

Table 12 The inner dependence matrix of the factors with respect to 'F_1'

F_1	F_3	F_5	Weight
F_3	(1, 1, 1)	(2/5, 1/2, 2/3)	0.3333
F_5	(3/2, 2, 5/2)	(1, 1, 1)	0.6667

$\lambda = 1.$

Table 13 The inner dependence matrix of the factors with respect to 'F_2'

F_2	F_3	F_5	Weight
F_3	(1, 1, 1)	(3/2, 2, 5/2)	0.6667
F_5	(2/5, 1/2, 2/3)	(1, 1, 1)	0.3333

$\lambda = 1.$

Table 14 Degree of relative impact for the factors

Factor	F_1	F_2	F_3	F_4	F_5
F_1	0.500	0.000	0.000	0.094	0.000
F_2	0.000	0.500	0.000	0.136	0.000
F_3	0.167	0.333	1.000	0.154	0.000
F_4	0.000	0.000	0.000	0.500	0.000
F_5	0.333	0.167	0.000	0.116	1.000

The position of S_1 is (2,1), and in this way, other positions are calculated as follows:

$$S_2 = (3,0), S_3 = (4,-2), S_4 = (-1,1) \text{ and } S_5 = (6,-5)$$

Definition of risk levels

Different points on SPACE matrix show the success possibility of each segment that is considered as risk amount of segments. The most possibility occurs when financial strength and industry strength get the most score as positive factors, and environmental stability and competitive advantage get the lowest score as negative factors. So, the pair (6, 6) has the most success possibility with lowest risk amount in the SPACE matrix, and the pair (−6,−6) has the most risk amount. Risk of other points is defined based on their distance from (6, 6). The surface of the SPACE matrix is separated into five areas according to the distance from the best point. These areas are defined by radiuses which have been calculated based on fuzzy approach. It means that the Euclidean distance from the worst and the best points has been separated into five sections according to the linguistic values and the mean of fuzzy numbers (Table 3 and Figure 5). Thus, the Euclidean distance of each segment from the best point towards the worst point can show a level of risk.

Let (x^j, y^j) show the position of segment j. Let (X, Y) show the best position, and (X', Y') show the worst

position in a SPACE graph. The risk amount of (x^j, y^j) is defined based on its Euclidean distance proportion that is showed as follows:

$$\text{Risk } (x^j, y^j) \sim \frac{\sqrt{(X-x^j)^2 + (Y-y^j)^2}}{\sqrt{(X-X')^2 + (Y-Y')^2}}. \tag{10}$$

Euclidean distance proportions of all positions are calculated by (10). It helps assign linguistic variables to each position according to Table 7.

Euclidean distances and proper risk of each segment was calculated as shown in Table 8 and Figure 6. ARL is defined to filter segments according to ability of risk acceptance of DMs. In this case, low risk level is considered as maximum ARL; it means that segments with risk level higher than low are rejected. Thus, S_1 and S_2 are selected to define proper strategies, and S_3, S_4, and S_5 are rejected because of their high risk levels.

Strategy definition

Strategy definition is done by SPACE matrix. Generic strategies of SPACE matrix are defensive, aggressive, conservative, and competitive which could be broken into the main strategies. In this case, aggressive and conservative strategies are suitable for S_2, and aggressive strategies for S_1. The two main strategies were defined for S_2 from two different views: the first one is 'putting decaffeinated coffee in old basket' as conservative strategy, and the second one is 'A new basket of decaffeinated coffee products with decaffeinated coffee stores development' as aggressive strategy. The aggressive strategy that was defined for S_1 is 'A new basket of decaffeinated coffee and decaffeinated coffee stores development'.

Select the best segment-strategy

The five competitive forces model is a common tool used in analyzing and supporting the competitive

Table 15 Pair-wise comparison matrix of F_1 sub-factors with local weight

F_1 S-F	F_{11}	F_{12}	F_{13}	F_{14}	F_{15}	F_{16}	Weight
F_{11}	(1, 1, 1)	(1/2, 2/3, 1)	(1, 3/2, 2)	(1/2, 1, 3/2)	(1, 3/2, 2)	(3/2, 2, 5/2)	0.1962
F_{12}	(1, 3/2, 2)	(1, 1, 1)	(3/2, 2, 5/2)	(1, 3/2, 2)	(3/2, 2, 5/2)	(3/2, 2, 5/2)	0.2385
F_{13}	(1/2, 2/3, 1)	(2/5, 1/2, 2/3)	(1, 1, 1)	(1/2, 2/3, 1)	(1/2, 1, 3/2)	(1,3/2, 2)	0.1351
F_{14}	(2/3, 1, 2)	(1/2, 2/3, 1)	(1, 3/2, 2)	(1, 1, 1)	(1, 3/2, 2)	(3/2, 2, 5/2)	0.1884
F_{15}	(1/2, 2/3, 1)	(2/5, 1/2, 2/3)	(2/3, 1, 2)	(1/2, 2/3, 1)	(1, 1, 1)	(1, 3/2, 2)	0.1351
F_{16}	(2/5, 1/2, 2/3)	(2/5, 1/2, 2/3)	(1/2, 2/3, 1)	(2/5, 1/2, 2/3)	(1/2, 2/3, 1)	(1, 1, 1)	0.1067

$\lambda = 0.5311$.

Table 16 Pair-wise comparison matrix of F_2 sub-factors with local weight

F_2 S-F	F_{21}	F_{22}	F_{23}	F_{24}	F_{25}	Weight
F_{21}	(1, 1, 1)	(1/2, 2/3, 1)	(1, 3/2, 2)	(1, 3/2, 2)	(1/2, 2/3, 1)	0.1964
F_{22}	(1, 3/2, 2)	(1, 1, 1)	(1, 3/2, 2)	(1, 3/2, 2)	(1/2, 2/3, 1)	0.2339
F_{23}	(1/2, 2/3, 1)	(1/2, 2/3, 1)	(1, 1, 1)	(1/2, 2/3, 1)	(2/5, 1/2, 2/3)	0.1329
F_{24}	(1/2, 2/3, 1)	(1/2, 2/3, 1)	(1, 3/2, 2)	(1, 1, 1)	(2/5, 1/2, 2/3)	0.1583
F_{25}	(1, 3/2, 2)	(1, 3/2, 2)	(3/2, 2, 5/2)	(3/2, 2, 5/2)	(1, 1, 1)	0.2785

$\lambda = 0.4810$.

strategic management in competitive markets. Porter (1980) developed these forces that model every single industry and market, and help DMs analyze industry competition for profitability and attractiveness. The five force factors and sub-factors in Porter's model, which are determined by the committee, are shown in Table 9.

Factors and sub-factors of Porter's (1980) five forces model are applied as decision criteria to select the best segment-strategy. FDNP is implemented to rank the segment-strategies. This method can consider all inner dependency effects among factors and sub-factors over time. Using the factors and sub-factors, a decision tree is made to rank the segment-strategies (Figure 7). The decision tree includes four levels. The first level is the decision making (choosing the best segment-strategy). The second comprise the factors and sub-factors; the third level includes the problem criteria. The fourth level consists of the alternatives.

The local weights of the factors are calculated by a useful method that Saaty and Takizawa (1986) and Saaty (1996) presented and developed in fuzzy prioritization approach. These are the fuzzy comparison values presented in Table 10.

The non-linear programming presented as follows resulted from pairwise comparisons and was solved using the LINGO 11 (2008) software (Lindo Systems Inc., Chicago). The other weights were calculated using the same approach for each pairwise comparison matrix.

Maximize $= \lambda$
Subject to:

$$1/6 \times \lambda \times w_2 - w_1 + 1/2 \times w_2 \leq 0$$
$$1/3 \times \lambda \times w_2 + w_1 - w_2 \leq 0$$
$$1/6 \times \lambda \times w_3 - w_1 + 1/2 \times w_3 \leq 0$$
$$1/3 \times \lambda \times w_3 + w_1 - w_3 \leq 0$$
$$1/10 \times \lambda \times w_4 - w_1 + 2/5 \times w_4 \leq 0$$
$$1/6 \times \lambda \times w_4 + w_1 - 2/3 \times w_4 \leq 0$$
$$1/2 \times \lambda \times w_5 - w_1 + w_5 \leq 0$$
$$1/2 \times \lambda \times w_5 + w_1 - 2 \times w_5 \leq 0$$
$$1/2 \times \lambda \times w_3 - w_2 + 1/2 \times w_3 \leq 0$$
$$1/2 \times \lambda \times w_3 + w_2 - 3/2 \times w_3 \leq 0$$
$$1/6 \times \lambda \times w_4 - w_2 + 1/2 \times w_4 \leq 0$$
$$1/3 \times \lambda \times w_4 + w_2 - w_4 \leq 0$$
$$1/2 \times \lambda \times w_5 - w_2 + 3/2 \times w_5 \leq 0$$
$$1/2 \times \lambda \times w_5 + w_2 - 5/2 \times w_5 \leq 0$$
$$1/6 \times \lambda \times w_4 - w_3 + 1/2 \times w_4 \leq 0$$
$$1/3 \times \lambda \times w_4 + w_3 - w_4 \leq 0$$
$$1/2 \times \lambda \times w_5 - w_3 + 3/2 \times w_5 \leq 0$$
$$1/2 \times \lambda \times w_5 + w_3 - 5/2 \times w_5 \leq 0$$
$$1/2 \times \lambda \times w_5 - w_4 + 2 \times w_5 \leq 0$$
$$1/2 \times \lambda \times w_5 + w_4 - 3 \times w_5 \leq 0$$
$$w_1 + w_2 + w_3 + w_4 + w_5 = 1$$
$$w_i > 0 \; i = 1, 2, ..., 5$$

The effects of the interdependency among the five force factors are shown in Figure 8. The inner dependency matrix is presented in Tables 11, 12, and 13, which was defined by the expert committee to obtain the local weights of the factors.

The vectors of the inner dependency weight of the factors (Tables 11, 12, and 13) are normalized to find the degree of relative impact matrix (Table 14). The final weights of the factors (w_{Factors}) are calculated by multiplying the normalized degree matrix (Table 14) with the local weight of the factors that had been calculated before in Table 10.

Table 17 Pair-wise comparison matrix of F_3 sub-factors with local weight

F_3 S-F	F_{31}	F_{32}	F_{33}	Weight
F_{31}	(1, 1, 1)	(2/5, 1/2, 2/3)	(1/2, 2/3, 1)	0.2239
F_{32}	(3/2, 2, 5/2)	(1, 1, 1)	(1, 3/2, 2)	0.4584
F_{33}	(1, 3/2, 2)	(1/2, 2/3, 1)	(1, 1, 1)	0.3177

$\lambda = 0.8855$.

Table 18 Pair-wise comparison matrix of F_4 sub-factors with local weight

F_4 S-F	F_{41}	F_{42}	F_{43}	F_{44}	F_{45}	Weight
F_{41}	(1, 1, 1)	(1, 3/2, 2)	(1, 3/2, 2)	(3/2, 2, 5/2)	(1, 3/2, 2)	0.2854
F_{42}	(1/2, 2/3, 1)	(1, 1, 1)	(1, 3/2, 2)	(1, 3/2, 2)	(1, 3/2, 2)	0.2327
F_{43}	(1/2, 2/3, 1)	(1/2, 2/3, 1)	(1, 1, 1)	(1, 3/2, 2)	(1/2, 2/3, 1)	0.1610
F_{44}	(2/5, 1/2, 2/3)	(1/2, 2/3, 1)	(1/2, 2/3, 1)	(1, 1, 1)	(1/2 ,2/3, 1)	0.1312
F_{45}	(1/2, 2/3, 1)	(1/2, 2/3, 1)	(1, 3/2, 2)	(1, 3/2, 2)	(1, 1, 1)	0.1897

$\lambda = 0.4536$.

$$
W_{\text{Factrs}} = \begin{bmatrix} F_1 \\ F_2 \\ F_3 \\ F_4 \\ F_5 \end{bmatrix} = \begin{bmatrix} 0.50 & 0.00 & 0.00 & 0.09 & 0.00 \\ 0.00 & 0.50 & 0.00 & 0.14 & 0.00 \\ 0.17 & 0.33 & 1.00 & 0.15 & 0.00 \\ 0.00 & 0.00 & 0.00 & 0.50 & 0.00 \\ 0.33 & 0.17 & 0.00 & 0.12 & 1.00 \end{bmatrix}
$$

$$
\times \begin{bmatrix} 0.16 \\ 0.22 \\ 0.22 \\ 0.29 \\ 0.11 \end{bmatrix} = \begin{bmatrix} 0.106 \\ 0.149 \\ 0.362 \\ 0.147 \\ 0.236 \end{bmatrix}.
$$

In the next step, the pairwise comparisons of the sub-factors should be done with respect to each factor to calculate the local weights and global weights. Tables 15, 16, 17, 18, and 19 show the pairwise comparisons of the sub-factors and their calculated weight for each sub-factor. To calculate the global weights of each sub-factor, their calculated local weights should be multiplied with the weight of each factor directly (Table 20).

To obtain the priority of the alternatives, the alternatives are compared with respect to each sub-factor. These comparisons should be done for each time period of the planning horizon. To make a better decision, considering the facts like 'changing future conditions' 'more preferred business in each time period' and 'changing the priority of each factor or its sub-factors' are very important. The certain planning horizon is dependent on the strategies that the company applied to launch a product in the market. It could be considered as the product life cycle that is planned for a certain time in a certain area. In this case, 5 years of planning horizon are considered to compare the alternatives

considering the sub-factors and future changes of alternative priorities. Table 21 shows the final priorities of the alternatives regarding the importance of each strategy that is planned for each segment on the specific periods. As shown in Figure 9, it is clear that the priority of segment$_1$-startegy$_1$ is preferred over the others at first, although its priority is decreased during the planning horizon. In the end, segment$_2$-startegy$_2$ becomes more interesting than the others. On the other hand, results show that the priority of segment$_2$-startegy$_2$ is preferred almost after 1 year; thus, it could be selected for a long-term strategic planning.

Conclusion

The purpose of the current study is to provide a modular decision support system to determine the best marketing strategy with an acceptable risk. This DSS helps companies to select appropriate segments to develop their business while they can care about their risk. Also, they can consider the effects of the strategies in their success based on priorities which may be changed over time. Two modules have been developed in this study: the first one used the SPACE matrix to allocate the risk to each segment, and the second one used FDNP method to monitor the segment-strategies over time and select the best one accordingly.

In the first module, segments have been evaluated based on the four main factors (and their sub-factors) of the SPACE matrix, and their risk have been calculated according to their success possibility. Then, the segments have been filtered with regard to their risk level which had been defined using the fuzzy approach. This method helps managers to take their acceptance

Table 19 Pair-wise comparison matrix of F_5 sub-factors with local weight

F_5 S-F	F_{51}	F_{52}	F_{53}	F_{54}	F_{55}	F_{56}	Weight
F_{51}	(1, 1, 1)	(2/5, 1/2, 2/3)	(2/5, 1/2, 2/3)	(2/5, 1/2, 2/3)	(1/2, 2/3, 1)	(1/2, 2/3, 1)	0.1030
F_{52}	(3/2, 2, 5/2)	(1, 1, 1)	(1/2, 2/3, 1)	(1/2, 2/3, 1)	(1, 3/2, 2)	(1, 3/2, 2)	0.1782
F_{53}	(3/2, 2, 5/2)	(1, 3/2, 2)	(1, 1, 1)	(1/2, 2/3, 1)	(1, 3/2, 2)	(1, 3/2, 2)	0.2046
F_{54}	(3/2, 2, 5/2)	(1, 3/2, 2)	(1, 3/2, 2)	(1, 1, 1)	(1, 3/2, 2)	(1, 3/2, 2)	0.2349
F_{55}	(1, 3/2, 2)	(1/2, 2/3, 1)	(1/2, 2/3, 1)	(1/2, 2/3, 1)	(1, 1, 1)	(1/2, 2/3, 1)	0.1300
F_{56}	(1, 3/2, 2)	(1/2, 2/3, 1)	(1/2, 2/3, 1)	(1/2, 2/3, 1)	(1, 3/2, 2)	(1, 1, 1)	0.1493

$\lambda = 0.3871$.

Table 20 Global weights for sub-factors and computed total weight of each

Factor	Sub-factors	Local weight	Global weights
F_1 (0.1065)	F_{11}	0.1962	0.0209
	F_{12}	0.2385	0.0254
	F_{13}	0.1351	0.0144
	F_{14}	0.1884	0.0201
	F_{15}	0.1351	0.0144
	F_{16}	0.1067	0.0114
F_2 (0.1487)	F_{21}	0.1964	0.0292
	F_{22}	0.2339	0.0348
	F_{23}	0.1329	0.0198
	F_{24}	0.1583	0.0235
	F_{25}	0.2785	0.0414
F_3 (0.3613)	F_{31}	0.2239	0.0809
	F_{32}	0.4584	0.1656
	F_{33}	0.3177	0.1148
F_4 (0.1447)	F_{41}	0.2854	0.0421
	F_{42}	0.2327	0.0343
	F_{43}	0.1610	0.0237
	F_{44}	0.1312	0.0193
	F_{45}	0.1897	0.0280
F_5 (0.2361)	F_{51}	0.1030	0.0243
	F_{52}	0.1782	0.0421
	F_{53}	0.2046	0.0482
	F_{54}	0.2349	0.0555
	F_{55}	0.1300	0.0306
	F_{56}	0.1493	0.0353

risk level into consideration and leads DMs to select segments with their reasonable risk levels. Moreover, the SPACE matrix helps managers define proper strategies, too. Filtered segments help them have more suitable alternatives in the decision-making process.

In the second module, the five forces model of Porter (1980) has been developed in a decision-making process to select the best segment-strategy. Because of the changing conditions in the market and the decreasing or increasing attractiveness of the alternatives, the alternative priorities are changed over time, so the FDNP is

Table 21 Global weights for each segment-strategy and computed total weight in each year

Segment-strategy	Year (total weight)				
	2013	2014	2015	2016	2017
Seg_1-Str_1	0.487	0.383	0.318	0.295	0.294
Seg_2-Str_1	0.312	0.285	0.261	0.236	0.213
Seg_2-Str_2	0.201	0.332	0.421	0.469	0.493

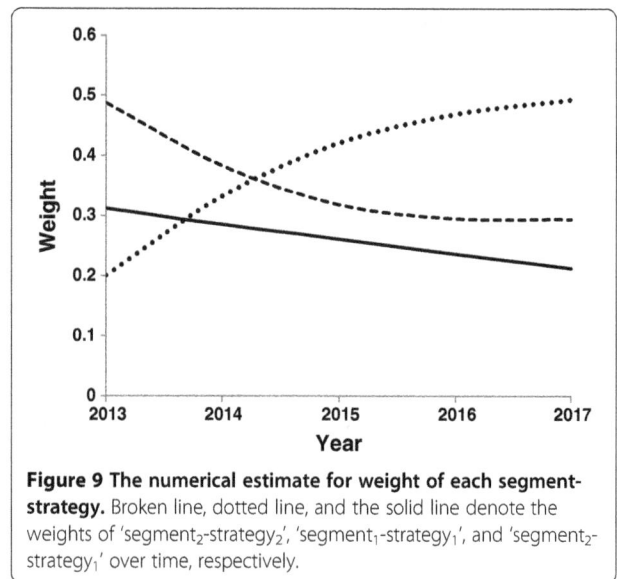

Figure 9 The numerical estimate for weight of each segment-strategy. Broken line, dotted line, and the solid line denote the weights of 'segment$_2$-strategy$_2$', 'segment$_1$-strategy$_1$', and 'segment$_2$-strategy$_1$' over time, respectively.

developed to consider the variation of segment priorities. As it is clear in the numerical results, time variation could affect the DMs' decision. The priority of segment$_1$-startegy$_1$ is more preferred over the others at first, although its priority decreased during the planning horizon. In the end, segment$_2$-startegy$_2$ becomes more interesting than the others. On the other hand, results show that the priority of segment$_2$-startegy$_2$ is more preferred almost after 1 year; thus, it could be selected for a long-term strategic plan.

Market segmentation is one of the most important issues in marketing process of industries such as food, dairy, beverage, home care, etc. In this process, risk consideration is very essential because it may have big effects on the expected results. The proposed method in this paper could mitigate this risk by bringing the risk into calculation, and it could be applied to mitigate risk consequences. Using this method, DM could filter its alternative and will not count on segments which are in high risk space. As a result, DM will not select strategies based on high risk segments, and the company could lead its investment to the most secure space. As shown in the results, segment 5 (S_5) has the maximum risk of selection because in this segment, environment stability is weaker than the other potential segments. Hence, disregarding risk factors and selecting S_5 as a potential segment, the company will enter an unstable market. In this way, the other steps of strategy definition such as distribution channels, pricing, and long-term and short-term strategies will undergo selected market instability. So, disregarding the risk effects could lead a business to the spaces which can decrease the possibility of success.

On the other hand, for each segment, a special strategy could be developed while the importance of each segment-strategy has its special trend over time. Practically, when a company is going to invest on segment-strategy, it should have a serious attention on the long-term results of its decision. In this condition, having a good view on the trend of segment-strategy importance could help DMs make effective decisions over time. In considering this issue, the developed FDNP method of this paper could be applied. The application of this method in industries will be more significant when they have marketing strategies such as pricing, distribution channels, and promotion in their appropriate segment.

Considering the risk amount and competitive factors with their effects on each other will drive the company to be more successful. Analysis effects of these strategies to decrease risk amount could be helpful in making a better and more complete decision merits future research.

Competing interest
The authors declare that they have no competing interests.

Authors' contributions
MHH managed the segment-strategies development process including holding the meeting of managers' committee to gather the market information. AMN, MHH, and SE proposed the decision making process including risk mitigation approach and developed the dynamic network process via five forces model. AMN developed the fuzzy DNP to calculate the weight of factors and alternative, and the calculations by Lingo11. SE managed and supported the team to provide the steps. He also modified the process. All authors read and approved the final manuscript.

Authors' information
AMN received his MSc from the Industrial Engineering Department of K. N. Toosi University of Technology of Iran. He has work experience in automotive and food production companies. Now, he is working in the PSIG Co. as Business Development Manager. His research interests are project portfolio selection, strategic management, market segmentation and multi-attribute decision making, supply chain management, and risk management. MHH received his MSc from the Industrial Engineering Department of Islamic Azad University of Qazvin. He has work experience in automotive, food and FMCG production companies. Now, he is working in Unilever Company as a demand planner. His research interests are project portfolio selection, strategic management, market segmentation, and multi-attribute decision making. SE is an assistant professor of operation research in the Islamic Azad University, Karaj Branch. He received his PhD in Operation Research and Operation Management from Science and Research Branch of the Islamic Azad University (SRBIAU), MSc from Amirkabir University of Technology (Tehran Polytechnic), and BS from Iran University of Science and Technology. His research interests include risk management, construction projects selection, fuzzy MADM, scheduling, mathematical programming, shortest path networks, and supply chain management.

Acknowledgments
The authors are thankful for the support of the Multicafe Co. for the research process of Iran coffee market and segment-strategies definition. The anonymous reviewers are acknowledged for their constructive comments as well as the editorial changes suggested by the language editor, which certainly improved the presentation of the paper.

Author details
[1]Department of Industrial Engineering, K. N. Toosi University of Technology, Tehran, Iran. [2]Department of Industrial and Mechanical Engineering, Islamic Azad University, Qazvin Branch, Qazvin, Iran. [3]Department of Industrial Engineering, Islamic Azad University, Karaj Branch, Alborz, Iran.

References
Aaker DA (1995) Strategic market management. Wiley, New York

Aghdaie MH, Zolfani SH, Rezaeinia N, Mehri-Tekmeh J (2011) A hybrid fuzzy MCDM approach for market segments evaluation and selection, Paper presented at the international conference on management and service science, Wuhan, China, 12–14 August 2011

Albert TC (2003) Need-based segmentation and customized communication strategies in a complex-commodity industry: a supply chain study. Ind Market Manag 32(4):281–290

Bharadwaj SG, Varadarajan PR (1993) Sustainable competitive advantage in service industries: a conceptual model and research propositions. J Market 57(4):83–99

Bock T, Uncles M (2002) A taxonomy of differences between consumers for market segmentation. IJRM 19(3):215–224

Bonoma TV, Shapiro BP (1983) Segmenting the industrial market. Lexington Books, Lexington

Brown CA, Bolton-Smith C, Woodward M, Tunstall-Pedoe H (1993) Coffee and tea consumption and the prevalence of coronary heart disease in men and women: results from the Scottish Heart Health Study. J Epidemiol Commun H 47(3):171–175

Buckley JJ (1985) Fuzzy hierarchical analysis. Fuzzy Set Syst 17(3):233–247

Chang DY (1996) Applications of the extent analysis method on fuzzy AHP. Eur J Oper Res 95(3):649–655

Cheng CH (1997) Evaluating naval tactical missile systems by fuzzy AHP based on the grade value of membership function. Eur J Oper Res 96(2):343–350

Clarke AH, Freytag PV (2008) An intra- and inter-organisational perspective on industrial segmentation: a segmentation classification framework. Eur J Marketing 42(9):1023–1038

Coughlan AT, Soberman DA (2005) Strategic segmentation using outlet malls. IJRM 22(1):61–86

Dağdeviren M, Yüksel I (2010) A fuzzy analytic network process (ANP) model for measurement of the sectoral competition level (SCL). Expert Syst Appl 37(2):1005–1014

Day GS, Wensley R (1988) Assessing advantage: a framework for diagnosing competitive superiority. J Market 52(2):1–20

Deng H (1999) Multicriteria analysis with fuzzy pairwise comparison. Int J Approx Reason 21(3):215–231

Dickson PR (1993) Marketing management. The Dryden Press, Orlando

Dincer O (2004) Strategy management and organization policy. Beta Publication, Istanbul

File KM, Prince RA (1996) A psychographic segmentation of industrial family businesses. Ind Market Manag 25(3):223–234

Freytag PV, Clarke AH (2001) Business to business market segmentation. Ind Market Manag 30(6):473–486

Goller S, Hogg A, Kalafatis S (2002) A new research agenda for business segmentation. Eur J Marketing 36(1):252–272

Hunt SD, Arnett DB (2004) Market segmentation strategy, competitive advantage, and public policy: grounding segmentation strategy in resource-advantage theory. AMJ 12(1):7–25

Jenkins M, McDonald M (1997) Market segmentation-organizational archetypes and research agendas. Eur J Marketing 31(1):17–32

Kahraman C, Ruan D, Tolga E (2002) Capital budgeting techniques using discounted fuzzy versus probabilistic cash flows. Inform Sci 142(1–4):57–76

Kahraman C, Ruan D, Doğan I (2003) Fuzzy group decision-making for facility location selection. Inform Sci 157:135–153

Kahraman C, Ertay T, Buyukozkan G (2006) A fuzzy optimization model for QFD planning process using analytic network approach. Eur J Oper Res 171(2):390–411

Keufmann A, Gupta MM (1991) Introduction to fuzzy arithmetic: theory and application. VanNostrand Reinhold, New York

Klir GJ, Yuan B (1995) Fuzzy sets and fuzzy logic: theory and applications. Prentice-Hall, New Jersey

Kotler P (1997) Marketing management: analysis, planning, implementation, and control, 9th edn. Prentice Hall International, Upper Saddle River, NJ

Kuo RJ, Ho LM, Hu CM (2002) Integration of self-organizing feature map and K-means algorithm for market segmentation. Compu Oper Res 29(11):1475–1493

Leung LC, Cao D (2000) On consistency and ranking of alternatives in fuzzy AHP. Eur J Oper Res 124(1):102–113

LINGO 11, (2008) Optimization modeling software for linear, nonlinear, and integer programming. Lindo Systems Inc., Chicago

Liu Q, Serfes K (2007) Market segmentation and collusive behavior. Int J Ind Organ 25(2):355–378

Liu Y, Ram S, Lusch RF, Brusco M (2010) Multicriterion market segmentation: a new model, implementation, and evaluation. Market Sci 29(5):880–894

Lu CS (2003) Market segment evaluation and international distribution centers. Transpor Res E-Log 39(1):49–6

Mazzafera P, Baumann TW, Shimizu MM, Silvarolla MB (2009) Decaf and the steeplechase towards decaffito-the coffee from caffeine-free Arabica plants. Trop Plant Biol 2(2):63–76

Meade LM, Sarkis J (1999) Analyzing organizational project alternatives for agile manufacturing processes: an analytical network approach. Int J Prod Res 37(2):241–261

Mikhailov L (2000) A fuzzy programming method for deriving priorities in the analytic hierarchy process. J Oper Res Soc 51(3):341–349

Mikhailov L (2003) Deriving priorities from fuzzy pairwise comparison judgments. Fuzzy Set Syst 134(3):365–385

Montoya-Weiss M, Calentone RJ (2001) Development and implementation of a segment selection procedure for industrial product markets. Market Sci 18(3):373–395

Nakip M (1999) Segmenting the global market by usage rate of industrial products: heavy-user countries are not necessary heavy users for all industrial products. Ind Market Manag 28(2):177–187

Ou CW, Chou SY, Chang YH (2009) Using a strategy-aligned fuzzy competitive analysis approach for market segment evaluation and selection. Expert Syst Appl 36(1):527–541

Pendergrast M (1999) Uncommon grounds: the history of coffee and how it transformed our world. Basic Books, New York

Phillips JM, Reynolds TJ, Reynolds K (2010) Decision-based voter segmentation: an application for campaign message development. Eur J Marketing 44(3–4):310–330

Porter M (1980) Competitive strategy: techniques for analyzing industries and competitors. The Free Press, New York

Ray S (2000) Strategic segmentation of a market. Int J Ind Organ 18(8):1279–1290

Rowe H, Mason R, Dichel K (1982) Strategic management and business policy: A methodological approach. Addison-Wesley, MA

Ren Y, Yang D, Diao X (2010) Market segmentation strategy in internet market. Physica A 389:1688–1698

Saaty TL (1980) The analytic hierarchy process. McGraw-Hill, New York

Saaty TL (1996) Decision making with dependence and feedback: the analytic network process. RWS Publications, Pittsburgh

Saaty TL (2007) Time dependent decision-making, dynamic priorities in the AHP/ANP: generalizing from points to functions and from real to complex variables. Math Comput Model 46(7–8):860–891

Saaty TL, Takizawa M (1986) Dependence and independence: from linear hierarchies to nonlinear networks. Eur J Oper Res 26(22):229–237

Sausen K, Tomczak T, Hermann A (2005) Development of a taxonomy of strategic market segmentation: a framework for bridging the implementation gap between normative segmentation and business practice. J Strat Market 13(3):151–173

Shani A, Reichel A, Croes R (2012) Evaluation of segment attractiveness by risk-adjusted market potential. J Trav Res 51(2):166–177

Söllner A, Rese M (2001) Market segmentation and the structure of competition: applicability of the strategic group concept for an improved market segmentation on industrial markets. J Bus Res 51(1):25–36

Tolga E, Demircan ML, Kahraman C (2005) Operating system selection using fuzzy replacement analysis and analytic hierarchy process. Int J Prod Econ 97(1):89–117

Tsai MC, Tsai YT, Lien CW (2011a) Generalized linear interactive model for market segmentation: the air freight market. Ind Market Manag 40(3):439–446

Tsai MC, Yang CW, Lee HC, Lien CW (2011b) Segmenting industrial competitive markets: an example from air freight. J Air Transp Manag 17:211–214

Varadarajan PR, Cunningham MH (1995) Strategic alliances: a synthesis of conceptual foundations. J Market 23(4):282–296

Xia Y (2011) Competitive strategies and market segmentation for suppliers with substitutable products. Eur J Oper Res 210:194–203

Zadeh LA (1965) Fuzzy sets. Inform Contr 8(3):338–353

Zadeh LA, Klir GJ, Yuan B (1996) Fuzzy sets, fuzzy logic, fuzzy systems. World Scientific Press, Hackensack

10

Customer involvement in greening the supply chain: an interpretive structural modeling methodology

Sanjay Kumar[1*], Sunil Luthra[2] and Abid Haleem[3]

Abstract

The role of customers in green supply chain management needs to be identified and recognized as an important research area. This paper is an attempt to explore the involvement aspect of customers towards greening of the supply chain (SC). An empirical research approach has been used to collect primary data to rank different variables for effective customer involvement in green concept implementation in SC. An interpretive structural-based model has been presented, and variables have been classified using *matrice d'impacts croises-multiplication appliqué a un classement* analysis. Contextual relationships among variables have been established using experts' opinions. The research may help practicing managers to understand the interaction among variables affecting customer involvement. Further, this understanding may be helpful in framing the policies and strategies to green SC. Analyzing interaction among variables for effective customer involvement in greening SC to develop the structural model in the Indian perspective is an effort towards promoting environment consciousness.

Keywords: Supply chain, Green supply chain management, Green distribution, Interpretive structural modeling (ISM), MICMAC analysis

Background

Green supply chain management (GSCM) is the integration of both environmental and supply chain managements and has been identified as a proven way to reduce an organization's impact on the environment while improving business performance (Torielli et al. 2011).

Some studies have suggested that the customer plays an important role in greening the supply chain (SC). The requirement for organizations to respond actively to the customers' need has increasingly been important (Christopher 2000; Zhu and Sarkis 2004; Zhu et al. 2007a, 2007b; Zhu et al. 2008a, 2008b; Eltayeb et al. 2011). Consumer demands have been realized as a powerful pressure for change within organizations offering products or services in those markets. Consumers demand more value and quality from products, and since environmental awareness has increased, this type of pressure creates market opportunities in the form of environmental

attributes and responsibility within the supply chain (Paquette 2005). The role of customers and environmental societies has been recognized for more environment-friendly products (Vachon and Klassen 2006a). The influence on green supply chain initiatives had been followed by regulations and customer pressures (Eltayeb et al. 2011). Environmentally responsible organizations make themselves more attractive to customers and investors. The human factor plays an important role on both levels (Hanna et al. 2000; Lazuraz et al. 2011).

To improve the environmental supply chain performance, organizations need to make interactions with customers (Carter and Ellram 1998). Organizations implementing green concepts may compete and export their products in international markets. Organizations using improved environmental performance may lower their costs by reducing waste, also reducing their environmental compliance costs, and lessening the threats of civil and criminal liability by preventing pollution (Mudgal et al. 2009, 2010). Environmental collaboration with upstream suppliers and downstream customers has been found

---

* Correspondence: skbhardwaj19711971@gmail.com
[1]International Institute of Technology and Management, Murthal, Haryana 131039, India
Full list of author information is available at the end of the article

useful for organizations to reap performance gains (Vachon and Klassen 2008; Yang et al. 2009; Zhu et al. 2010).

Due to governmental legislations, environmental concerns, and customer awareness, more and more industries are giving attention to green practices (Thomas and Bijulal 2011; Wang and Gupta 2011). Customers having the choice of purchase and persuading organizations to act pro-environmentally may be possible by creating environmental consumer demand, when consumers request only environment-friendly products and refuse to buy products not meeting this requirement (H'Mida 2009).

GSCM practices have been about developing policies and practices protecting the environment along the supply chain and involve as many people as possible in this process, including manufacturers and suppliers, retailers, and customers (Zhu and Sarkis 2006; Lazuraz et al. 2011). Green supply chain programs may be initiated to position manufacturers to their customers or investors and to facilitate environmental compliance (IFS 2012).

Hence, the need arises to identify variables for effective customer involvement in the implementation of the green concept in SC. The main objectives of the research are the following:

- To identify variables for effective customer involvement in the implementation of the green concept in the supply chain.
- To prioritize the identified variables.
- To establish contextual relationships among the identified variables.
- To develop structural model using interpretive structural modeling (ISM) technique.
- To carry out *matrice d'impacts croises-multiplication appliqué a un classement* (MICMAC) analysis to classify the variables.

The next section discusses literature review of the related work. In the 'Methods' section, subsections 'Identification of the variables for effective customer involvement in implementation of green concept in SC', 'Questionnaire-based survey', 'Interpretive structural modeling', and 'MICMAC analysis' of the problem are presented. Finally, the 'Results and discussion' section of this research are presented, followed by the 'Conclusions' section.

Literature review

Some researchers have attempted work related to customer perspective in greening the supply chain. The research work done by various researchers may be highlighted in chronological order as follows:

(Tan et al. 1998) explored the relationship among supplier management, customer relations, and organizational performance and used purchasing, quality, customer relations, and supplier closeness to evaluate the suitability of a supplier selection model. Hall (2000) suggested that large firms have to meet stakeholder pressure beyond legal environmental responsibilities, and many suppliers are under considerable pressures from their customers. (Hanna et al. 2000) suggested that operational improvements can be made through techniques as just-in-time manufacturing, total quality management, concurrent engineering, and employee involvement. Employee involvement in operational and environmental activities may cause cost reduction, process improvement, reduced process waste, improved morale, enhanced customer satisfaction, improved process safety, improved community relations, and an enhanced public perception as a green firm. Chan (2003) proposed SC performance measurement system which includes qualitative measures such as customer satisfaction, flexibility, information and material flow integration and effective risk management, and quantitative measures to evaluate the SC performance in terms of strategic planning, order planning, suppliers, production, and delivery. Theyel (2006) studied customer and supplier relations for achieving environmental performance in the US chemical industry and indicated that relations between customers and suppliers affect environmental performance by waste reduction. Vachon and Klasson (2006b) noted that environmental collaboration with suppliers and customers leads to improved manufacturing performance, such as improved quality, delivery, and flexibility.

(Simpson et al. 2007) explored the moderating impact of relationship conditions existing between a customer and suppliers and the effectiveness of the customer's environmental performance requirements. Zhu et al. (2007a) found five GSCM components or constructs including eco-design, green purchasing, internal environmental management, cooperation with customers, and investment recovery. Walker et al. (2008) stated that customer pressures on organizations vary depending on the size of the organization. Zhu et al. (2008a) studied GSCM practices in China's manufacturing industry, and GSCM practices have been divided into five primary dimensions: internal environmental management, green purchasing, eco-design, cooperation with customers, and investment recovery. These five dimensions of green SCM practices distinguish it from the traditional definitions of SCM. Zhu et al. (2008b) suggested that cooperation with suppliers and customers has become extremely critical for the organizations to close the supply chain loop. Wang et al. (2009) determined a methodology to identify customers who are satisfied and willing to pay more for green products in the consumer-oriented market where the consumers' behavior towards green product has become an important issue for business.

(Azevedo et al. 2011) examined the relationships between green practices of supply chain management and supply chain performance in the context of the Portuguese automotive supply chain and obtained the conceptual model from data analysis providing evidence as to which green practices have positive effects on quality, customer satisfaction, and efficiency. Further, the practices having negative effects on supply chain performance were also identified in this study. Lazuras et al. (2011) suggested that the human factor plays an important role in the adoption and effective implementation of GSCM practices and presented a psychological perspective to GSCM adoption. Wu et al. (2011) evaluated supplier performance to find key factor criteria to improve performance. Results showed that the satisfied customer needs a criteria factor that has the greatest influence among the criteria for selecting suppliers. Green et al. (2012) suggested the model of green supply chain practices that link manufacturers with suppliers and customers to support environmental sustainability throughout the supply chain. The result suggested that organizations working with suppliers and customers achieve better environmental sustainability in the supply chain. Cooperation with customers is strongly associated with environmental performance. (Shi et al. 2012) suggested a structural model of natural resource-based GSCM and found intra- and interorganizational environmental practices, performance measures, and institutional drivers as important constructs. Causal relationships within and between the constructs were proposed in the form of hypotheses. Cooperation among customers has been found an important construct.

Methods

We have identified various variables for effective customer involvement in green concept implementation in the supply chain from the literature review and experts' opinions. Literature was reviewed to identify various variables for effective customer involvement in green concept implementation in supply chain. Experts from academia and industry were invited to idea engineering workshop, and brainstorming session was conducted where 13 variables relevant to effective customer involvement in green concept implementation in the supply chain were identified. A questionnaire-based survey was conducted to rank these variables. Using the ISM approach and MICMAC analysis technique, it had been identified that three variables ('Education Level of Customers,' 'Customer Income,' and 'Customer Intelligence') were falling into the category of autonomous variables, hence discarded, and research work had been continued with the remaining ten variables. Autonomous variables have weak driver power and weak dependence. These variables are relatively disconnected from the system, with which they have only few

links, which may not be strong, and hence, may be discarded (Ravi and Shankar 2005; Mudgal et al. 2009, 2010; Luthra et al. 2011).

Identification of variables for effective customer involvement in green concept implementation in the supply chain

These variables identified for effective customer involvement in green concept implementation in the supply chain are as follows: 'Awareness Level of Customers,' 'Encouragement and Support of Customers,' 'Motivation by Organization Sales Network,' 'Positive Perception about Top Management Commitment and Openness in Policy towards Greening,' 'Effective Advertisement and Marketing Campaign towards Green Efforts of Organization,' 'IT Enablement and Effective Communication,' 'Environment-Friendly Distribution,' 'Effective Training Program Schedule for Customer,' 'Green Labeling and Use of Green Packaging Material,' and 'Recycling and Reuse Efforts of Organization.' The abovesaid identified variables are explained in the following subsections.

Awareness level of customers

Customers have been reported as strong drivers for greening activities in the literature (Green et al. 1996). Producing environment-friendly products and creating awareness among consumers are some of the ways through which companies can contribute towards nature conservation. Customer demands have a strong influence on the decisions that companies take towards eco-design (Alhola 2008). To obtain the most sustainable solution, the environment consideration of properties of products and services must meet customer requirement (Zhu et al. 2008a, 2008b).

Encouragement and support of customers

Some studies have found that ultimate individual consumer interest in the environment and environmentally sound products is quite substantial, even though there has been a slight decline (Reijonen 2011). Implementation of environmental technology may build a positive brand image, mitigate environmental liabilities associated with a firm's products and services, and influence the mindset of customers and investors (Rao and Holt 2005). In the USA, an estimated 75% of consumers claim that their purchases are influenced by reputation, and 80% would be willing to pay more for environment-friendly products (Lamming and Hamapson 1996; Chien and Shih 2007a, 2007b).

Motivation by organization sales network

Customers are in direct contact with the organizations' sales personnel in most of the cases and may be informed, influenced, and convinced about the green products and services offered by the organizations. As one of the results of the brainstorming session, this factor was strongly recommended by the participants of the session.

Positive perception about top management commitment and openness in policy towards greening

Top management may be held responsible directly and indirectly for each activity at all the levels of the organizations (Singh and Kant 2008). Top management commitment is necessary for supporting GSCM ideas, practices, and cooperation across organizational functions (Sarkis et al. 2007; Zhu et al. 2007a, 2007b), and success of any strategic program needs to be derived from top management (Yu and Hui 2008). Top management has a significant ability to support actual formation and implementation of green initiatives across the organization. Top management may provide continuous support for GSCM in the strategic and action plans for successful implementation (Ravi and Shankar 2005; Mudgal et al. 2009). Positive perception about top management commitment and openness in policy towards greening may be achieved by publishing sincere green efforts of organization.

Effective advertisement and marketing campaign towards green efforts of organization

Organizations may advertise environment-friendly products and services to create awareness among customers. Customers aware of green products may prefer to purchase green products, which may further increase an organization's reputation and sales volumes (Luthra et al. 2011). Newspapers, hording, magazines, printed material distribution (leaflets, booklets, etc.), and various audiovisual media (e.g., radio, television,cinema) may be a few media for advertisement and marketing campaign for making the customers more aware of green efforts of the organization. In India, few retail organizations are providing recycled paper and jute bags (with green slogans) for carrying their products.

IT enablement and effective communication

IT enablement may be required for processing and updating accurate information of products, materials, and other resources (Sarkis et al. 2007) and for supporting various GSCM activities (Ravi and Shankar 2005). Informal linkages and improved communication may help the organization to adopt green practices (Yu 2007; Yu and Hui 2008), and increased environmental performance in GSCM may be achieved by information sharing of improved quality (Wu et al. 2010).

Environment-friendly distribution

Environment-friendly or green distribution is the process of moving a product from its manufacturing source to its customers with a low impact on the environment.

Reverse logistics is identified as the process of planning, implementing, and controlling flows of raw materials, in-process inventory, and finished goods from a manufacturing, distribution, or use point to a point of recovery or point of proper disposal (Ilgin and Gupta 2010; Srivastva 2007). The use of green fuel-like compressed natural gas-driven vehicles may exhibit seriousness about green efforts of an organization.

Effective training program schedule for customers

Training and education are the prime requirements for achieving successful implementation of GSCM in any organization (Ravi and Shankar 2005; Sarkis et al. 2007; Wu et al. 2010). Trained personnel may contribute in training the customers, leading to better customer involvement in GSCM implementation.

Green labeling and use of green packing material

Environment-friendly packing refers to use of recyclable or dissolvable materials for packing and has a clear objective of encouraging business to market greener products (Fielding 2001), and it may be a good way to make the customers better informed about environmental choices while purchasing. Eco-labeling is a voluntary scheme designed to encourage businesses to market environment-friendly products and services (Mudgal et al. 2009).

Recycling and reuse efforts of organization

Recycling is the process of collecting used products, components, and materials from the field and separating them into categories of like materials (recyclable and nonrecyclable), and recyclable materials may be processed into recycled products, components, and materials. Reuse is the process of collecting used materials, products, or components from the field, and distributing or selling them as used. Waste management and recyclability evaluation methods may help in managing and minimizing waste and improving the environment (Ilgin and Gupta 2010; Srivastva 2007). Lean is a competitive practice that reduces costs, improves the environment, and improves quality (Bhetja et al. 2011). The use of lean or flexible manufacturing may help in the continuous improvement and elimination of waste in all forms and has great potential for reciprocal benefits to firm environmental management practices (Mudgal et al. 2009).

Questionnaire-based survey

Based upon the abovesaid variables for effective customer involvement in green concept implementation in the supply chain, a research questionnaire was designed. The questionnaire was developed taking into account the experts' opinions. A first draft was reviewed by experts from the academia and industry. Their feedback was used to improve the questions and eliminate redundancies. A second version was developed. These variables were tested for content validity and reliability through the pretesting of the questionnaire. Content validity is the technique used to ensure that the measures adequately quantify the concepts that they are supposed to be tested (Sekaran 2003). Reliability concerns the extent to which an experience, test, or any measuring procedure yields the same results on repeated trials (Carmines and Zeller 1979). Reliability evaluates the accuracy of measures through assessing the internal stability and consistency of items in each variable (Hair et al. 2009). Validity of the variables was pretested among selected experts from the academia and industry. The reliability of measures was also pretested by applying Cronbach's alpha coefficients on the responses from experts. All values of the coefficients fall within the range of 0.60 to 0.80, ensuring an acceptable level of reliability (Nunnally 1987). The results from this pretest were used to further improve the questionnaire. After a discussion with the experts and the pretest, we kept our questionnaire very simple and short due to low rate of responses from respondents. After pretesting, the final version will be used in the survey. The population of this study consists of academicians, manufacturing firms, and valuable customers from North Indiasince the population size was very large. After identifying the target population, it was necessary to determine the sample size. The sample size was taken using the following mathematical relationship for proportions (Israel 1996; Rea and Parker 2005; Sanchez Gomez 2011).

$$\text{Sample size} = Z^2 \times p \times (1 - p)/H2 \qquad (1)$$

where $Z=$ Z value in normal distribution tables (1.96 for 95% confidence level), $p=$ the estimated proportion of the population that presents the characteristics (0.5 is used as a conservative value, higher or lower values yield a smaller required sample size), and $H=$ the precision level or margin of error, expressed as decimal (10%= 0.1). Then, sample size = $(1.96)^2 \times 0.5(1-0.5)/(0.1)^2$ =96.04 or sample size= 96.

Therefore, approximately 96 complete questionnaires were needed. A questionnaire-based study was carried out, and respondents were asked to rank the variables on a five-point Likert scale (where '1' means not important and '5' means most important). This research was conducted from August 2011 to November 2012 at the North India zone. We used convenience sampling as well as random sampling. Due to difficulties of mail surveys and the possibility of respondents to misunderstand the questionnaire items, we used convenience sampling through interviewing various academicians, top/middle level managers/engineers of industries, and customers. One hundred seventy-eight completed questionnaires

were collected via interviews. Further, to test the convenience sampling bias, we carried out random surveys through e-mail; 643 questionnaires were sent to various academicians, top/middle level managers/engineers of various industries, and customers. After reminder emails in addition to telephonic calls, 171 questionnaires were received. Twenty-seven questionnaires were incomplete and were discarded. This gives an overall response rate of 22.4%. A response rate of 20% is considered for positive assessment of the surveys (Malhotra and Grover 1998). A total of 322 questionnaires were considered for further research work.

Interpretive structural modeling

The mathematical foundations of the ISM methodology can be found in reference works (Harary et al. 1965), while the philosophical basis for the development of this approach has been presented by Warfield (1974). ISM has been used for policy analysis (Sage 1977) and, in recent years, for management research (Mandal and Deshmukh 1994; Jharkharia and Shankar 2005; Ravi and Shankar 2005; Sushil 2005; Sarkis et al. 2007; Mudgal et al. 2009, 2010; Diabat and Kannan 2011; Luthra et al. 2011). ISM was first proposed by J. Warfield in 1974 to analyze the complex socioeconomic systems. Its basic idea is to use the experts' practical experience and knowledge to decompose a complicated system into several subsystems and construct a multilevel structural model. The ISM is interpretive as the judgment of the selected group for the study decides whether and how the factors are interrelated. ISM generally has the following steps (Ravi and Shankar 2005; Sage 1977; Warfield 1974):

Step 1: Variables affecting the system are listed.
Step 2: From the variables identified in step 1, the contextual relationships among the variables are found.
Step 3: A structural self-interaction matrix (SSIM) is developed for variables, which indicated pairwise relationships among variables of the system.
Step 4: A reachability matrix is developed from the SSIM, and the matrix is checked for transitivity. The transitivity of the contextual relation is a basic assumption made in ISM. It states that if variable A is related to variable B and variable B is related to variable C, then variable A is necessarily related to variable C.
Step 5: The reachability matrix obtained in step 4 is partitioned into different levels.
Step 6: Based on the contextual relationships given above in the reachability matrix, a directed graph is drawn and the transitive links are removed.

Step 7: The resultant diagraph is converted into an ISM by replacing variable nodes with statements.

Step 8: The ISM model developed in step 7 is reviewed to check for conceptual inconsistency, and necessary modifications are made.

The flow chart for the ISM methodology is shown in Figure 1.

MICMAC analysis

Matrice d'impacts croises-multipication applique' an classment (cross-impact matrix multiplication applied to classification) is abbreviated as MICMAC (Mudgal et al. 2009). In the MICMAC analysis, the dependence power and driver power of the variables are analyzed. Variables will be classified into four clusters. The four clusters are autonomous, dependent, linkage, and driver/independent. In the final reachability matrix, the driving power and dependence power of each of the variables will be plotted. Autonomous variables (first cluster) have weak driving power and weak dependence power. These variables can be disconnected from the system. The second

clusters named dependent variables have weak driving power and strong dependence power. The third cluster named linkage variables has strong driving power and strong dependence power. The fourth cluster named independent variables has strong driving power and weak dependence power.

Results and discussion

The results and discussions of questionnaire-based survey, interpretive structural modeling technique, and MICMAC analysis have been discussed in the following subsections, respectively.

Questionnaire-based survey

Questionnaire-based data have been further processed with the help of the software package Minitab version 16. The values of mean, standard error of mean, trimmed mean, standard deviation, variance, and rank for each variable have been tabulated in Table 1. Five percent of the highest and five percent of the lowest data values are excluded for the calculation the trimmed

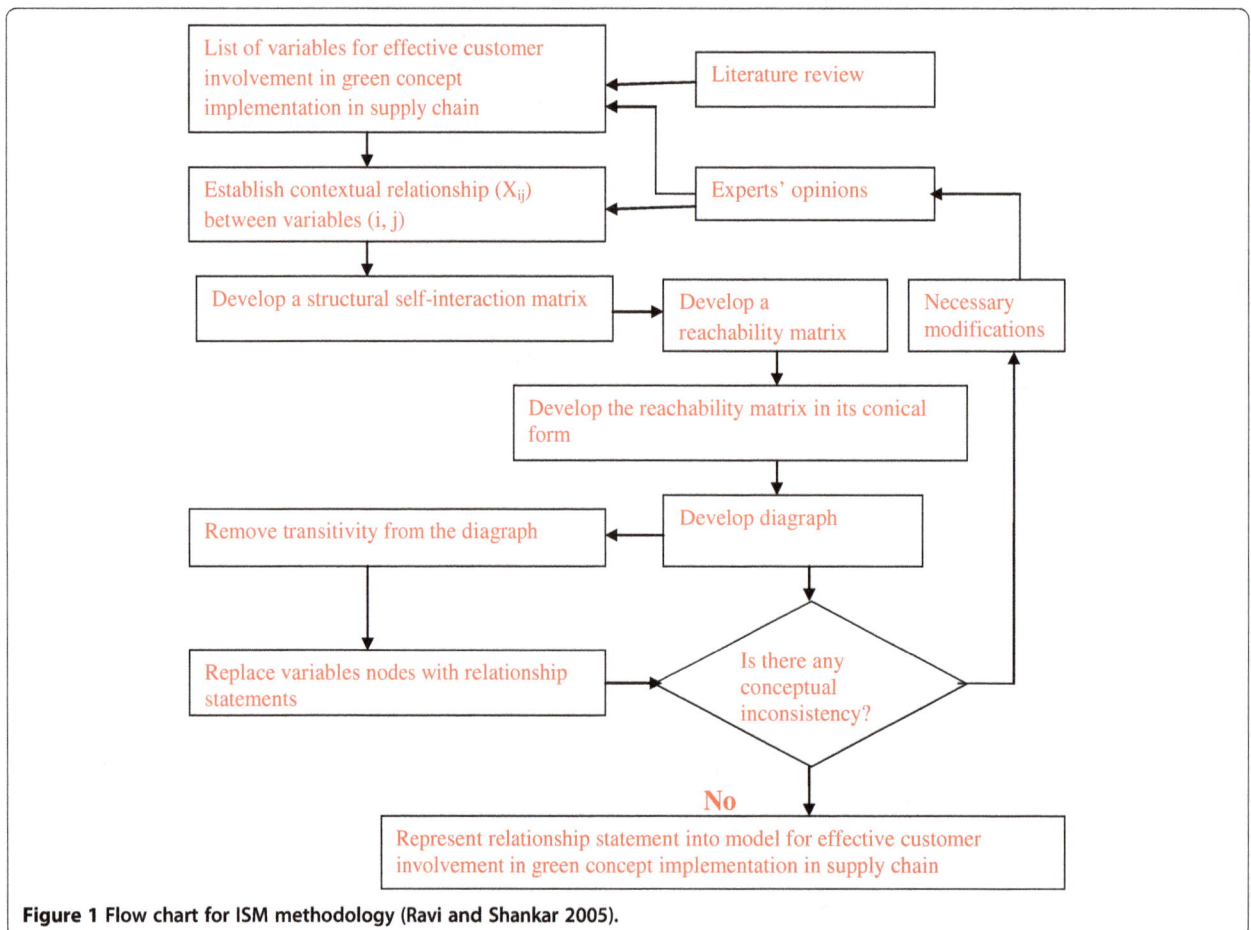

Figure 1 Flow chart for ISM methodology (Ravi and Shankar 2005).

Table 1 Statistical data analysis result

Number	Variable	Mean	Standard error mean	Trimmed mean	Standard deviation	Variance	Rank
1	Awareness level of customers	3.2688	0.0476	3.2425	0.9187	0.8440	X
2	Encouragement and support of customers	3.2124	0.0454	3.1886	0.8753	0.7661	IX
3	Motivation by organization sales network	3.6371	0.0330	3.6078	0.6359	0.4043	VIII
4	Positive perception about top management commitment and openness in policy towards greening	3.6882	0.0329	3.6766	0.6355	0.4038	VII
5	Effective advertisement and marketing campaign towards green efforts of organization	4.0838	0.0297	4.0994	0.5712	0.3263	V
6	IT enablement and effective communication	4.0538	0.0320	4.0689	0.6163	0.3799	VI
7	Environment-friendly distribution	4.4785	0.0288	4.5090	0.5563	0.3095	II
8	Effective training program schedule for customers	4.7339	0.0233	4.7635	0.4486	0.2012	I
9	Green labeling and use of green packing material	4.3925	0.0308	4.4371	0.5935	0.3523	III
10	Recycling and reuse efforts of organization	4.3172	0.0300	4.3533	0.5795	0.3358	IV

mean. Ranking of variables has been done on the basis of the trimmed mean.

'Effective training program schedule for customers,' 'Environment-friendly distribution,' and 'Green labeling and use of green packaging material' have been ranked by respondents as the top three variables.

'Effective training program schedule for customers' has also been observed having the lowest standard deviation and lowest variance. 'Awareness level of customers' has been observed having the highest standard deviation and highest variance. Variance and standard deviation measure variability within a distribution. Standard deviation indicates how much, on average, each of the values in the distribution deviates from the mean of the distribution. Higher standard deviation means having large variations in the data, and lower standard deviation means having small variations in the data. Lower standard deviation will indicate the reliability of data. Variance is the average squared deviations of the mean.

The coefficients of correlation among variables have been tabulated in Table 2. Maximum value of coefficient of correlation has been indentified between 'Awareness level of customers' and 'Encouragement and support of customers.'

These correlation coefficients of different variables have been classified according to the rules of thumb about the strength of correlation coefficient (Hair et al. 2003, p.568). These variables have been classified into five categories (Table 3). 'Awareness level of customer' and 'Encouragement and support of customers,' 'Motivation by organization sales network,' and 'Positive perception about top management commitment and openness in policy towards greening,' and 'Effective advertisement and Marketing campaign towards green efforts of organization' and 'IT enablement and effective communication' have been classified as strongly correlated variables.

Interpretive structural modeling

ISM can be used for identifying and summarizing relationships among specific variables, which define a problem or an issue (Warfield 1974; Sage 1977). It provides means by which order can be imposed on the complexity of such variables (Mandal and Deshmukh 1994; Jharkharia and Shankar 2005; Luthra et al. 2011).

Structural self-interaction matrix

ISM model suggests the use of experts' opinions in identifying the contextual relationship among variables. Thus, in this research for identifying the contextual relationship among the variables for effective customer involvement in green concept implementation in SC, three experts from the academia and four experts from the industry were consulted. Four symbols (V, A, X, and O) are used to denote the direction of the relationship between the variables (i and j):V- variable i will lead to variable j, A- variable j will lead to variable i, X- variable i and j will lead to each other, and O- variable i and j are unrelated.

Based on the contextual relationships, SSIM has been developed (Table 4). Variable 9 leads to variable 10, so symbol 'V' has been given in the cell (9, 10); variable 3 leads to variable 1, so symbol 'A' has been given in the cell (1, 3); variables 1 and 2 lead to each other, so symbol 'X' has been given in the cell (1, 2); variables 6 and 7 do not lead to each other, so symbol 'O' has been given in the cell (6, 7), and so on. The number of pairwise comparison question addressed for developing the SSIM are $((N) \times (N-1)/2)$, where N is the number of variables.

Table 2 Coefficient of correlation of variables

Variable	1	2	3	4	5	6	7	8	9	10
1	1.000	0.776	0.226	0.276	0.159	0.139	0.069	0.060	0.069	0.086
2	0.776	1.000	0.149	0.169	0.186	0.165	0.101	0.097	0.012	0.117
3	0.226	0.149	1.000	0.606	0.084	0.124	0.004	−0.011	−0.082	−0.002
4	0.276	0.169	0.606	1.000	0.132	0.145	0.088	0.009	−0.092	0.012
5	0.159	0.186	0.084	0.132	1.000	0.612	0.351	0.193	0.183	0.255
6	0.139	0.165	0.124	0.145	0.612	1.000	0.236	0.118	0.042	0.099
7	0.069	0.101	0.004	0.088	0.351	0.236	1.000	0.437	0.288	0.387
8	0.060	0.097	−0.011	0.009	0.193	0.118	0.437	1.000	0.320	0.284
9	0.069	0.012	−0.082	−0.092	0.183	0.042	0.288	0.320	1.000	0.287
10	0.086	0.117	−0.002	0.012	0.255	0.099	0.387	0.284	0.287	1.000

The use of symbols V, A, X, and O may be clearly understood from SSIM made after reaching to a consensus in the opinions of experts (Table 4):

Reachability matrix

SSIM has been converted into a binary matrix, named initial reachability matrix, by substituting V, A, X, and O by 1 or 0 applying the following rules:

- If (i, j) value in the SSIM is V, (i, j) value in the reachability matrix will be 1 and (j, i) value will be 0; for V (9, 10) in SSIM, '1' has been given in cell (9, 10) and '0' in cell (10, 9) in the initial reachability matrix.
- If (i, j) value in the SSIM is A, (i, j) value in the reachability matrix will be 0 and (j, i) value will be 1; for A (1, 3) in SSIM, '0' has been given in cell(1, 3) and '1' in cell (3, 1) in the initial reachability matrix.
- If (i, j) value in the SSIM is X, (i, j) value in the reachability matrix will be 1 and (j, i) value will also be 1; for X (1, 2) in SSIM, '1' has been given in cell (1, 2) and '1' in cell (2, 1) also in the initial reachability matrix.

- If (i, j) value in the SSIM is O, (i, j) value in the reachability matrix will be 0 and (j, i) value will also be 0; for O (6, 7) in SSIM, '0' has been given in cell (6, 7) and '0' in cell (7, 6) also in initial reachability matrix.

By applying these rules, an initial reachability matrix for variables for effective customer involvement in green concept implementation in the supply chain has been obtained (Table 5).

The final reachability matrix (Table 6) is constructed from the initial reachability matrix taking into account the transitivity rule, which states that if a variable 'A' is related to 'B' and 'B' is related to 'C', then 'A' is necessarily related to 'C'. Variable 5 leads to variables 1, 2, 3, 4, and 6. Variable 7 leads to variables 1, 2, 3, 4, and 5, and then variable 6 must be added to variable 7 as transitive element and so on.

In Table 6, the driving power and the dependence of each enabler are also shown. The driving power for each variable is the total number of variables (including itself), intowhich it may impact. Dependence is the total

Table 3 Classification of variables based upon significance of correlation

Variable number	Very strongly correlated[a]	Strongly correlated[b]	Moderately correlated[c]	Weakly correlated[d]	Not correlated[e]
1	1	2		3, 4	5,6,7,8,9,10
2	2	1			3,4,5,6,7,8,9,10
3	3	4		1	2,5,6,7,8,9,10
4	4	3		1	2,5,6,7,8,9,10
5	5	6		7,10	1,2,3,4,8,9
6	6	5		7	1,2,3,4,8,9,10
7	7		8	5,6,9,10	1,2,3,4
8	8		7	9,10	1,2,3,4,5,6
9	9			7,8,10	1,2,3,4,5,6
10	10			5,7,8,9	1,2,3,4,6

[a]Variable numbers having a correlation coefficient between 0.801 and 1.000; [b]variable numbers having a correlation coefficient between 0.601 and0.800; [c]variable numbers having a correlation coefficient between 0.401 and0.600; [d]variable numbers having a correlation coefficient between 0.201 and0.400; [e]variable numbers having a correlation coefficient less than or equal to 0.200.

Table 4 Structural self interaction matrix

Variable	10	9	8	7	6	5	4	3	2
1	A	A	A	A	A	A	A	A	X
2	A	A	A	A	A	A	A	A	
3	A	A	A	A	A	A	X		
4	A	A	A	A	A	A			
5	A	A	A	A	X				
6	O	O	O	O					
7	X	A	O						
8	O	A							
9	V								

number of variables (including itself), which may be impacting it. These driving power and dependencies will be used in the MICMAC analysis, where the variables will be classified into four groups of autonomous, dependent, linkage, and independent (driver) variables.

Level partitioning

The reachability set and antecedent set (Warfield 1974) for each variable have been found out from final reachability matrix. Subsequently, the intersection set of these sets has been derived for all variables. The variable, for which the reachability and the intersection sets are the same, has been given the top-level variable in the ISM hierarchy. From Table 7, it is seen that the 'Awareness Level of Customers' and 'Encouragement and Support of Customers' have been found at Level I. The iteration is continued until the level of each variable is found out. The identified levels aid in building the diagraph and the final model of the ISM.

From the final reachability matrix, the structural model generated is known as a diagraph. After removing the transitivity links and replacing the node numbers by statements, the ISM model has been generated, which has been shown in Figure 2. It is observed from the figure that 'Green labeling and use of green packaging

Table 5 Initial reachability matrix

Variable	1	2	3	4	5	6	7	8	9	10
1	1	1	0	0	0	0	0	0	0	0
2	1	1	0	0	0	0	0	0	0	0
3	1	1	1	1	0	0	0	0	0	0
4	1	1	1	1	0	0	0	0	0	0
5	1	1	1	1	1	1	0	0	0	0
6	1	1	1	1	1	1	0	0	0	0
7	1	1	1	1	1	0	1	0	0	1
8	1	1	1	1	1	0	0	1	0	0
9	1	1	1	1	1	0	1	1	1	1
10	1	1	1	1	1	0	1	0	0	1

material' is a very significant variable for effective customer involvement in green concept implementation in SC as it comes at the base of the ISM hierarchy. 'Awareness level of customers' and 'Encouragement and support of customers' are the top-level variables in the model.

The structural model for effective customer involvement in greening the supply chain, presented in the study, addresses the importance of customer role in the process of making the SC sustainable. 'Use of green packing material' supports 'Recycling and reuse efforts of the organization' and 'Environment-friendly distribution.' Also, 'Environment-friendly distribution' may be utilized for recollecting used and/or discarded (due to improper quality or because of any other reason) products/components. In India, an unorganized sector involves collection of junk/waste/scrap material by ragpickers and selling to big dealers. Here, authors want to draw the attention towards this area ('Kabaadi-Wala' SC) for future research having green potential for research focusing upon the 'recycling and reuse' aspect. Green labeling of products with appropriate display of information about greening may be proven as an important help to train the customers, which may further facilitate effective communication among various actors in the greening process. 'Environment-friendly distribution' system may itself be utilized as a strong way of advertising and publicized in the organization's marketing campaign. IT enablement, effective communication, advertisement, and marketing campaign may complement and help each other towards green efforts of organization and, further, may help in motivating customers by sales personnel and having a positive perception about top management efforts towards greening. However, this may lead to developing the trust of customers on top management sincerity for greening. All the above factors in totality may lead to a high level of customer awareness, encouragement, and support.

MICMAC analysis

MICMAC analysis is done with the help of driving power and dependence power of variables. The driver power-dependence diagram has been constructed and is shown in Figure 3.

The first cluster consists of the autonomous variables that have weak driver power and weak dependence. No variable has been identified as an autonomous variable. The second cluster consists of the dependent variables that have weak driver power but strong dependence. 'Awareness level of customers', 'Encouragement and support of customers', 'Motivation by organization sales network', and 'Positive perception about top management commitment and openness in policy towards greening' have been identified as dependent variables. The third cluster has the linkage variables that have strong driver

Table 6 Final reachability matrix

Variables	1	2	3	4	5	6	7	8	9	10	Driver power	Driver rank
1	1	1	0	0	0	0	0	0	0	0	2	VI
2	1	1	0	0	0	0	0	0	0	0	2	VI
3	1	1	1	1	0	0	0	0	0	0	4	V
4	1	1	1	1	0	0	0	0	0	0	4	V
5	1	1	1	1	1	1	0	0	0	0	6	IV
6	1	1	1	1	1	1	0	0	0	0	6	IV
7	1	1	1	1	1	1[a]	1	0	0	1	8	II
8	1	1	1	1	1	1[a]	0	1	0	0	6	III
9	1	1	1	1	1	1[a]	1	1	1	1	10	I
10	1	1	1	1	1	1[a]	1	0	0	1	8	II
Dependence	10	10	8	8	6	6	3	2	1	3		
Dependence rank	I	I	II	II	III	III	IV	V	VI	IV		

[a]Means transitive.

power and also strong dependence. 'Effective advertisement and marketing campaign towards green efforts of organization,' and 'IT enablement and effective communication' have been found out as linkage variables in our study. The fourth cluster includes the independent variables having strong driving power but weak dependence. 'Environment-friendly distribution,' 'Effective training program schedule for customers,' 'Green labeling and use of green packaging material,' and 'Recycling and reuse efforts of organization' have been identified as the driver variables.

Conclusions

Greening of SC has been identified as an important approach for improving the environmental performance of processes and products. Ten variables for effective customer involvement in green concept implementation

Table 7 Partitioning of variables

Variables	Reachability set	Antecedent set	Intersection set	Level
1	1,2	1,2,3,4,5,6,7,8,9,10	1,2	I
2	1,2	1,2,3,4,5,6,7,8,9,10	1,2	I
3	1,2,3,4	3,4,5,6,7,8,9,10	3,4	II
4	1,2,3,4	3,4,5,6,7,8,9,10	3,4	II
5	1,2,3,4,5,6	5,6,7,8,9,10	5,6	III
6	1,2,3,4,5,6	5,6,7,8,9,10	5,6	III
7	1,2,3,4,7,5,6,10	7,9,10	7,10	IV
8	1,2,3,4,5,6,8	8,9	8	IV
9	1,2,3,4,5,6,7,9,10,8	9	9	V
10	1,2,3,4,5,6, 7,10	7,9,10	7,10	VI

in SC have been identified from the literature. These variables have been validated from a questionnaire-based survey.

ISM methodology has been used in finding contextual relationships among various variables. A model has been developed from ISM methodology. 'Awareness level of customers' and 'Encouragement and support of customers' have been identified as top-level variables and 'Green labeling and use of green packing material' as the most important bottom-level variable. MICMAC analysis has also been carried out. The driving power-dependence power diagram helps to categorize various variables for effective customer involvement in the implementation of green concept in the supply chain. Four variables have been identified as driver variables and four variables as dependent variables. Two variables have been identified as linkage variables and no variable as autonomous variable. Dependent variables represent desired objectives for customer involvement in greening the supply chain. Driver variables will play an important driving role in customer involvement in greening the supply chain. Management needs to address these variables more carefully. Linkage variables are unstable that any action on these variables will have an effect on others and also a feedback on themselves.

Human factor plays a major role in 'greening' the supply chain. The legislative framework and regulations promoting environmental practices cannot guarantee GSCM adoption, unless the dynamic interplay between the people involved throughout the SC is taken into consideration (Lazuras et al. 2011).

The customer has been reported as repeated enabler in the literature. Some researchers identified cooperation of customers with eco-design, cleaner production, and using less energy during product transportation and green design as an important variable (Zhu and Sarkis 2004; Zhu et al. 2007a, 2007b; Zhu et al. 2008a, 2008b; Green et al. 2012). Our study identifies ten variables for effective customer involvement in the implementation of green concept in supply chain.

Consumer demand for eco-friendly goods is increasing. A large number of customers prefer organizations that have superior environmental records and greener products and are ready to pay a premium for it (Lakshmi and Visalakshmi 2012). Organizations have begun to implement GSCM practices in response to customer demand (Green et al. 2012). Customers have impacted on the decisions of manufacturing enterprises to employ GSCM practices (Huang 2012). With increasing customer awareness and regulatory norms, organizations with greener supply chain may have a competitive advantage (Cognizant 2009).

Earlier studies did not provide a structural model of customer involvement in greening the supply chain in

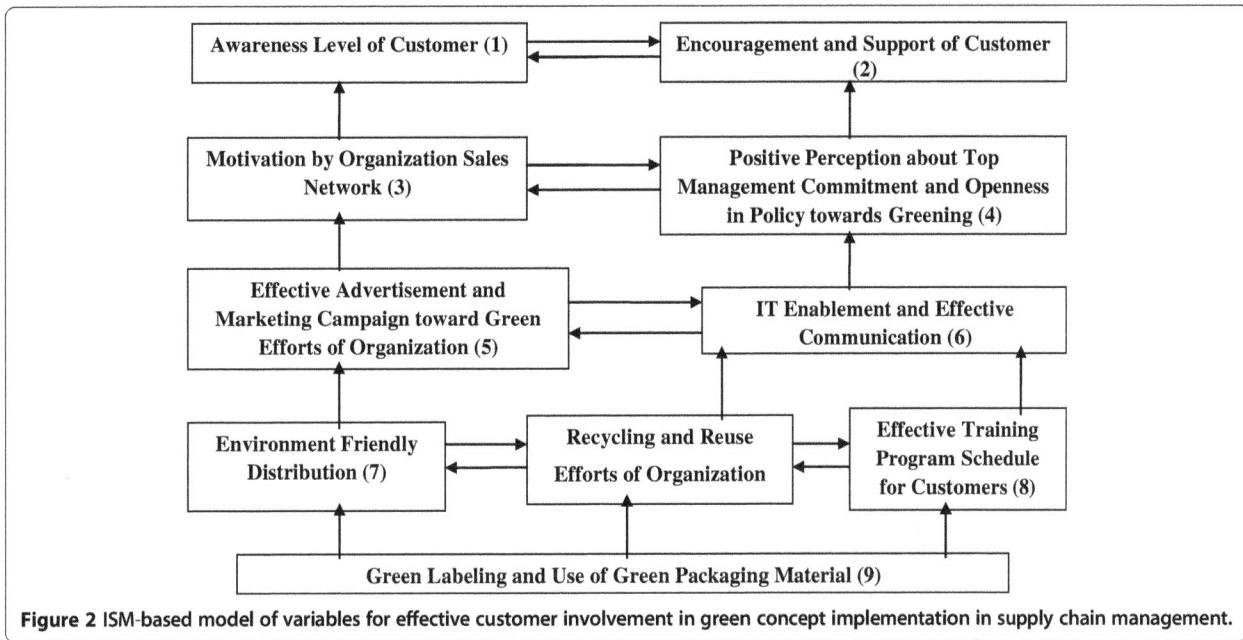

Figure 2 ISM-based model of variables for effective customer involvement in green concept implementation in supply chain management.

the Indian context. Further, this study provided hierarchy of variables for effective customer involvement in green concept implementation in the supply chain; this hierarchy may help managers/supply chain practitioners in strategic and tactical decisions towards customer involvement in green practices. Customers aware of green products would like to purchase green products, which may increase an organization's reputation and sales volumes (Mudgal et al. 2009; Luthra et al. 2011).

The model developed in this research is based upon experts' opinions. The experts' opinion may be biased. The results of model analysis may vary in real world setting. We have considered ten variables. In case a model needs to be developed for some specific industry, some variables may be deleted and/or added. Hypothesis testing may be further used to test the validity of this hypothetical model. The ragpickers/junk dealers/members involved in recycling/reuse, i.e., 'Kabaadi-Wala' SC, have been purposed as an important area for research in context with greening.

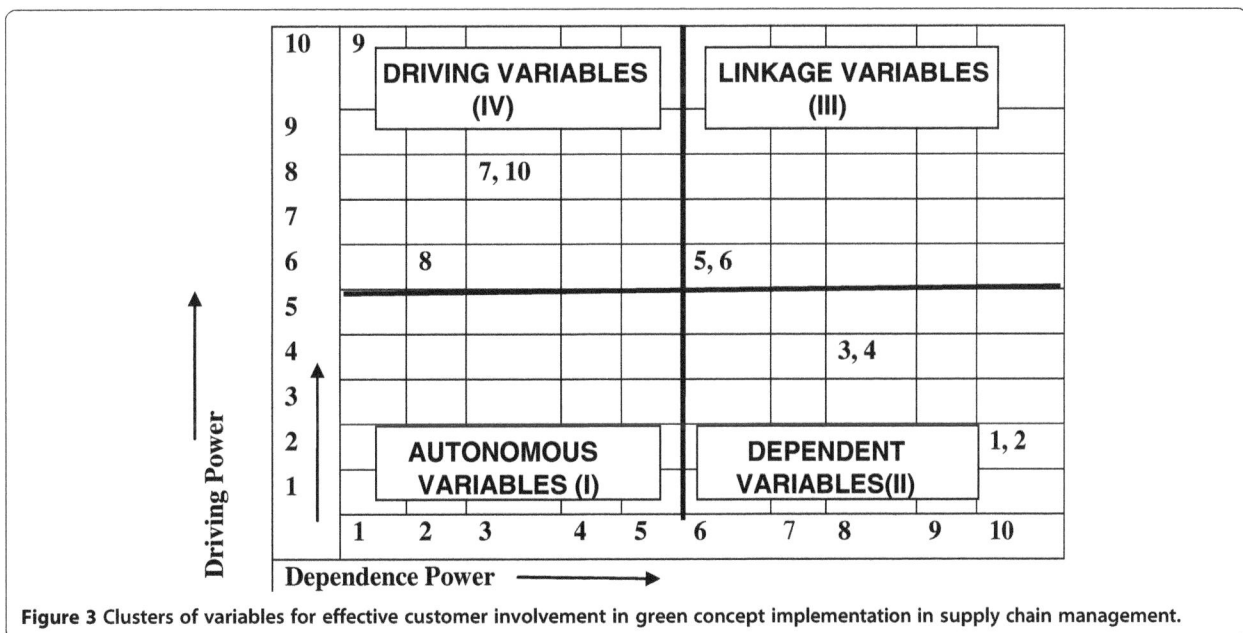

Figure 3 Clusters of variables for effective customer involvement in green concept implementation in supply chain management.

Competing interests

The authors declare that they have no competing interests.

Authors' contributions

All authors read and approved the final manuscript.

Authors' information

Dr. Sanjay Kumar is working as a Professor in Mechanical Engineering Department at the International Institute of Technology and Management, Murthal-131039, Haryana, India. He completed his Ph.D. degree in Mechanical Engineering from Jamia Millia Islamia University, New Delhi, India. He accomplished his Master's Degree in Mechanical Engineering with specialization in Production Engineering from the Delhi Technological University, Delhi, India and Bachelor's Degree in Mechanical Engineering from the National Institute of Technology, Kurushetra, Haryana, India. He has been associated with industry and teaching in the various fields of Mechanical Engineering and operation management over 15 years. He has contributed over 20 research papers in international referred and national journals and conferences at international and national level. His specific areas of interest are industrial engineering, supply chain management, green supply chain management, etc.

Sunil Luthra is working as a lecturer in Government Polytechnic, Jhajjar. He is a research scholar (part-time) in the Mechanical Engineering Department at the National Institute of Technology, Kurukshetra, Haryana, India. He has been associated with teaching for the last 12 years. His specific areas of interest are operation management, optimization techniques, green supply chain management, etc.

Dr Abid Haleem is working as a professor and head of Mechanical Engineering at the Faculty of Engineering and Technology, Jamia Millia Islamia, New Delhi, India. He has more than 110 research papers to his credit, published in international and national journals. His research interests are e-governance, technology management, supply chain management and systems modeling, flexibility, etc. He is an editor of the Asia Pacific of Global Journal of Flexible Systems Management. He is also a member of an editorial board of contemporary management research.

Acknowledgments

The authors are very much thankful to Dr. Dixit Garg, Professor at the Department of Mechanical Engineering, Kurukshetra, India for his valuable support and guidance. The authors are also very much thankful to the unanimous reviewers, to Sherlyn C Machica, and to the language editors of the paper for their constructive and helpful comments that improved the quality of the paper.

Author details

[1]International Institute of Technology and Management, Murthal, Haryana 131039, India. [2]National Institute of Technology, Kurukshetra, Haryana 136119, India. [3]Faculty of Engineering and Technology, Jamia Mallia Islamia, New Delhi 110025, India.

References

Alhola KP (2008) Promoting environmentally sound furniture by green public procurement. Ecological Econ 68(1–2):472–485

Azevedo SG, Carvalho H, Machado VC (2011) The influence of green practices on supply chain performance: a case study approach. Transportation Research Part E 47(6):850–871

Bhetja AK, Babbar R, Singh S, Sachdeva A (2011) Study of green supply chain management in the Indian manufacturing industries: a literature review cum an analytical approach for the measurement of performance. IJCEM Int J of Computational Engineering & Manage 13:84–99

Carmines E, Zeller R (1979) Reliability and validity assessment, series: quantitative applications in social science. Sage, Newbury Park, CA

Carter C, Ellram L (1998) Reverse logistics: a review of the literature and framework for future investigation. J of Business Logistics 19(1):85–102

Chan F (2003) Performance measurement in a supply chain. The Int J of Advanced Manufacturing Technology 21(7):534–548

Chien MK, Shih LH (2007a) An empirical study of the implementation of green supply chain management practices in the electrical and electronics industries and their relation to organizational performances. Int J of Science and Technology 4(3):383–394

Chien MK, Shih LH (2007b) Relationship between management practice and organization performance under European Union directives such as ROHS: a case study on the electrical and electronics industry in Taiwan. African J of Environmental Science and Technology 1(1):37–48

Christopher M (2000) The agile supply chain: competing in volatile markets. Industrial Marketing Manage 29(1):37–44

Cognizant (2009) Creating a green supply chain information technology as an enabler for a green supply chain. http://www.cognizant.com/InsightsWhitepapers/Creating_a_Green%20Supply_Chain_WP.pdf. Accessed 9 July 2012

Diabat A, Govidan K (2011) An analysis of the drivers affecting the implementation of green supply chain management. Resources, Conservation and Recycling 55(6):659–667

Eltayeb TK, Zailani S, Ramayah T (2011) Green supply chain initiatives among certified companies in Malaysia and environmental sustainability: investigating the outcomes. Resource, Conservation and Recycling 55(5):495–506

Fielding S (2001) ISO 14001: a plan for environmental excellence. Industrial Maintenance & Plant Oper 62(8):11–15

Gomez LSS (2011) Identifying success factors in the wood pallet supply chain. Virginia Polytechnic Institute and State University, Virginia, Master of Science Thesis

Green K, Morton B, New S (1996) Purchasing and environmental management: interactions policies and opportunities. Business Strategy and the Environment 5(3):188–197

Green KW Jr, Zelbst PJ, Meacham J, Bhadauria VS (2012) Green supply chain management practices: impact on performance. Supply Chain Manage: An Int J 17(3):290–305

H'Mida S (2009) Factors contributing in the formation of consumers' environmental consciousness and shaping green purchasing decisions. Proceedings of the Int conference on computers & industrial engineering 2009:957–962

Hair JF, Bush RP, Ortinau DJ (2003) Marketing Research, 2nd edn. McGraw-Hill, New Delhi

Hair JF, Black WC, Babin JB, Anderson RE, Tatham RL (2009) Multivariate data analysis(6th edition). Pearson Education, India

Hall J (2000) Environmental supply chain dynamics. J of Cleaner Prod 8(6):206–225

Hanna MD, Newman WR, Johnson P (2000) Linking operational and environmental improvement through employee involvement. Int J of Oper & Prod Manage 20(2):148–165

Harary F, Norman R, Cartwright Z (1965) Structural models: an introduction to the theory of directed graphs. Wiley, New York

Huang X, Tan BL, Li D (2012) Pressures on green supply chain management: a study on manufacturing small and medium-sized enterprises in China. Int Business and Manage 4(1):76–82

IFS (2013) ERP for green supply chain management in manufacturing. http://www.ifsworld.com/en/search/?searchParam=ERP%20for%20green%20supply%20chain%20management%20in%20manufacturing. Accessed 9 January 2013

Ilgin MA, Gupta SM (2010) Environmentally conscious manufacturing and product recovery: a review of the state of the art. J of Environmental Manage 91(3):563–591

Israel GD (1996) Determining sample size. IFAS Extension, PEOD-6, University of Florida, edis.ifas.ufl.edu/pd006. Accessed 16 January 2013

Jharkharia S, Shankar R (2005) IT enablement of supply chains: understanding the barriers. J of Enterprise Information Manage 18(1):11–27

Lakshmi P, Visalakshmi S (2012) Managing green supply chain: initiatives and outcomes. Int J of Managing Value and Supply Chains 3(4):55–63

Lamming R, Hamapson J (1996) The environmental as a supply chain management issue. British J of Manage 7:45–62

Lazuraz L, Ketikidis PH, Bofinger AB (2011) Promoting green supply chain management: the role of the human factor.In: 15th Panhellenic Logistics Conference and 1st Southeast European Congress on Supply Chain Management. Greek Association of Supply Chain Management (EEL of Northern Greece),Thessaloniki, Greece, 11–12 November 2011. SSRN, New York, pp:1–13

Luthra S, Kumar V, Kumar S, Haleem A (2011) Barriers to implement green supply chain management in automobile industry using interpretive structural modeling technique-An Indian perspective. J of Industrial Engineering and Manage 4(2):231–257

Malhotra MK, Grover V (1998) An assessment of survey research in POM: from constructs to theory. J of Oper Manage 16(4):407–425

Mandal A, Deshmukh SG (1994) Vendor selection using interpretive structural modeling. Int J of Oper & Prod Manage 14(6):52–59

Mudgal RK, Shankar R, Talib P, Raj T (2009) Greening the supply chain practices: an Indian perspective of enablers' relationship. Int J of Advanced Oper Manage 1(2–3):151–176

Mudgal RK, Shankar R, Talib P, Raj T (2010) Modeling the barriers of green supply chain practices: an Indian perspective. Int J of Logistics Systems and Manage 7(1):81–107

Nunnally JC (1987) Psychometric theory. McGraw Hill, New York

Paquette J (2005) The supply chain response to Environmental Pressures Discussion., Massachusetts Institute of Technology, http://hdl.handle.net/1721.1/34530. Accessed 12 January 2013

Rao P, Holt D (2005) Do green supply chains lead to competitiveness and economic performance? Int J of Oper & Prod Manage 25(9):898–916

Ravi V, Shankar R (2005) Analysis of interactions among the barriers of reverse logistics. Int J of Technological Forecasting and Social change 72:1011–1029

Rea LM, Parker RA (2005) Designing and conducting survey research: a comprehensive guide, 3rd edn. Jossey-Bass, San Francisco

Reijonen S (2011) Environmentally friendly consumer: from determinism to emergence. Int J of Consumer Studies 35(4):403–409

Sage A (1977) Interpretive structural modeling: methodology for large scale systems. McGraw-Hill, New York, pp 91–164

Sarkis J, Hasan MA, Shankar R (2007) Evaluating environmentally conscious manufacturing barriers with interpretive structural modeling. http://dx.doi.org/10.2139/ssrn.956954

Sekaran U (2003) Research methods for business: askill building approach, 5th edn. Wiley, Singapore

Shi VG, Lenny Koh SC, Baldwin J, Cucchiella F (2012) Natural resource based green supply chain management. Supply Chain Manage: An Int J 17(1):54–67

Simpson D, Power D, Samson D (2007) Greening the automotive supply chain: a relationship perspective. Int J of Oper & Prod Manage 27(1):28–48

Singh MD, Kant R (2008) Knowledge management barriers: an interpretive structural modeling approach. Int J of Manage Science and Engineering Manage 3(2):141–150

Srivastva S (2007) Green supply chain management: a state of the art literature review. Int J of Manage Review 9(1):53–80

Sushil (2005) Interpretive matrix: a tool to aid interpretation of management and social research. Global J of Flexible Systems Manage 6(2):27–30

Tan KC, Kannan VR, Handfield RB (1998) Supply chain management: supplier performance and firm performance. Int J of Purchasing and Material Manage 34(3):2–9

Theyel G (2006) Customer and Supplier Relations for Environmental Performance. In: Sarkis J (ed) Greening the Supply Chain. Springer, Berlin, pp 139–150

Thomas AVand Bijulal D (2011) Closed-loop supply chains: the green practices in supply chains.In: National technological congress, College of Engineering Trivandrum, Kerala, 28–29 January 2011. NATCON 2011, Ottawa, pp. 196–199Torielli RM, Abrahams RA, Smillie RW, Voigt RC (2011) Using lean methodologies for economically and environmentally sustainable foundries. China Foundry 8(1):74–88

Vachon S, Klassen RD (2006a) Extending green practices across the supply chain: the impact of upstream and downstream integration. Int J of Oper & Prod Manage 26(7):795–821

Vachon S, Klassen RD (2006b) Green project partnership in the supply chain: the case of the package printing industry. J of Cleaner Prod 14(6–7):661–71

Vachon S, Klassen RD (2008) Environmental management and manufacturing performance: the role of collaboration in the supply chain. Int J of Prod Econ 111(2):295–308

Walker H, Sisto L, McBain D (2008) Drivers and barriers to environmental supply chain management practices: lessons from the public and private sectors. J of Purchasing & Supply Manage 14(1):69–85

Wang HF, Gupta SM (2011) Green supply chain management: product life cycle approach, 1stedition, McGraw Hill, New York. Wang ML, Kuo TC, Wen LJ (2009) Identifying target green 3C customers in Taiwan using multiattribute utility theory. Expert Systems with Applications 36(10)):12562–12569

Warfield JW (1974) Developing interconnected matrices in structural modeling. IEEE Transcript on Systems, Men and Cybernetics 4(1):51–81

Wu GC, Cheng YH, Hang SY (2010) The study of knowledge transfer and green management performance in green supply chain management. African J of Business Manage 4(1):44–48

Wu KL, Tseng ML, Vy T (2011) Evaluation the drivers of green supply chain management practices in uncertainty. Procedia-Social and Behavioral Sciences 25:384–397

Yang J, Wong CWY, Lai KH, Ntoko AN (2009) The antecedents of dyadic quality of performance and its effect on buyer–supplier relationship improvement. Int J of Prod Econ 102(1):243–51

Yu LC (2007) J of Manage Study August, Adoption of green supply in Taiwan logistic industry, pp 90–98

Yu LC, Hui HY (2008) An empirical study on logistics services provider, intention to adopt green innovations. J of Technology, Manage and Innovation 3(1):17–26

Zhu Q, Sarkis J (2004) Relationships between operational practices and performance among early adopters of green supply chain management practices in Chinese manufacturing enterprises. J of Oper Manage 22(3):265–289

Zhu Q, Sarkis J (2006) An inter-sectoral comparison of green supply chain management in China: drivers and practices. J of Cleaner Prod 14(5):71–74

Zhu Q, Sarkis J, Lai KH (2007a) Green supply management: pressures, practices and performance within the Chinese automobile industry. J of Cleaner Prod 15(11–12):1041–1052

Zhu Q, Sarkis J, Lai KH (2007b) Initiatives and outcomes of green supply chain management implementation by Chinese manufacturers. J of Environmental Manage 85(1):179–189

Zhu Q, Sarkis J, Lai KH (2008a) Confirmation of a measurement model for green supply chain management practices implementation. Int J of Prod Econ 111(2):261–273

Zhu Q, Sarkis J, Lai KH (2008b) Green supply chain management implications for "closing the loop". Transportation Research Part E 44(1):1–18

Zhu Q, Geng Y, Lai KH (2010) Circular economy practices among Chinese manufacturers varying in environmental-oriented supply chain cooperation and the performance implications. J of Environmental Manage 91(6):1324–1331

11

Building a maintenance policy through a multi-criterion decision-making model

Elahe Faghihinia[1][*] and Naser Mollaverdi[2]

Abstract

A major competitive advantage of production and service systems is establishing a proper maintenance policy. Therefore, maintenance managers should make maintenance decisions that best fit their systems. Multi-criterion decision-making methods can take into account a number of aspects associated with the competitiveness factors of a system. This paper presents a multi-criterion decision-aided maintenance model with three criteria that have more influence on decision making: reliability, maintenance cost, and maintenance downtime. The Bayesian approach has been applied to confront maintenance failure data shortage. Therefore, the model seeks to make the best compromise between these three criteria and establish replacement intervals using Preference Ranking Organization Method for Enrichment Evaluation (PROMETHEE II), integrating the Bayesian approach with regard to the preference of the decision maker to the problem. Finally, using a numerical application, the model has been illustrated, and for a visual realization and an illustrative sensitivity analysis, PROMETHEE GAIA (the visual interactive module) has been used. Use of PROMETHEE II and PROMETHEE GAIA has been made with Decision Lab software. A sensitivity analysis has been made to verify the robustness of certain parameters of the model.

Keywords: Preventive maintenance, Age-dependent PM policy, PROMETHEE II, Bayesian approach, PROMETHEE GAIA

Background

Global trade, higher levels of automation, and the desire to apply lean production are some factors that increase the demand for effective maintenance (Salonen and Deleryd 2011). In recent decades, industrial and service systems have realized that establishing a proper maintenance policy plays an essential role in achieving their objectives (Cholasuke et al. 2004; van der Meulen et al. 2008). It can also lead to maximizing their profits (Alsyouf 2009). One of the most important reasons of considering maintenance as a crucial concept can be its large contribution of operating budget in organizations with heavy investments in machinery and equipment (Tsang et al. 1999). Moreover, because of the development of technology, competitive industrial and service systems should make use of more advanced machinery which need higher levels of maintenance because they are more complex and more difficult to control (Alsyouf 2009). The role of maintenance in modern

manufacturing systems is becoming even more important with companies adopting maintenance as a profit-generating business element (Sharma and Yadava 2011). In order to avoid failures at random times and the effect of such failures on the performance of systems that appear as a reducing production rate and loss of quality of the products, maintenance management is required to reduce the loss of system operating time and the number of defective parts produced (Tsarouhas 2011).

Maintenance is becoming a critical functional area in most types of organizations and systems including construction, manufacturing, transportation, etc. This increasing role of maintenance is reflected in its high cost, which is estimated to be around 30% of the total running cost of modern manufacturing and construction businesses. As such, planning for maintenance is becoming an essential part of planning for the whole organization (Al-Turky 2011). Therefore, common practices like repairing a system when there is a problem have to be substituted by monitoring the system condition and planning the maintenance intervals (Cavalcante and De Almedia 2007). Also, effective maintenance can extend the equipment life,

* Correspondence: elahe_faghihinia60@yahoo.com
[1]Department of Industrial Engineering, Islamic Azad University of Najafabad, Isfahan 8514143131, Iran
Full list of author information is available at the end of the article

improve equipment availability, and restore the equipment to a good condition (Swanson 2001). Well-defined maintenance system will ensure optimal performance of the machineries (Oberschmidt et al. 2010). Therefore, it can not only improve the quality of goods and services but also satisfy and rather exceed customers' demands especially in service sectors (Oke and Charles-Owaba 2006). The importance of running proper maintenance policies in organizations has led researchers to define maintenance in several ways. For example, Tsarouhas (2011) defines maintenance as a tool whose objectives are to increase the time to failure and reduce the repair time of equipment. Al-Turky (2011) defines it as the activities related to maintaining a certain level of availability and reliability of the system and its components and the system ability to perform with a standard level of quality. Still, maintenance can be defined as the combination of all technical efforts which can retain an item or equipment, or restore it to an acceptable operating condition (Dhillon 2002; British Standards Institute Staff 1993). With regard to the critical role of maintenance in improving reliability, preventing unexpected system failures and reducing maintenance costs, maintenance and replacement problems have been widely studied from different perspectives in the literature, and several models have been proposed (Wang 2002). All of them seek to elaborate on different maintenance problems and propose more rational solutions. This paper proposes a multi-criterion decision-aided maintenance model with regard to three criteria important to selecting the best maintenance policy. They are maintenance costs, reliability, and maintenance downtime criteria. This model not only considers the various aspects of a maintenance problem but also attends to the preference of a decision maker. Furthermore, Bayesian approach has been applied to overcome failure data shortage. Finally, a sensitivity analysis has been made to verify the robustness of certain parameters of the model.

Preventive maintenance

Complex equipment and machinery systems used in the production of goods and delivery of services constitute the vast majority of capital invested in industry (Savsar 2011). As time passes, the machines age and unplanned failures occur, causing the system performance to drift away from its initial state. In fact, no piece of equipment or system can continue to function without failure forever; however, carefully it might have been designed and manufactured (Samar Ali and Kannan 2011). System deterioration is often reflected in higher production costs and lower product quality. Therefore, the function of the system must be periodically restored to the desired level; this is practically achieved by maintenance operations. Proper maintenance can increase the reliability of a piece of

equipment or a system at regular intervals (Samar Ali and Kannan 2011). Such maintenance is known as preventive maintenance (PM); it is done periodically before the failure of the system; hence, it is different from corrective or repair maintenance, which is carried out only after the failure of the item or the system (Savsar 2011). To keep production costs down while maintaining good product quality, PM is often performed on systems subject to deterioration (Savsar 2011). The probability of failure would increase as a machine is aged, and it would sharply decrease after a planned PM is implemented (Savsar 2011).

It should be pointed out quickly that the maintenance actions which are normally classified as corrective maintenance (CM) include all actions performed as a result of a failure to restore an item to a specified working condition, while PM includes all actions performed on an operating equipment to restore it to a better condition (Oberschmidt et al. 2010). Moreover, making use of CM could be costly for organizations because most of the time, CM takes a long time to have an acceptable effect on a failed system or component (Nakagawa 2005). Thus, it can be disastrous for some systems where failures and interruptions could be dangerous. For example, we can consider military systems, aircraft, and health systems where a small mistake can lead to a horrible disaster (Cavalcante and De Almedia 2008).

Also, the costs of applying CM in organizations are usually three or four times bigger than applying PM (Chitra 2003). So, it would be more rational to study PM models as a basic concept for the purpose of proposing an optimum maintenance model. In addition, PM policies are used for contexts where the component failure rate increases by age and usage (Cavalcante and De Almedia 2008).

PM models

Although a lot of maintenance models have been created during the past decades, there are few maintenance policies on which all the other maintenance models can be based (Wang 2002). There is a categorization proposed by Wang (2002). According to him, there are seven categories of maintenance policies, of which five are preventive. They are age-dependent PM, periodic PM, failure limit, sequential PM, and repair limit.

According to age-dependent PM policy, a unit is replaced at the predetermined time T or in the case of failure, whichever occurs first, where T is a constant (Barlow and Hunter 1960). The given time T is measured from the time of the last replacement (Wang 2002). According to periodic PM policy, a unit is preventively maintained at fixed time intervals independent of the failure history of the unit and repaired at intervening failures where T is a constant.

According to failure limit policy, PM is performed only when the failure rate or other reliability indices of a unit reach a predetermined level, and intervening failures are corrected by repairs. According to sequential PM policy, a unit is preventively maintained at unequal time intervals under the sequential PM policy. Usually, the time intervals become shorter and shorter as time passes, considering that most units need more frequent maintenance with increased ages.

According to repair limit PM policy, when a unit fails, the repair cost is estimated and repair is undertaken if the estimated cost is less than a predetermined limit; otherwise, the unit is replaced. He also indicates that the age-dependent policy can be the most common and popular PM. In several recent works, age-replacement policy was extensively studied. The age-replacement policy and its extensions belong to the age-dependant policy (Wang 2002).

Therefore, by taking a look at PM models, we can realize that there are a large variety of preventive maintenance models and their extensions, so it would be necessary to specify a given problem to resolve in this context. Therefore, the age-replacement policy has been chosen as the basis for this research. Also in this paper, it is assumed that the replacement of a piece of equipment or part gives the system a good-as-new performance.

In addition, there are two requisites for PM implementation in each system where (Cavalcante and De Almedia 2008):

1. The replacement cost of a component (c_p) before failures should be less than the cost of replacement due to failures (c_f).
2. The component failure rate should increase by age and usage.

This paper proposes a multi-criterion decision-aided model with three criteria which deals with the problem of the replacement times. It determines the best timing and frequency for replacing components by taking into account three criteria, which are the total cost of maintenance per unit of time, the reliability, and the total maintenance downtime per unit of time. Thus, after choosing the policy followed by this research, it is important to describe the importance of these three criteria in maintenance decision making.

Maintenance has become one of the most important issues in the manufacturing industry due to high costs involved (Savsar 2011). In production systems, maintenance managers concentrate on reducing maintenance costs (Cavalcante and De Almedia 2008). In manufacturing organizations, maintenance-related costs are estimated to be 25% of the overall operating cost (Cross 1988).

According to Maggard and Rhyne (1992), the maintenance can represent between 10% and 40% of the production cost in a company. Coetzee (2004) means that the numbers should be 15% to 50%. Bevilacqua and Braglia (2000) state that maintenance costs can represent as much as 15% to 70% of the total production cost. So, it seems plausible that the maintenance costs may very well represent over 15% of the total production cost in industry (Salonen and Deleryd 2011). These findings show that maintenance cost cannot be ignored by maintenance managers.

But there are several situations in some organizations where other criteria like reliability, availability, downtime, etc., play critical roles in systems. Earlier, researchers were using the optimization criteria as minimizing system maintenance cost rate, ignoring the reliability performance. In fact, minimizing system maintenance cost rate may not imply maximizing the system reliability measures. Sometimes, when the maintenance cost rate is minimized, the system reliability measures are also so low that they are not acceptable in practice (Sharma and Yadava 2011).

According to Cavalcante and De Almedia (2008), in the services sector, the decision maker can show a preference for minimizing undesirable consequences which are difficult to measure in financial units. Because in this context, the customer is in direct contact with the production, and frequent interruption in the service can negatively affect the desire of the customer to enter into a new contract with that supplier or can lead the customer to cancel the current contract, which is unacceptable in competitive markets today.

Generally, managers would like to see their system run as planned, and an unscheduled event such as a machine failure will disrupt the smooth running of the plant. Sometimes, the marketing department brings emergency product orders for important customers, and a system failure may result in severe losses (Chareonsuk et al. 1997).

Therefore, looking at the cost criterion as the most important factor to establish an optimum maintenance model is a very dangerous perspective for industrial and service systems, especially for systems where failures and interruptions could be disastrous. It is moreover impossible to capture all of a system's effects in a cost function.

In some systems, the reliability criterion plays an essential role and which must be taken into account when an optimum maintenance model is to be established. Therefore, in a number of situations, maintenance managers mean to consider reliability as a separate criterion (Chareonsuk et al. 1997).

Reliability, $R(t)$, is the probability that a component or system will perform its design function for a specified mission time, given the operating conditions.

Table 1 Preference functions (adapted from Brans and Mareschal 2005)

Generalized criterion	Definition	Parameters to fix
Type 1: Usual Criterion	$P(d) = \begin{cases} 0 & d \le 0 \\ 1 & d > 0 \end{cases}$	–
Type 2: U-shape Criterion	$P(d) = \begin{cases} 0 & d \le q \\ 1 & d > q \end{cases}$	Q
Type 3: V-shape Criterion	$P(d) = \begin{cases} 0 & d \le 0 \\ \dfrac{d}{p} & 0 \le d \le p \\ 1 & d > p \end{cases}$	P
Type 4: Level Criterion	$P(d) = \begin{cases} 1 & d \le q \\ \dfrac{1}{2} & q < d \le p \\ 1 & d > p \end{cases}$	p,q
Type 5: V-shape with indifference Criterion	$P(d) = \begin{cases} 0 & d \le q \\ \dfrac{d-q}{p-q} & q < d \le p \\ 1 & d > p \end{cases}$	p,q
Type 6: Gaussian Criterion	$P(d) = \begin{cases} 0 & d \le 0 \\ 1 - e^{-\frac{d^2}{2e^2}} & d > 0 \end{cases}$	S

$s = (p+q)/2.$

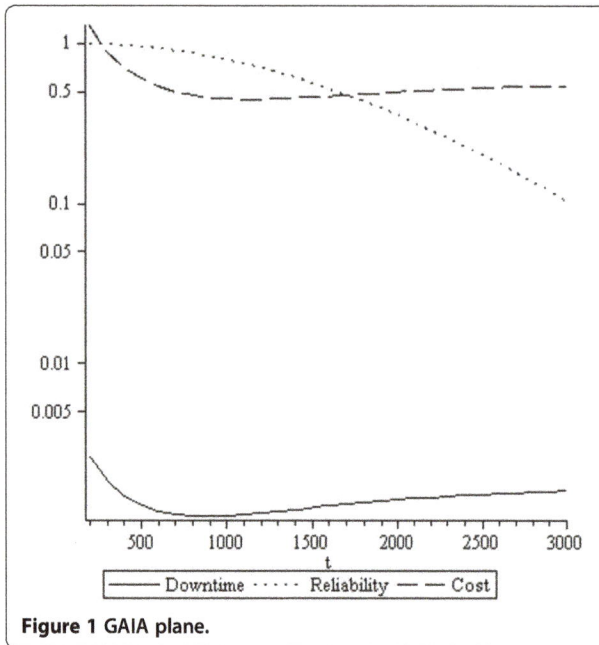

Figure 1 GAIA plane.

Table 2 Model's parameters

c_a ($)	c_b ($)	T_f (days)	T_p (days)	β_1	η_1	β_2	η_2
1,000	250	3	0.5	3.40	4.15	2.80	2200

per unit of time, the reliability criterion and the maintenance downtime in making a proper maintenance decision need to be integrated in considering maintenance scheduling in a multi-criterion environment. This paper seeks to determine PM intervals during which the three criteria are in their best compromise with each other.

Bayesian approach

Mathematics has had an important role to extend maintenance models. Stochastic mathematical models have been developed to improve system reliability, prevent unexpected failures, and reduce maintenance costs (Zhang 2005). The use of mathematical modeling for this purpose is well established in the literature (Sortrakul and Cassady 2007). A number of surveys have been published by some authors in this area.

McCall (1965) proposes a survey of researches on maintenance policies subject to stochastic failure. Pierskalla and Voelker (1976) also present a survey on maintenance models for deteriorating systems. Sherif and Smith (1981) review various maintenance models subject to failure and propose a classification.

Sharma and Yavada (2011) present a survey on maintenance optimization models. More surveys can be found in the studies of Valdez-Flores and Feldman (1989), Cho and Parlar (1991), Dekker (1996), and Wang (2002).

This part explains the mathematical requirements. In order to plan a maintenance program in this research, a failure distribution is needed which has a wear-out characteristic, namely the failure rate should increase with age. The Weibull model is a most prevalent distribution that satisfies this perquisite. It can be shown to be of the form:

There is another factor added to the model maintenance downtime. A question might be in order here. How important is this factor in maintenance decision making?

Chareonsuk et. al (1997) considered a situation where preventive maintenance cost is not high, so in order to keep the system at a high level of reliability, the maintenance department decides to run maintenance programs very frequently. This can lead to very frequent shutdowns where the production department will be reluctant to attend to the preventive maintenance programs. Therefore, they will either cause forced postponed preventive maintenance or schedule when there is no production (for example, at night when it would be inconvenient for maintenance people). This problem can postpone maintenance programs. Therefore, in practice, maintenance programs cannot be fully maintained (Chareonsuk et al. 1997).

Besides, downtime is very important and must not be neglected because the minimum downtime for a piece of equipment could result in undesirable consequences (Cavalcante and De Almedia 2008). With respect to the importance of taking into account the criterion of cost

$$f(t) = \frac{\beta}{\eta} \cdot \left(\frac{t}{\eta}\right)^{\beta-1} \quad \beta, \eta > 0, t > 0 \tag{1}$$

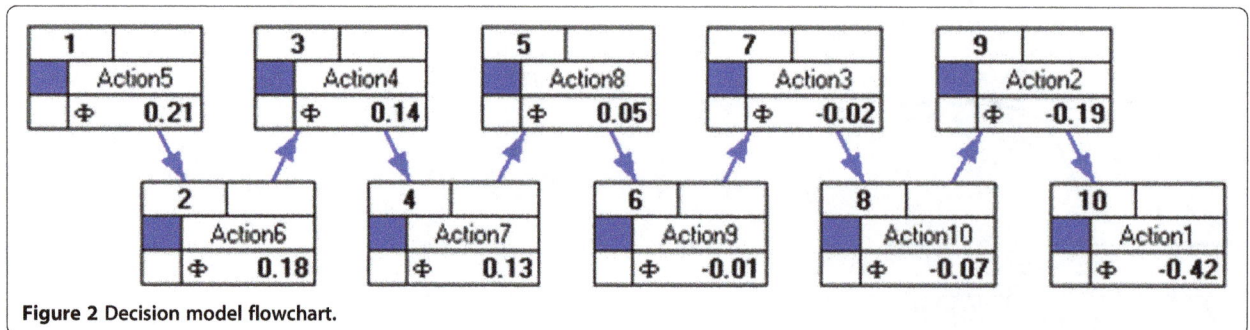

Figure 2 Decision model flowchart.

Table 3 Performances of alternatives (g$_i$(t))

T (days)	R(t)	C(t)	D(t)
200	0.9904	1.2874	0.0030
300	0.9797	0.8890	0.0018
400	0.9643	0.6996	0.0014
500	0.9441	0.5944	0.0013
600	0.9192	0.5315	0.0012
700	0.8899	0.4928	0.0011
800	0.8566	0.4690	0.0011
900	0.8198	0.4551	0.0011
1,000	0.7802	0.4478	0.0011
1,100	0.7383	0.4452	0.0012
1,200	0.6947	0.4458	0.0012
1,300	0.6502	0.4489	0.0012
1,400	0.6051	0.4534	0.0012
1,500	0.5604	0.4591	0.0013
1,600	0.5162	0.4654	0.0013
1,700	0.4730	0.4722	0.0013
1,800	0.4314	0.4792	0.0014
1,900	0.3916	0.4861	0.0014
2,000	0.3538	0.4930	0.0014
2,100	0.3182	0.4996	0.0014
2,200	0.2849	0.5060	0.0015
2,300	0.2541	0.5120	0.0015
2,400	0.2258	0.5179	0.0015
2,500	0.1998	0.5230	0.0015
2,600	0.1762	0.5274	0.0016
2,700	0.1549	0.5318	0.0016
2,800	0.1358	0.5357	0.0016
2,900	0.1187	0.5393	0.0016
3,000	0.1034	0.5425	0.0016

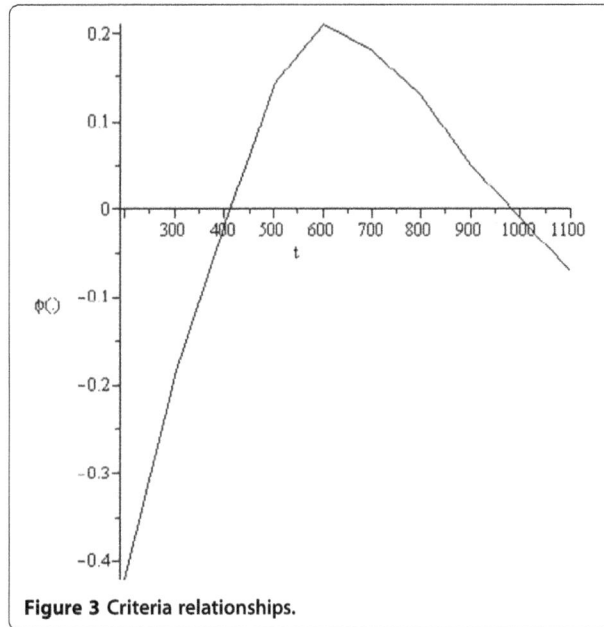

Figure 3 Criteria relationships.

where η is called the scale parameter; β, the shape parameter. In order to establish optimum maintenance intervals, we need to recognize the failure behavior of system or component. Thus, the parameters of the failure distribution of system or component should be estimated. In order to estimate the distribution function parameters, historical data are often used; therefore, a large quantity of data is needed to obtain reliable estimates. But because of the rapid growth of industry, often sufficient historical information about the components or system failures is not available (Chen and Popova 2002). Often, only a few failure data are available, and in some cases where there are enough data, they are not reliable (Scarf 1997). Therefore, estimation parameters from failure data is another difficulty in maintenance program (Cavalcante and De Almedia 2008).

However, during the process of the system production and its operating time, reliability engineers and specialists find out by intuition about its failure behavior (Chen and Popova 2002). Combined with actual observations, this information can provide better assessment of the failure rate parameters. Bayesian analysis is one way to enter this information into the decision-making process in order to make a more objective decision. Therefore, a major advantage of Bayesian analysis is when only a few data are available. Bayesian statistics provides a way to incorporate specialist advice about a system into the maintenance model. The Bayesian maintenance models have been used frequently to establish maintenance policies in recent decades (Cavalcante and De Almedia 2008).

Some authors such as Jorgenson et al. (1967), McCall (1965), Dayanlk and Gurler (2002), Wilson and Benmerzouga (1995), Sheu et al. (2001), Juang and Anderson (2004), Kallen and Van Noortwijk (2005), Makis and Jardine (1992), McNaught and Chan (2011), and many others have used this approach in different maintenance models (Oberschmidt et al. 2010).

Finally, using a Weibull distribution to model failure in cases of incomplete data, specialist knowledge can be used. Therefore, the Weibull distribution parameters are considered random variables with a priori distributions representing specialist knowledge: $\mu(\eta)$ and $\mu(\beta)$.

Table 4 Criteria thresholds

Criteria	Thresholds	
	p	Q
Reliability	0.0300	0.1000
Cost	0.0150	0.2000
Downtime	0.0011	0.0001

Table 5 Criteria weights

Criteria	Weights (%)
Reliability	35
Cost	40
Downtime	25

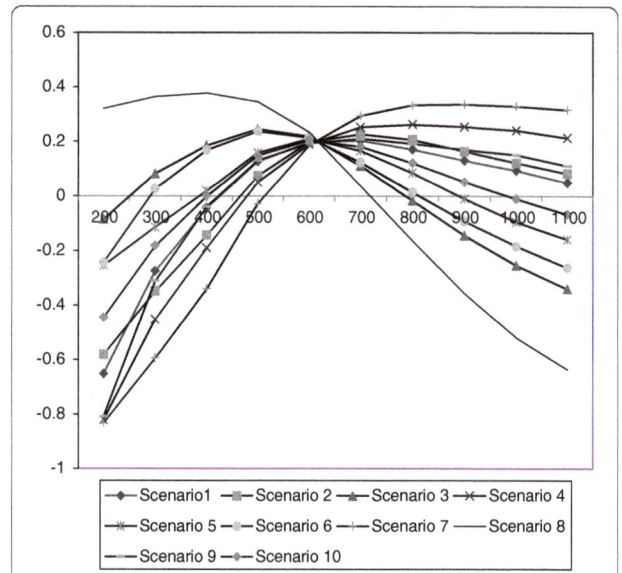

Figure 5 Net flow values.

After estimating the Weibull parameters, evaluation of the cost, reliability, and maintenance downtime criteria can be obtained using the model, and then a multi-criterion decision with PROMETHEE methods can be made.

PROMETHEE: one of the multi-criterion decision-making methods

By taking a look at decision-making problems in the real world, it can be seen that most of them are multi-criterion. Decision making in many contexts depends on several criteria not just on one criterion. This can be seen in many fields such as industries, economics, finance, or politics. Making decisions in maintenance programs can be a multi-criterion decision problem. According to Shyjith et al. (2008), selecting a maintenance policy based on a few factors makes it unrealistic. There is a need to consider maintenance problems as multi-criterion. This outlook can give a comprehensive view to maintenance management. So, it can be critical to consider maintenance problems as multi-criterion especially in systems that take into account only the cost criterion for making a maintenance decision because in some systems with special conditions, it could result in

disasters. If the maintenance department only wants to look at the cost criterion, it could lead it to ignoring other criteria like reliability or maintenance downtime.

A multi-criterion problem is mathematically defined as (Brans and Mareschal 1994 a,b; Brans et al. 1984):

$$\text{Max}\{g_1(a), g_1(a), \ldots, g_i(a), \ldots, g_k(a) \cdot | \cdot a\epsilon A\}, \quad (8)$$

where A is a finite set of n possible alternatives $\{a_1, a_2, \ldots, a_n\}$, and $\{g_1(.), g_2(.), \ldots, g_k(.)\}$ is a set of evaluation

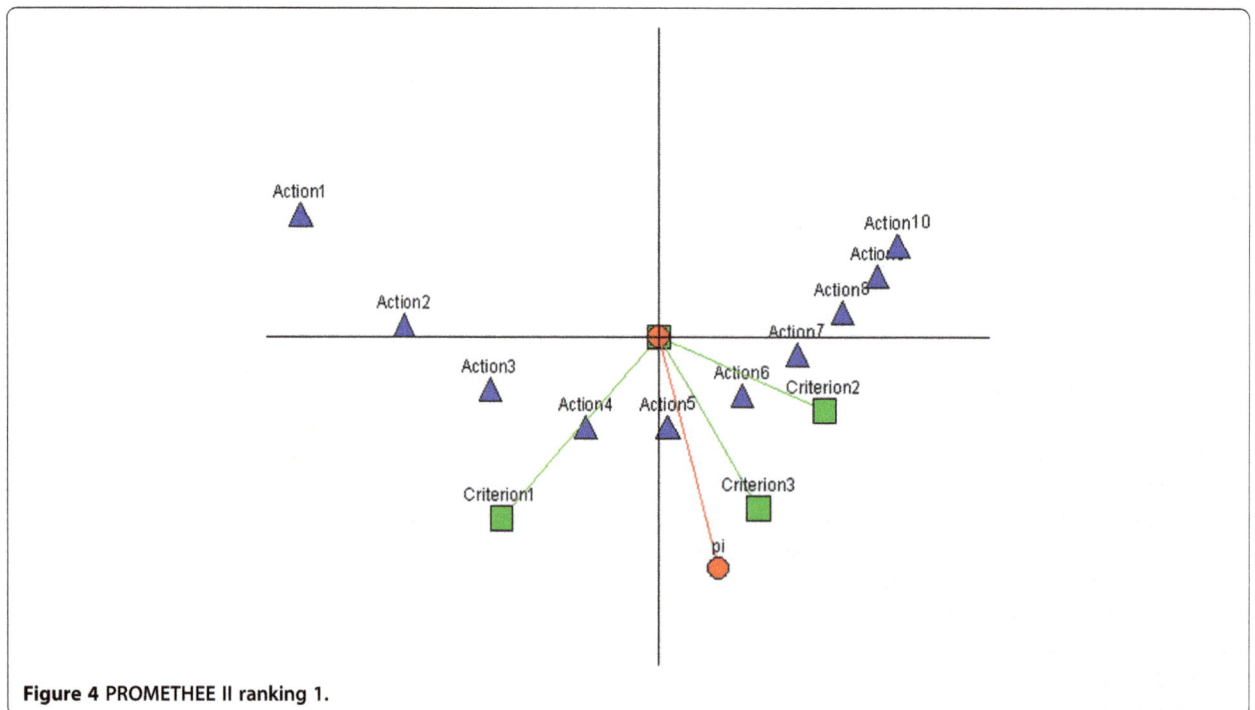

Figure 4 PROMETHEE II ranking 1.

Table 6 Scenarios

Scenarios	Reliability weight (%)	Cost weight (%)	Downtime weight (%)
1	20	25	55
2	25	55	20
3	55	20	25
4	10	45	45
5	45	45	10
6	45	10	45
7	10	80	10
8	80	10	10
9	10	10	80
10	33	33	34

criteria. They are called the basic data of a multi-criterion problem (Brans and Mareschal 1994a,b; Brans et al. 1984).

In recent years, many decision-aid methods have been proposed. The PROMETHEE methods are one group of these methods consisting of seven. PROMETHEE methods developed by Brans are one of the best known and most widely used outranking approaches in many applications (Makis and Jardine 1992). A comprehensive overview of applications can be found in Behzadian et al. (2010). In general, outranking approaches are based on comparisons of pairs of alternatives (Oberschmidt et al. 2010). The PROMETHEE methods have been frequently used in many fields, and their success is due to their mathematical processes and the fact that they are easy to use by decision makers (Brans and Mareschal 1994a,b; Brans et al. 1984).

The input required concerns the evaluation of the criteria for all of the alternatives considered as well as the weightings needed to reflect their relative importance. In order to apply PROMETHEE, first, the performance of the alternatives regarding all criteria needs to be determined. Then, alternatives are compared in pairs for each criterion based on generalized preference functions. Based on the weighted sum of single criterion preferences, positive and negative outranking flows are calculated as a measure of dominance of alternatives. Criteria weights reflect the subjective relative importance of the criteria. Based on positive and negative outranking flows, a partial preorder of alternatives can be defined according to PROMETHEE I. The net outranking flow can also be calculated to avoid incomparabilities and define a complete preorder on the set of alternatives according to PROMETHEE II (Oberschmidt et al. 2010). After that, PROMETHEE III that ranks alternatives based on intervals and PROMETHEE IV, the continuous case, were developed by Brans and Mareschal.

They also proposed the visual interactive module GAIA in 1988, which provides an interesting graphical view to support the PROMETHEE methodology. In 1992 and 1994, Brans and Mareschal extended these two types: PROMETHEE V, an extension of PROMETHEE I and II where a subset of alternatives has to be selected by considering a set of constraints, and PROMETHEE VI, an extension of the results from PROMETHEE I and II that provides the decision maker with the freedom to think of the weight in terms of intervals, rather than of exact values (Brans and Mareschal 1994a,b; Brans et al. 1984; Cavalcante and De Almedia 2008).

The PROMETHEE II method has been chosen for outranking results in this research, and the PROMETHEE GAIA has been chosen for a visual realization and sensitivity analysis of the results in this research. The reasons for selecting these methods are fast use, easy-to-analyze results, and a flexible comparison process (Cavalcante and De Almedia 2008). Moreover, the information which needed to use PROMETHEE and GAIA is easy and clear to define for decision makers (Brans et al. 1984; Brans and Mareschal 1994b).

To make use of PROMETHEE methods, first, the two following phases should be passed (Brans and Mareschal 1994a,b; Brans et al. 1984; Cavalcante and De Almedia 2008):

1. Calculating the evaluation of each alternative for each criterion : $g_i(a)$; and
2. Calculating the differences between the evaluations of the alternatives within each criterion:

$$d_i(a, b) = g_i(a) - g_i(b) \qquad (9)$$

We also need two types of additional information to run PROMETHEE (Brans and Mareschal 1994a):

1. The information between the criteria that consists of the relative importance of the different criteria and which depends on the preferences of a decision maker. They are shown by $w_j, j = 1, 2, \ldots, k$. They are considered as norm weights.
2. The information within the criteria is referred to assign a preference function to each criterion. After calculating the differences between each two alternatives for a criterion, $d_i(a, b)$, the decision maker's preferences are needed to identify the indifference threshold(q) that is the largest deviation to ignore and the preference threshold (p), i.e., the smallest deviation considered to be a

Table 7 Scenario 1

t(days)	200	300	400	500	600	700	800	900	1,000	1,100
φ(.)	−0.6523	−0.2751	−0.0442	0.1267	0.198	0.2029	0.1712	0.1305	0.0937	0.0486

Table 8 Scenario 2

t(days)	200	300	400	500	600	700	800	900	1,000	1,100
$\varphi(.)$	−0.5816	−0.3483	−0.1435	0.0738	0.1979	0.2268	0.2065	0.1637	0.1217	0.0831

preference relation between two alternatives. So in this phase, decision maker selects a generalized criterion, F_i (d_i (a, b)), to model his preferences for every criterion. After specifying the function parameters by the decision maker, the preference function can be obtained.

$$P_i(a, b) = F_i(d_i(a, b)) \quad d_i(a, b) > 0 \tag{10}$$

$$P_i(a, b) = 0 \quad d_i(a, b) < 0 \tag{11}$$

There are six preference functions suggested for decision makers. These functions have satisfied the conditions of many real-world problems. They are shown in Table 1.

The PROMETHEE II method has been chosen to rank the alternatives. This ranking is based on net flow $\phi(a)$ (Brans and Mareschal 1994a; Cavalcante and De Almedia 2008):

$$\phi(\text{a}) = \frac{1}{n-1} \sum x \epsilon A \sum_i^k [Pi(a,x) - Pi(x,a)]wi. \tag{12}$$

Therefore, each alternative can get the highest score on the net flow which is the best compromise solution. Because of considering the differences d_i to rank the alternatives in PROMETHEE II, more information get lost (Brans and Mareschal 1994a,b; Brans et al. 1984). In order to overcome and have a better realization of the situations of alternatives and criteria, this paper has made use of PROMETHEE GAIA. The GAIA plane will be shown and interpreted in Figure 1. The GAIA plane can also provide a powerful graphical visualization tool for a decision maker (Brans et al. 1984; Brans and Mareschal 1994b). The specific power of each criterion, the conflicting aspects, and the attributes of each alternative on the different criteria are GAIA plane's specific qualities (Brans et al. 1984). Finally, a sensitivity analysis usually serves to demonstrate the influence of different weightings on the results of the assessment (Oberschmidt et al. 2010).

Results and discussions
Mathematical configurations of the criteria
The goal of this paper is to evaluate the different time alternatives and rank them according to the decision maker's preferences and to the values of the three criteria in each alternative. Therefore, the mathematical configurations of these criteria are needed. To achieve this goal, the mathematical equations of these three criteria have been illustrated.

The first equation is the reliability formula which is based on reliability definition. It is the probability of lack of a component or system failure before time t. Thus, it is mathematically defined as follows, where $f(t)$ is the density function of the component failure behavior:

$$R(t) = \int_t^\infty f(t)dt \tag{2}$$

The second and third equations are the maintenance cost and maintenance downtime formulas, respectively. They were created by Jardine (1973). The details of these formulas are defined as follows:

c_p: Replacement cost before failure
c_f: Replacement cost due to failure
T_p: The time taken to make a preventive replacement
T_f: The time taken to make a replacement due to failure

$$C(t) = \frac{Cp \times R(t) + C_f \times [1 - R(t)]}{(t + T_p) \times R(t) + \left(\left(\int_{-\infty}^t xf(x)dx \middle/ {[1-R(t)]} \right) + T_f \right) \times [1 - R(t)]} \tag{3}$$

$$D(t) = \frac{Tp \times R(t) + T_f \times [1 - R(t)]}{(t + T_p) \times R(t) + \left(\left(\int_{-\infty}^t xf(x)dx \middle/ {[1-R(t)]} \right) + T_f \right) \times [1 - R(t)]} \tag{4}$$

Because of the absence of the maintenance data and uncertainty in both parameters of the component failure distribution function, specialist information can be used to estimate them. Hence, by making use of Bayesian approach, the parameters of Weibull distribution are

Table 9 Scenario 3

t(days)	200	300	400	500	600	700	800	900	1,000	1,100
$\varphi(.)$	−0.0859	0.0824	0.1855	0.2452	0.2185	0.1102	−0.0167	−0.1452	−0.2545	−0.3396

Table 10 Scenario 4

t(days)	200	300	400	500	600	700	800	900	1,000	1,100
$\varphi(.)$	−0.8203	−0.4524	−0.1897	0.0506	0.1903	0.2518	0.2624	0.2531	0.2403	0.2139

considered random variables, and their distributions should be assessed from the specialist information on these variables $\pi(\eta)$ and $\pi(\beta)$. Because of uncertainty on parameters η and β, the computations of reliability, cost, and downtime criteria should incorporate the distributions $\pi(\eta)$ and $\pi(\beta)$ that follow the Weibull distribution. Hence, according to Cavalcante and De Almedia (2008), reliability and cost criteria formulas follow these equations:

$$E(R(t;\eta,\beta)) = \int_t^\infty \int_{-\infty}^\infty \int_{-\infty}^\infty \pi(\eta)\pi(\beta)f(x;\eta,\beta)d\eta d\beta dx \tag{5}$$

$$C(t) = \frac{Cp \times R(t) + C_f \times [1 - R(t)]}{(t + T_p) \times R(t) + \left(\int_{-\infty}^t \int_{-\infty}^\infty \int_{-\infty}^\infty x\pi(\eta)\pi(\beta)\right.} $$
$$\left. f(x;\eta,\beta)d\eta d\beta dx\right)$$
$$\times T_f \times [1 - R(t)] \tag{6}$$

At the end the third equation, which calculates the maintenance downtime, this equation follows:

$$D(t) = \frac{Tp \times R(t) + T_f \times [1 - R(t)]}{(t + T_p) \times R(t) + \left(\int_{-\infty}^t \int_{-\infty}^\infty \int_{-\infty}^\infty x\pi(\eta)\pi(\beta)\right.}$$
$$\left. f(x;\eta,\beta)d\eta d\beta dx\right)$$
$$\times T_f \times [1 - R(t)] \tag{7}$$

The decision model process
Up to this point, all the details of the model have been described. In order to describe the process of the model, the following 14 steps need to be completed. The flowchart of the model has been brought in Figure 2.

1. Determine the time alternatives T_i by decision makers, those which are applicable for doing PM;
2. Determine Bayesian parameters $\beta_1, \eta_1, \beta_2, \eta_2$;
3. Determine c_p, c_f, T_p, T_f;
4. Calculate $g_i(a)$ where $g_1 = R(t)$, $g_2 = C(t)$, $g_3 = D(t)$ for all of the time alternatives T;

5. Analyze the values of the three criteria and determine the acceptable alternatives;
6. Calculate $d_i(a, b) = g_i(a) - g_i(b)$ for each two time alternatives;
7. Determine w_j, $j = 1..3$;
8. Determine $F_j (d_i (a,b))$, $j = 1..3$ and their thresholds;
9. Calculate $\varphi(a_i)$ for each time alternatives;
10. Rank the time alternatives by the values of $\varphi(a_i)$;
11. Draw the GAIA plane;
12. Analyze the sensitivity of the results into variation of w_j $j = 1..3$;
13. Choose the best time alternative;
14. Stop

Numerical application
In order to evaluate the practical aspects of the model and to see the model's value in practice, a numerical application will be needed. In fact, by a numerical example, the effectiveness of the model can be observed, and decision makers can get a better idea of it. Therefore, this section presents a hypothetical example which is closer to the real situation of a component.

The data consist of information about the prior distributions of η and β. Therefore, there are β_1 and η_1 which are the parameters of the Weibull distribution that belongs to β and β_2 and η_2, the parameters of the Weibull distribution that belongs to η. They have been obtained from specialist information. Also, the replacement costs before (c_p) and after failure (c_f), the time taken to make a replacement before (Tp) and after failure (T_f) are needed. These values are shown in Table 2.

Also, the time alternatives incorporate the interval between 200 and 3,000 days with an interval of 100 days between the alternatives. The performances of the alternatives are calculated for the three criteria and are shown in Table 3. The calculations in relation to Table 3 have been done by making use of Maple 13 software.

In order to see the relationships between these three criteria in the time alternatives whose values have been plugged in Table 3, they have been drawn as three curves in Figure 3. The horizontal axes show the time alternatives from 200 to 3,000 days and the vertical axes show the values of three criteria in each the time alternative.

It is obvious that the cost criterion should be minimized. As seen in Table 3 and Figure 3, this criterion

Table 11 Scenario 5

t(days)	200	300	400	500	600	700	800	900	1,000	1,100
$\varphi(.)$	−0.2556	−0.1151	0.0184	0.1578	0.2104	0.1652	0.0847	−0.0107	−0.0954	−0.1597

Table 12 Scenario 6

t(days)	200	300	400	500	600	700	800	900	1,000	1,100
$\varphi(.)$	−0.2439	0.0265	0.1691	0.2373	0.2138	0.1229	0.0139	−0.0934	−0.184	−0.2621

gets its best value in the time alternative 1,100 days. Also, the reliability criterion should be maximized. As seen in Table 3 and Figure 3, it is descending during the time alternatives. In fact, it has its best value at point zero. The downtime criterion also needs to be minimized, and as can be seen in Table 3 and Figure 3, it gets its minimum value in 1,200 days. Therefore, for the time alternatives greater than 1,100 days, the cost criterion increases during the time and simultaneously the reliability criterion is descending and the downtime criterion is increasing during the time. Therefore, evaluating the alternatives greater than 1,100 days is not useful. They cannot result in the best compromise response between these three criteria. Hence, they can be neglected. Finally, there are ten alternatives which need to be ranked. There are the time alternatives from 200 to 1,100 days.

In order to make use of the PROMETHEE II ranking, the preference function is determined as the linear function, the fifth among the Decision Lab functions (according to Table 1), and it has been used for all of the three criteria. Their thresholds have been determined according to the preference of the decision maker, and they are shown in Table 4.

Moreover, the PROMETHEE methods need criteria weights which are chosen by the decision maker. The weights which are assumed for the three criteria in this paper are shown in Table 5.

After determining the whole data needed in order to use the Decision Lab software and rank the alternatives, they can be put in the software. Decision Lab 2000 is a multi-criterion analysis and decision-making software. Decision Lab 2000 was designed to be applied to various multi-criterion decision problems and designed for all Windows platforms. After putting the required data in the software, it ranks the time alternatives immediately. The PROMETHEE II ranking is used for the ten-time alternatives chosen as shown in Figure 4.

As said before, the PROMETHEE II method ranks the alternatives by calculating the $\phi(.)$ values. Therefore, each alternative capable of getting the highest score in ϕ (.) is the best compromise solution. Figure 5 shows that the best compromise solution is Action 5 which presents the time alternative 600 days. In order to see the variations between the alternatives in the value of the net flow, Figure 5 can be illustrative. It shows that alternative

600 days has made the highest score in net flow; therefore, it is the best compromise solution that PROMETHEE II has determined.

GAIA plane

GAIA plane is a useful tool to evaluate a decision-making problem. It can show the relationship between criteria and alternatives. In order to get a better understanding of the problem, the GAIA plane of the problem has been shown in Figure 1. The alternatives are shown by triangle-shaped points, and the criteria are shown by square-shaped points. There is an axis named Pi. It is called the PROMETHEE decision axis, namely, each alternative which is closer to this axis than the others is better to choose.

Criteria 2 and 3 show similar preferences because they are approximately in the same direction. Moreover, in GAIA plane, each alternative which is closer to a criterion should be good at the criterion. It can be seen about alternative 4 at criterion 1, alternative 5 at criterion 3, and alternative 6 at criterion 2.

It is so obvious that alternatives 1, 2, 3, and 4 are not good at criterion 2 or alternatives 6,7,8,9 and 10 are not good at criterion 1. Alternative 6 is between criteria 2 and 3; therefore, it is good at both criteria. Alternative 5 is between criteria 1 and 3; therefore, it is good at both criteria. But seen in Figure 1, alternative 5 is closer to Pi than others. Therefore, it should be the best compromise solution.

As seen in Table 5, these results are obtained from specific weights. Therefore, if the weights change, the ranking will change. Changing the weights will only change the situation of axis Pi and the situations of the criteria, and the alternatives will remain unchanged. Therefore, it can be seen how this ranking is sensitive to the variation of the weights. In order to answer this question, a sensitivity analysis has been made.

Sensitivity analysis

With a good sensitivity analysis, it is possible to obtain more interpretative results which enhance the decision-maker understanding of the maintenance problem and to evaluate whether solutions proposed by the model are sensitive to parameter change. Therefore, in this section,

Table 13 Scenario 7

t(days)	200	300	400	500	600	700	800	900	1,000	1,100
$\varphi(.)$	−0.832	−0.594	−0.3403	−0.0289	0.1869	0.2941	0.3332	0.3358	0.3289	0.3163

Table 14 Scenario 8

t(days)	200	300	400	500	600	700	800	900	1,000	1,100
$\varphi(.)$	0.3209	0.3638	*0.3772*	0.3445	0.2339	0.0362	−0.1639	−0.3573	−0.5196	−0.6357

the research has tried to test different weights to get a better idea about the problem.

For the purpose of analyzing the sensitivity of the results to the changing weights, different weights need to be chosen and tested. In order to choose some weights, ten scenarios have been defined in Table 6:

For each scenario, the net flow values of the alternative have been calculated by Decision Lab software and are brought in Tables 7, 8, 9, 10, 11, 12, 13, 14, 15, and 16.

It should be noticed that the values in italics show that they are the best answers of those scenarios based on the maximum values of $\phi(.)$

Moreover, to achieve a visual realization of the results of net flows ϕ for the scenarios, they are shown in Figure 6.

At first, it is obvious that the variation of the weights changes the ranking of the alternatives. Therefore, it is important to properly determine the weights. The decision maker should study the condition of the system carefully and then define the weights regarding their preferences. With regard to the tables, it is obvious that alternatives 600 days and 700 days are more frequently used as the best compromise solutions.

For the purpose of studying the behavior of each criterion, the weight of the criterion can be changed at a constant rate, and the weights of the other two criteria need to be change simultaneously equally. For this purpose, the following eight scenarios have been defined for each criterion, and the movement of the decision axis Pi has been in mind:

Scenario 1 = {20, 25, 55}
Scenario 2 = {25, 55, 20}
Scenario 3 = {55, 20, 25}
Scenario 4 = {10, 45, 45}
Scenario 5 = {45, 45, 10}
Scenario 6 = {45, 10, 45}
Scenario 7 = {10,10,80}
Scenario 8 = {80, 10, 10}
Scenario 9 = {10, 10, 80}
Scenario 10 = {33, 33, 34}

So, the observation of the behavior of the reliability criterion shows that by moving from scenario 1 to scenario 8, the decision axis moves from alternative 800 days to alternative 400 days. The observation of the

behavior of the cost criterion shows that by moving from scenario 1 to scenario 8, the decision axis moves from alternative 500 days to alternative 900 days.

The observation of the behavior of the downtime criterion shows that by moving from scenario 1 to scenario 8, the decision axis moves from alternative 600 days to alternative 700 days. These observations show that the two criteria of reliability and cost are conflicting. Also, the alternative downtime does not change widely. It only moves between two alternatives.

Gap of research

It is obvious that researchers in the field of maintenance try to make the best policy that is most compatible with their systems. Thus, there are many researches in this field with different focuses. Tsarouhas (2011) performs a comparative study of performance evaluation between four pizza production lines. He estimates the reliability and maintainability of the lines, focusing on the maintenance and repair strategies necessary for maintenance staff to keep equipment operating at the required level of reliability, which leads to the situation where lines are operating more profitably through reduced maintenance costs and increased productivity and efficiency (Tsang et al. 1999). Savsar (2011) presents a practical application of modeling and analysis procedures for maintenance operations in the context of an oil filling plant. System is analyzed under the current and a proposed PM policy, which reduced the equipment down time due to CMs (Samar Ali and Kannan 2011). Mohideen et al. (2011) presents a proposal to minimize the recovery time and the breakdown cost in the system in a construction plant (McNaught and Chan 2011). Sharma and Yadava (2011) review the literature on maintenance optimization models and associated case studies (Scarf 1997). They conclude that a good research work has been reported on optimization to bring down maintenance cost. The maintenance cost optimization work has been done on selecting maintenance policies, equipment availability, spare parts management, workforce scheduling, and interval of inspection frequency based on different simulation model. This finding shows that in most cases other criteria, like reliability, availability, down time, have been ignored in the choice of a maintenance policy. Sharma and Yadava (2011) also show that the applications

Table 15 Scenario 9

t(days)	200	300	400	500	600	700	800	900	1,000	1,100
$\varphi(.)$	−0.8086	−0.3108	−0.0391	0.1301	0.1937	*0.2095*	0.1917	0.1705	0.1516	0.1114

Table 16 Scenario 10

t(days)	200	300	400	500	600	700	800	900	1,000	1,100
$\varphi(.)$	−0.4452	−0.1822	−0.0013	0.1483	0.2046	0.1804	0.1213	0.0514	−0.0107	−0.0667

on DSS of optimization models are very limited in industry and not much has been from literature. As seen before, the application of PROMETHEE in this research is one answer to this lack.

A point of strength in this model is specifying periodic frequency for PM operations by the analysts and those involved in the system. In those model where optimum time for PM operations is calculated by the model itself, ordinarily, the obtained answers and the output of the model are not immediately applicable and would need adaptations and modifications, since the maintenance

operation is not performed independently and would require coordination with the production and operations sections, with production scheduling section, and even at times other engineering sections. In fact, the answer produced is not applicable and would generally require concurrence with the conditions at the workshop, and this is while modification would distance our answer from optimal conditions. In the present model, frequency is determined beforehand by the analyst and those involved in the system, and their opinions are thoroughly incorporated into the system; thus, the obtained

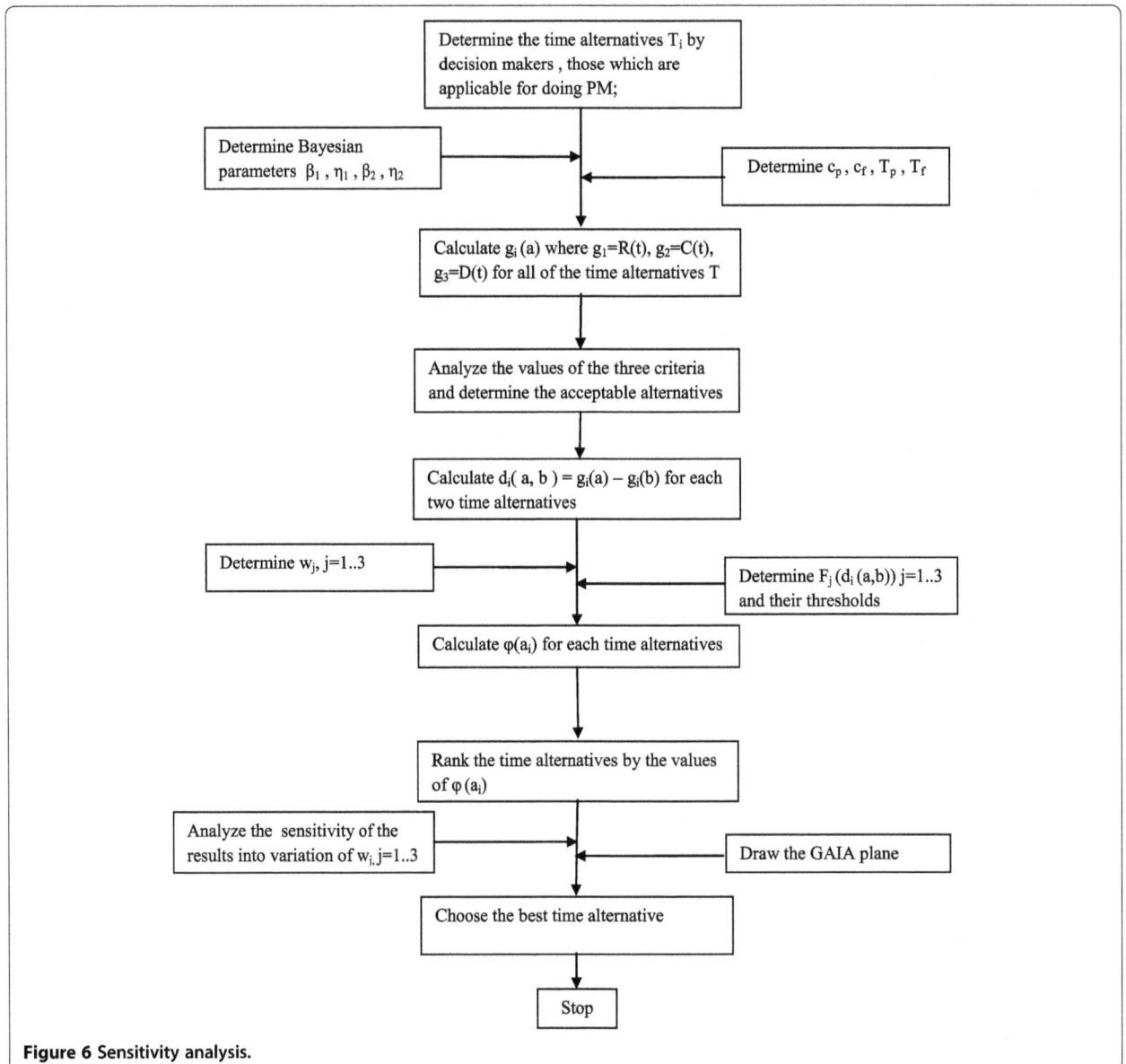

Figure 6 Sensitivity analysis.

answers and, in fact, the output of the model are immediately and without any alterations applicable and can be utilized.

Conclusion

This paper presents a multi-criterion decision-making model for preventive maintenance planning which determines the best compromise time for replacement of a certain item based on more than one criterion. This model also envisions the difficulty with the shortage of maintenance failure data by making use of Bayesian approach and PROMETHEE II for decision making.

In most cases, when maintenance managers try to determine the best policy for their systems, they only consider the cost criterion as the most important and the only criterion to be taken into account. This is a very dangerous point of view. Therefore, one of the most important goals that this paper seeks to reach is to give a broader view of the maintenance managers by considering more than one criterion in making an appropriate decision for replacement of an item in PM problems. Taking these three criteria into consideration, this paper does not imply that they are the most important criteria that need to be considered for replacement of an item in PM planning. It implies that in order to make a complete and timely PM planning which considers many aspects of the problem, decision makers have to study the problem completely and consider the factors which affect a PM planning for replacement of item because ignoring the influential factors in different situations can lead to disastrous results. Therefore, it is not true to say that there are some factors which are important for all the systems. Moreover, changing the weights shows that for different preferences of decision makers and different conditions of the systems, different weights are needed. Therefore, the structure of the model can be applied to different systems and situations.

Methods

In this section, the methods used in this research have been reviewed. In this research, multi-criterion decision making methods have been used to model a maintenance planning. Three criteria as reliability, maintenance cost, and maintenance downtime have been considered to make the best replacement intervals for preventive maintenance. In order to compensate the loss of historical data, Bayesian analysis has been used. This research has chosen PROMETHEE II method to outrank the results because of fast use, easy-to-analyze results, and a flexible comparison process. This method requires criteria weights reflecting the subjective relative importance of the criteria by decision makers. In this research, in order to analyze sensitivity and graphical visualization of results, PROMETHEE GAIA has been used.

Competing interests
The authors declare that they have no competing interests.

Authors' contributions
Both authors have participated in completing every section of the paper equally. All authors read and approved the final manuscript.

Author details
[1]Department of Industrial Engineering, Islamic Azad University of Najafabad, Isfahan 8514143131, Iran. [2]Department of Industrial Engineering, Isfahan University of Technology, Isfahan 8415683111, Iran.

References
Alsyouf I (2009) Maintenance practices in Swedish industries: survey results. Int J Prod Econ 121:212–223
Al-Turky U (2011) A framework for strategic planning in maintenance. J Qual Mainten Eng 17(2):150–162
Barlow RE, Hunter LC (1960) Optimum preventive maintenance polices. Oper Res 8:90–100
Behzadian M, Kazemzadeh RB, Albadvi A, Aghdasi M (2010) PROMETHEE: a comprehensive literature review on methodologies and applications. Eur J Oper Res 200(1):198–215
Bevilacqua M, Braglia M (2000) The analytic hierarchy process applied to maintenance strategy selection. Reliab Eng Syst Saf 70(1):71–83
Brans JP, Mareschal B (1994a) The PROMCALC and GAIA decision-support system for multicriteria decision aid. Decis Support Syst 12(4–5):297–310
Brans JP, Mareschal B (1994b) The PROMETHEE-GAIA decision support system for multicriteria investigations. Investigation Operativa 4(2):107–117
Brans JP, Mareschal B (2005) PROMETHEE methods. In: Figueira J, Greco S, Ehrgott M (eds) Multiple criteria decision analysis: state of the art surveys. Springer, Boston
Brans JP, Mareschal B, Vincke P (1984) PROMETHEE: a new family of outranking methods in multicriteria analysis. Oper Res 84:477–490
British Standards Institute Staff (1993) Glossary of terms used in terotechnology. BSI, London
Cavalcante CAV, De Almedia AT (2007) A multicriteria decision aiding model using PROMETHEE III for preventive maintenance planning under uncertain conditions. J Qual Mainten Eng 13:385–397
Cavalcante CAV, De Almedia AT (2008) A preventive maintenance decision model based on multicriteria method PROMETHEE II integrated with Bayesian approach. IMA J Manag Math 16:1–16
Chareonsuk C, Nagarur N, Tabucanon MT (1997) A multicriteria approach to the selection of preventive maintenance intervals. Int J Prod Econ 49:55–64
Cholasuke C, Bhardwa R, Antony J (2004) The status of maintenance management in UK manufacturing organisations: results from a pilot survey. J Qual Mainten Eng 10(1):5–15
Chitra T (2003) Life based maintenance policy for minimum cost, In: Annual reliability and maintainability (RAMS). IEEE Press, NY, Bangalore, India, February 28, 2003, pp 470–pp 474
Cho D, Parlar M (1991) A survey of maintenance models for multi-unit systems. Eur J Oper Res 51:1–23
Chen T, Popova E (2002) Maintenance policies with two-dimensional warranty. Reliability Engineering and Systems Safety 77:61–69
Coetzee JL (2004) Maintenance. Trafford Publishing, Victoria
Cross M (1988) Raising the value of maintenance in the corporate environment. Manag Res News 1(3):8–11
Dayanlk S, Gurler U (2002) An adaptive Bayesian replacement policy with minimal repair. Oper Res 50:552–558
Dekker R (1996) Applications of maintenance optimization models: a review and analysis. Reliab Eng Syst Saf 51:229–240
Dhillon BS (2002) Engineering maintenance. CRC Press, Boca Raton, USA
Jardine AKS (1973) Maintenance, replacement and reliability. John Wiley & Sons, New York
Jorgenson DW, Mccall JJ, Radner R (1967) Optimal replacement policy. North-Holland, Amsterdam
Juang MG, Anderson G (2004) A Bayesian method on adaptive preventive maintenance problem. Eur J Oper Res 155:455–473

Kallen MJ, Van Noortwijk JM (2005) Optimal maintenance decisions under imperfect inspection. Reliab Eng Syst Saf 90:177–185

Maggard B, Rhyne D (1992) Total productive maintenance: a timely integration of production and maintenance. Prod Inventory Manag J 33(4):6–10

van der Meulen P, Petraitis M, Pannese P (2008) Advanced semiconductor manufacturing conference. In: Design for maintenance. IEEE Press, NY, Washington DC, USA, May 5–7, 2008, pp 278–281

McCall JJ (1965) Maintenance policies for stochastically failing equipment: a survey. Manag Sci 11:493–624

McNaught K, Chan A (2011) Bayesian networks in manufacturing. J Manuf Tech Manag 22(6):734–747

Makis V, Jardine AKS (1992) Optimal replacement in the proportional hazards model. INFOR 30:172

Mohideen APB, Ramachandran M, Narasimmalu RR (2011) Construction plant breakdown criticality analysis – part 1: UAE perspective. Benchmark Int J 18(4):472–489

Nakagawa T (2005) Maintenance theory of reliability. Springer, London

Oberschmidt J, Geldermann J, Ludwig J, Schmehl M (2010) Modified PROMETHEE approach for assessing energy technologies. Int J En Sect Manag 4(2):183–212

Oke SA, Charles-Owaba OE (2006) Application of fuzzy logic control model to Gantt charting preventive maintenance scheduling. Int J Qual Reliab Manag 23(4):441–459

Pierskalla WP, Voelker JA (1976) A survey of maintenance models: the control and surveillance of deteriorating systems. Nav Res Logistics Q 23:353–388

Salonen A, Deleryd M (2011) Cost of poor maintenance: a concept for maintenance performance improvement. J Qual Mainten Eng 17(1):63–73

Samar Ali S, Kannan S (2011) A diagnostic approach to Weibull-Weibull stress-strength model and its generalization. Int J Qual Reliab Manag 28(4):451–463

Savsar M (2011) Analysis and modeling of maintenance operations in the context of an oil filling plant. J Manuf Tech Manag 22(5):679–697

Scarf PA (1997) On the application of mathematical models in maintenance. Eur J Oper Res 99:493–506

Sharma A, Yadava GS (2011) A literature review and future perspectives on maintenance optimization. J Qual Mainten Eng 17(1):5–25

Sherif SML (1981) Optimal maintenance models for systems subject to failure—a review. Nav Res Logistics Q 28:47–74

Shyjith K, Ilangkumaran M, Kumanan S (2008) Multi-criteria decision-making approach to evaluate optimum maintenance strategy in textile industry. J Qual Mainten Eng 14(4):375–386

Sheu SH, Yeh RH, Lin YB, Juang MG (2001) A Bayesian approach to an adaptive preventive maintenance model. Reliab Eng Syst Saf 71:33–44

Sortrakul N, Cassady CR (2007) Genetic algorithms for total weighted expected tardiness integrated preventive maintenance planning and production scheduling for a single machine. J Qual Mainten Eng 13(1):49–61

Swanson L (2001) Linking maintenance strategies to performance. Int J Prod Econ 70(3):237–244

Tsang AHC, Jardine AKS, Kolodny H (1999) Measuring maintenance performance a holistic approach. J Qual Mainten Eng 19(7):691–715

Tsarouhas PH (2011) A comparative study of performance evaluation based on field failure data for food production lines. J Qual Mainten Eng 17(1):26–39

Valdez-Flores C, Feldman RM (1989) A survey of preventive maintenance models for stochastic deteriorating single-unit systems. Nav Logistics Q 36:419–446

Wang H (2002) A survey of maintenance policies of deteriorating systems. Eur J Oper Res 139:469–489

Wilson JG, Benmerzouga A (1995) Bayesian group replacement policies. Oper Res 43:471–476

Zhang J (2005) Maintenance planning and cost effective replacement strategies. University of Alberta, Canada

12

Allocation models for DMUs with negative data

Ghasem Tohidi[†] and Maryam Khodadadi[*†]

Abstract

The formulas of cost and allocative efficiencies of decision making units (DMUs) with positive data cannot be used for DMUs with negative data. On the other hand, these formulas are needed to analyze the productivity and performance of DMUs with negative data. To this end, this study introduces the cost and allocative efficiencies of DMUs with negative data and demonstrates that the introduced cost efficiency is equal to the product of allocative and range directional measure efficiencies. The study then intends to extend the definition of the above efficiencies to DMUs with negative data and different unit costs. Finally, two numerical examples are given to illustrate the proposed methods.

JEL classification: C6, D2

Keywords: DEA, Cost efficiency, Negative data, Allocative efficiency, RDM

Introduction

Data envelopment analysis (DEA) is a nonparametric method for computing and assessing the relative efficiency of homogeneous decision making units (DMUs) with multiple inputs and outputs. The traditional DEA models assume that all of the inputs and outputs are nonnegative, while in many situations and applications, the negative values in data might exist, which is a weakness of the traditional DEA models. To overcome the shortcoming, in recent years, different DEA models have been proposed in the literature about DMUs with negative data.

Portela et al. (2004) provided the range directional measure (RDM) approach to measure the efficiency of DMUs with negative data based on a directional distance function without the need to transform the data. The efficiency measurement of DMUs with negative data is also much debated. For example, in order to overcome the shortcomings of the slack-based measure model (Tone 2001) in dealing with negative inputs and outputs, Sharp et al. (2006) presented the modified slack-based measure model. Emrouznejad et al. (2010) proposed the semi-oriented radial measure model for dealing with negative inputs and outputs. Also, recently, a two-phase approach

model has been proposed by Kazemi Matin and Azizi (2011) based on a modified version of the additive model to achieve a target with nonnegative components for each DMU with negative inputs and outputs.

One of the most significant types of efficiency is cost efficiency. This type of efficiency is used to identify the different kinds of inefficiencies when information on costs is available. DMUs can achieve the best cost efficiency score with a combination of inputs which allow them to produce the desired outputs at minimum costs.

In many DEA literatures, the economic concepts of cost efficiency have been considered. The primary discussion of cost efficiency can be traced back to Farrell (1957) and Debreu (1951), from whom many of the ideas about DEA are derived. Farrell offered a measure of cost efficiency under fixed and known prices. His method extended to situations with different prices of inputs for DMUs (Tone 2002).

So far, all of the previous studies have explored cost efficiency of DMUs with nonnegative data, and there is no discussion concerning cost efficiency in the presence of negative data. However, in some cases, the inputs of DMUs have negative values with positive costs. This paper defines the cost and allocative efficiencies for DMUs with negative data, then demonstrates that the

---
* Correspondence: maryam8khodadadi@yahoo.com
[†]Equal contributors
Department of Mathematics, Islamic Azad University of Central Tehran Branch, Simaye Iran Ave., Tehran 14676-86831, Iran

defined cost efficiency with negative data is equal to the product of allocative and RDM efficiencies, and then extends the definition of efficiencies to DMUs with negative data and different unit costs.

The rest of this paper is organized as follows: the Section 'Background models' explains the RDM efficiency and cost efficiency models. 'Cost efficiency in the presence of negative data' briefly introduces cost efficiency under common and different prices in the presence of negative data. The Section 'Illustrative examples' provides two numerical examples, and in the last section, the conclusion is given.

Background models

In productive activities, we assume that there are n homogeneous DMUs. Each DMU produces s different outputs from m different inputs. Input and output vectors for the DMU which is under evaluation are denoted by x_o and y_o, and for DMU$_j$, are denoted by x_j and y_j. The next section explains the RDM and cost efficiency models.

RDM

The RDM model, introduced by Portela et al. (2004), can be used for comparing DMUs when some inputs and/or outputs are negative. Consider a point with maximum outputs and minimum inputs as an ideal point (i.e., the ith ($i = 1, ..., m$) input x_{iI} as min $_j\{x_{ij}\}$ and the rth ($r = 1, ..., s,$) output y_{rI} as max $_j\{y_{rj}\}$). In RDM, a directional vector is considered as $R_{ro}^+ = \max_j \{y_{rj}\} - y_{ro}; r = 1, ..., s$ and $R_{io}^- = x_{io} - \max_j\{x_{ij}\}; \ i = 1, ..., m$. The RDM model for DMU$_o$ is as follows (Portela et al. 2004):

$$\max \beta$$
$$\text{s.t.} \quad \sum_{j=1}^{n} \lambda_j x_{ij} \leq x_{io} - \beta R_{io}^-, \quad i = 1, ..., m$$
$$\sum_{j=1}^{n} \lambda_j y_{rj} \geq y_{ro} + \beta R_{ro}^+, \quad r = 1, ..., s \quad (1)$$
$$\sum_{j=1}^{n} \lambda_j = 1 \ , \ \lambda_j \geq 0, \quad j = 1, ..., n.$$

The optimal value of model (1), β^*, represents the inefficiency measurement for DMU$_o$, while $1 - \beta^*$ represents the efficiency measurement for DMU$_o$. Unit invariance and translation invariance are the two important properties of the RDM model.

Cost efficiency

For measuring the cost efficiency of the DMUs with multiple inputs and outputs under common unit input prices, the following linear program is solved (Farrell 1957):

$$cx^* = \min \sum_{i=1}^{m} c_i x_i$$
$$\text{s.t.} \quad \sum_{j=1}^{n} \lambda_j x_{ij} \leq x_i, \quad i = 1, ..., m$$
$$\sum_{j=1}^{n} \lambda_j y_{rj} \geq y_{ro}, \quad r = 1, ..., s \quad (2)$$
$$\sum_{j=1}^{n} \lambda_j = 1 \ , \ \lambda_j \geq 0, \ j = 1, ..., n.$$

The cost efficiency is obtained as the following ratio:

$$\text{CE} = \frac{cx^*}{cx_o}, \quad (3)$$

where the nominator represents the minimum cost (i.e., the optimal value of model (2)) and the denominator shows the current cost at DMU$_o$.

Cost efficiency in the presence of negative data

In this section, we define cost and allocative efficiencies in the presence of negative data under common and different prices, and then it is shown that cost efficiency is the product of allocative and RDM efficiencies. Finally, the above subjects are extended to new cost, allocative, and RDM efficiencies.

Cost and allocative efficiencies with negative data under common unit prices

Definition 1 Under common unit input prices, we define cost (overall) efficiency in the presence of negative data as follows:

$$\text{CE} = \frac{cx^* - cx_I}{cx_o - cx_I}. \quad (4)$$

In the above mentioned ratio, the nominator represents the difference between cx^* (the optimal value of model (2)) and the cost of the ideal point, i.e., cx_I. In addition, the denominator depicts the difference between the observed cost of DMU$_o$, i.e., cx_o and cx_I. It is clear that the value of CE is equal to or less than 1. Cost efficiency might be less than 1 for one of the following two reasons: excessive input usage in production or production with a wrong input mix in light of input prices or both. In a particular case, when the ideal point is one of the observed DMUs, we define CE = 1.

Figure 1 illustrates the concepts dealing with allocative efficiency, RDM efficiency, and cost efficiency, using the units A, B, D, E, F, H, G, and P in the presence of

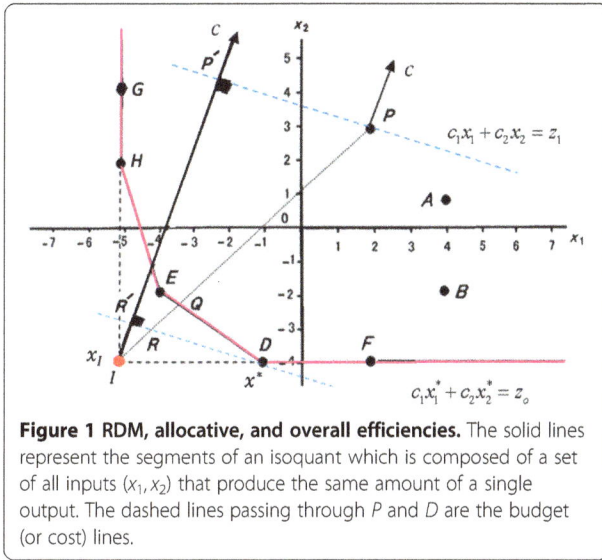

Figure 1 RDM, allocative, and overall efficiencies. The solid lines represent the segments of an isoquant which is composed of a set of all inputs (x_1, x_2) that produce the same amount of a single output. The dashed lines passing through P and D are the budget (or cost) lines.

negative data. In Figure 1, the RDM inefficiency of DMU$_P$ can be evaluated by

$$\beta = \frac{d(Q,P)}{d(I,P)}. \tag{5}$$

Hence,

$$1-\beta = \frac{d(I,P)-d(Q,P)}{d(I,P)} = \frac{d(I,Q)}{d(I,P)} \tag{6}$$

Equation 6 represents the RDM efficiency of DMU$_P$ which is between 0 and 1. $d(I, P)$ and $d(I, Q)$ denote the distance from the ideal point I to P and the distance from the ideal point I to Q, respectively, and $d(Q, P)$ is the distance from Q to P.

In order to illustrate allocative efficiency in Figure 1, we consider the budget (cost) line $c_1x_1 + c_2x_2 = z_1$ passing through the point P. We move this cost line in parallel form until it crosses the isoquant at D.

By moving the budget line in parallel form, cost can be reduced. The lowest cost is associated with the budget (cost) line $c_1x_1^* + c_2x_2^* = z_o$, where $z_o < z_1$ and z_o can be obtained by substituting the coordinates of DMU$_D$ in the budget (cost) line $c_1x_1 + c_2x_2 = z_1$. The best point D is achieved as the optimal solution x^* of the linear program (2) (Farrell 1957).

Now, we define $\frac{d(I,R)}{d(I,Q)}$ as a measure of allocative efficiency, where $d(I, R)$ and $d(I, Q)$ denote the distance from the ideal point I to R and the distance from the ideal point I to Q, respectively. It shows that the minimum cost is not reached since we have failed to make the replacements which are involved in moving from point Q to D along the efficiency frontier.

According to Definition 1, we have

$$\text{CE} = \frac{cx^* - cx_I}{cx_o - cx_I}$$

$$\text{CE} = \frac{cx^* - cx_I}{cx_o - cx_I} = \frac{|c||x^* - x_I|\cos\theta}{|c||x_o - x_I|\cos\phi} = \frac{|x^* - x_I|\cos\theta}{|x_o - x_I|\cos\phi} \tag{7}$$

where $|.|$ in Equation 7 represents the norm function of a vector, and θ and ϕ are the angles between the vector c and vectors $(x^* - x_I)$ and $(x_o - x_I)$, respectively. The numerator in Equation 7 represents the projection of vector $(x^* - x_I)$ on vector c which is equal to $d(I, R')$, and the denominator in Equation 7 demonstrates the projection of vector $(x_o - x_I)$ on vector c which is equal to $d(I, P')$. We can represent the defined cost efficiency in the presence of negative data by means of the following ratio:

$$\text{CE} = \frac{d(I,R')}{d(I,P')} \tag{8}$$

where $d(I, R')$ and $d(I, P')$ denote the distance from the ideal point I to R' and P', respectively. According to Thales theorem, we have:

$$\text{CE} = \frac{d(I,R')}{d(I,P')} = \frac{d(I,R)}{d(I,P)}. \tag{9}$$

Therefore, the three above mentioned efficiencies in the presence of negative data have the following relationship:

$$\frac{d(I,R)}{d(I,Q)} \times \frac{d(I,Q)}{d(I,P)} = \frac{d(I,R)}{d(I,P)}. \tag{10}$$

That is, the product of allocative and RDM efficiencies is equal to cost efficiency.

Cost efficiency with negative data under different prices
In some situations, the unit prices of input are not the same among DMUs. Therefore, the above definitions of cost and allocative efficiencies have shortcomings, and they are not applicable to these cases (Farrell 1957). In this case, we define new cost and allocative efficiencies, which are extensions of the prior definitions and are applicable to situations in which the input prices are not the same among DMUs.

In order to discuss cost efficiency in the presence of different unit input costs, we consider the cost-based production possibility set, P_C, as follows (Tone 2002):

$$P_C = \left\{ (\bar{x}, y) \mid \bar{x} \geq \sum_{j=1}^{n} \lambda_j \bar{x}_j, \right.$$

$$\left. y \leq \sum_{j=1}^{n} \lambda_j y_j, \sum_{j=1}^{n} \lambda_j = 1, \lambda_j \geq 0, j = 1, ..., n \right\}$$

$$(11)$$

where $\bar{x}_j = (\bar{x}_{1j}, ..., \bar{x}_{mj})^T = (c_{1j}x_{1j}, ..., c_{mj}x_{mj})^T$, in which $c_j = (c_{1j}, ..., c_{mj})^T$ is the positive cost vector of DMU$_j$.

Definition 2 The ideal point, by using the new production possibility set, is a point with maximum outputs and minimum inputs, i.e., $\bar{x}_I = (\bar{x}_{1I}, ..., \bar{x}_{mI})^T$ where $\bar{x}_{iI} = \min_j \bar{x}_{ij}$, $i = 1, ..., m$, and $y_I = (y_{1I}, ..., y_{sI})^T$ where $y_{rI} = \max_j y_{rj}$, $r = 1, ..., s$. In the new production possibility set, P_C, the new RDM inefficiency, $\bar{\beta}^*$, is obtained by solving the following linear program:

[NRDM] $\quad \bar{\beta}^* = \max \bar{\beta}$

$$\text{s.t.} \quad \sum_{j=1}^{n} \lambda_j \bar{x}_{ij} \leq \bar{x}_{io} - \bar{\beta}\bar{R}_{io}^-, \quad i = 1, ..., m$$

$$\sum_{j=1}^{n} \lambda_j y_{rj} \geq y_{ro} + \bar{\beta}R_{ro}^+, \quad r = 1, ..., s \qquad (12)$$

$$\sum_{j=1}^{n} \lambda_j = 1, \quad \lambda_j \geq 0, \quad j = 1, ..., n,$$

where $\bar{R}_{io}^- = \bar{x}_{io} - \min_j \{\bar{x}_{ij}\}$; $i = 1, ..., m$ and $R_{ro}^+ = \max_j \{y_{rj}\} - y_{ro}$; $r = 1, ..., s$. To interpret $\bar{\beta}^*$ (the optimal value of model (12)) and $1 - \bar{\beta}^*$, it can be said that $\bar{\beta}^*$ demonstrates the new RDM inefficiency, and $1 - \bar{\beta}^*$ demonstrates the new RDM efficiency of DMU$_o$.

Definition 3 The new cost efficiency in the presence of negative data under different unit prices is defined as follows:

$$\bar{CE} = \frac{e\bar{x}^* - e\bar{x}_I}{e\bar{x}_o - e\bar{x}_I} \qquad (13)$$

where e is a vector in R_m with each component equal to 1, and $\bar{x}^*$ is an optimal solution of the following model (Tone 2002):

[NCost] $\quad e\bar{x}^* = \min e\bar{x}$

$$\text{s.t.} \quad \sum_{j=1}^{n} \lambda_j \bar{x}_j \leq \bar{x}$$

$$\sum_{j=1}^{n} \lambda_j y_j \geq y_o \qquad (14)$$

$$\sum_{j=1}^{n} \lambda_j = 1,$$

$$\lambda_j \geq 0, \quad j = 1, ..., n.$$

It is obvious that $0 \leq \bar{CE} \leq 1$. The following theorem states a relationship between new RDM efficiency and new cost efficiency scores.

Theorem 1 $\bar{CE} \leq 1 - \bar{\beta}^*$.

Proof Let $(\bar{\beta}^*, \lambda^*)$ be an optimal solution of model (12); then $(\bar{x}_o - \bar{\beta}^*\bar{R}_o^-, \lambda^*)$ is a feasible solution for model (14). This shows that $e\bar{x}^* \leq e\bar{x}_o - e\bar{\beta}^*\bar{R}_o^-$. Using Equation 13 we have:

$$\bar{CE} = \frac{e\bar{x}^* - e\bar{x}_I}{e\bar{x}_o - e\bar{x}_I} \leq \frac{e\bar{x}_o - e\bar{\beta}^*\bar{R}_o^- - e\bar{x}_I}{e\bar{x}_o - e\bar{x}_I} \leq \frac{e\bar{x}_o - e\bar{\beta}^*(\bar{x}_o - \bar{x}_I) - e\bar{x}_I}{e\bar{x}_o - e\bar{x}_I}$$

$$= \frac{e\bar{x}_o(1 - \bar{\beta}^*) - e\bar{x}_I(1 - \bar{\beta}^*)}{e\bar{x}_o - e\bar{x}_I} = (1 - \bar{\beta}^*).$$

$$(15)$$

This fact completes the proof.

Definition 4 The new allocative efficiency under the different unit prices in the presence of negative data is defined as follows:

$$\bar{AE} = \frac{\bar{CE}}{1 - \bar{\beta}^*}. \qquad (16)$$

It is obvious that $0 \leq \bar{AE} \leq 1$.

Illustrative examples

In this section, two numerical examples are used to illustrate the concepts of what was mentioned earlier.

Table 1 The inputs, outputs, and costs for eight DMUs

DMUs	x_1	c_1	x_2	c_2	y
A	4	1	1	2	1
B	4	1	−2	2	1
D	−1	1	−4	2	1
E	−4	1	−2	2	1
F	2	1	−4	2	1
G	−5	1	4	2	1
H	−5	1	2	2	1
P	2	1	3	2	1

Table 2 RDM, cost, and allocative efficiencies

DMUs	β^*	$1 - \beta^*$	cx^*	cx_o	cx_I	CE	AL
A	0.76	0.24	−9	6	−13	0.21	0.87
B	0.67	0.33	−9	0	−13	0.30	0.90
D	0	1	−9	−9	−13	1	1
E	0	1	−9	−8	−13	0.8	0.8
F	0.43	0.57	−9	−6	−13	0.57	1
G	0.25	0.75	−9	3	−13	0.25	0.33
H	0	1	−9	−1	−13	0.33	0.33
P	0.77	0.23	−9	8	−13	0.19	0.82

Table 4 New data set

DMUs	$\bar{x}_1$	$\bar{x}_2$	y
A	12	2	1
B	4	−4	1
D	−2	−8	1
E	−4	−6	1
F	6	−4	1
G	−20	20	1
H	−15	8	1
P	6	6	1
Ideal	−20	−8	1

Example 1 In Table 1, we have eight DMUs with two inputs and one output. Figure 1 depicts the production possibility set composed of these input and output data. The values of x_1 and x_2 and their relative unit costs are exhibited in the columns of Table 1.

Table 2 reports the obtained results for the data of Table 1. To compare the results of Table 2 with Equations 5, 6, 9, and 10 as an example, we select DMU$_P$. As it can be seen in Figure 1, the coordinates of Q are $\left(\frac{-17}{5}, \frac{-12}{5}\right)$, which are obtained from the intersection of the line passing through E and D and the line passing through I and P.

The RDM efficiency for P is obtained as follows:

$$1 - \beta^* = \frac{d(I,Q)}{d(I,P)} = \frac{\sqrt{\left(-5 + \frac{17}{5}\right)^2 + \left(-4 + \frac{12}{5}\right)^2}}{\sqrt{(-5-2)^2 + (-4-3)^2}}$$
$$= \frac{8}{35} \approx 0.23$$

where $(-5, -4)$ are the coordinates of the ideal point. The coordinates of R are $\left(\frac{-11}{3}, \frac{-8}{3}\right)$, which were achieved from the intersection of the line passing through D and R' and the line passing through I and P. Hence, the cost and the allocative efficiencies of DMU$_P$ are obtained as follows:

$$CE = \frac{d(I,R)}{d(I,P)} = \frac{\sqrt{\left(-5 + \frac{11}{3}\right)^2 + \left(-4 + \frac{8}{3}\right)^2}}{\sqrt{(-5-2)^2 + (-4-3)^2}} = \frac{4}{21} \approx 0.19$$

$$AL = \frac{d(I,R)}{d(I,Q)} = \frac{\sqrt{\left(-5 + \frac{11}{3}\right)^2 + \left(-4 + \frac{8}{3}\right)^2}}{\sqrt{\left(-5 + \frac{17}{5}\right)^2 + \left(-4 + \frac{12}{5}\right)^2}} = \frac{5}{6} \approx 0.82.$$

Hence,

$$\frac{d(I,R)}{d(I,Q)} \times \frac{d(I,Q)}{d(I,P)} = \frac{d(I,R)}{d(I,P)}.$$

The above results are the same as the results which are exhibited in Table 2.

Example 2 Table 3 represents the data set of Table 1 under the different unit costs. The coordinates of inputs and output in the PPS P_C and the ideal point are represented in Table 4. These data were obtained by multiplying the relevant unit costs of x_1 and x_2 by the values of x_1 and x_2.

Table 5 shows the new RDM inefficiency, new RDM, cost, and allocative efficiencies under the different unit prices in the presence of the negative data of Table 4. The results of Table 5 indicate that the DMUs D and E have

Table 3 Data for eight DMUs

DMUs	x_1	c_1	x_2	c_2	y
A	4	3	1	2	1
B	4	1	−2	2	1
D	−1	2	−4	2	1
E	−4	1	−2	3	1
F	2	3	−4	1	1
G	−5	4	4	5	1
H	−5	3	2	4	1
P	2	3	3	2	1

Table 5 New RDM, new cost, and new allocative efficiencies

DMU	$\bar{\beta}^*$	$1 - \bar{\beta}^*$	$e\bar{x}^*$	$e\bar{x}_o$	$e\bar{x}_I$	$\bar{CE}$	$\bar{AL}$
A	0.56	0.44	−10	14	−13	0.11	0.25
B	0.35	0.65	−10	0	−13	0.23	0.35
D	0	1	−10	−10	−13	1	1
E	0	1	−10	−10	−13	1	1
F	0.45	0.55	−10	2	−13	0.2	0.36
G	0	1	−10	0	−13	0.23	0.23
H	0	1	−10	−7	−13	0.5	0.5
P	0.53	0.47	−10	12	−13	0.12	0.25

the best performance. The DMUs G and H are new RDM efficient, in spite of the fact that these DMUs fell short in their new cost and new allocative efficiency scores.

Conclusions

This paper introduced cost and allocative efficiencies under common and different unit prices in the presence of negative data. It was shown that under common and different unit prices, the defined cost efficiency is the product of allocative efficiency and RDM efficiency. Finally, to illustrate the mentioned concepts, two numerical examples were used.

Competing interests
Both authors declare that they have no competing interests.

Authors' contributions
GT proposed the definition of cost efficiency of DMUs in the presence of negative data. Then MKH and GT extended this definition to DMUs with negative data and different unit costs and then demonstrated that the introduced cost efficiency is equal to the product of allocative and range directional measure efficiencies. Both authors read and approved the final manuscript.

Authors' information
GT is currently an assistant professor in the Department of Mathematics of Islamic Azad University, Central Tehran Branch. His research interests include data envelopment analysis (DEA) and fuzzy and multi objective programming (MOP). He has published many articles in a number of peer-reviewed academic journals and conferences. MKH is a doctoral student in applied mathematics (operational research) at Islamic Azad University, Central Tehran Branch in Iran. She holds a M.Sc. and B.Sc. in Applied Mathematics from Kharazmi University and Urmia University in Iran. Her research interests include applied mathematics, operation research, and data envelopment analysis.

Acknowledgements
The financial support for this research from Islamic Azad University, Central Tehran Branch, is acknowledged.

References
Debreu G (1951) The coefficient of resource utilization. Econometrica 19:273–292
Emrouznejad A, Anouze AL, Thanassoulis E (2010) A semi-oriented radial measure for measuring the efficiency of decision making units with negative data, using DEA. Eur J Oper Res 200:297–304
Farrell MJ (1957) The measurement of productive efficiency. J Roy Stat Soc 120:253–281
Kazemi Matin R, Azizi R (2011) A two-phase approach for setting targets in DEA with negative data. Appl Math Model 35(12):5794–5803
Portela MCAS, Thanassoulis E, Simpson GPM (2004) Negative data in DEA: a directional distance approach applied to bank branches. J Oper Res Soc 55:1111–1121
Sharp JA, Liu WB, Meng WA (2006) Modified slacks-based measure model for data envelopment analysis with 'natural' negative outputs and inputs. J Oper Res Soc 57:1–6
Tone K (2001) Slacks-based measure of efficiency in data envelopment analysis. Eur J Oper Res 130:498–509
Tone K (2002) A strange case of the cost and allocative efficiencies in DEA. J Oper Res Soc 53:1225–1231

13

Contemporary methods for evaluating complex project proposals

Hans J Thamhain

Abstract

The ability to evaluate project proposals, assessing future success, and organizational value is critical to overall business performance for most enterprises. Yet, predicting project success is difficult and often unreliable. A four-year field study shows that the effectiveness of available methods for evaluating and selecting large, complex project depends on the specific project type, organizational culture, and managerial skills. This paper examines the strength and limitations of various evaluation methods. It also shows that, especially in complex project situations, the decision-making process has to go beyond the application of just analytical methods, but has to incorporate both quantitative and qualitative measures into a combined rational judgmental evaluation process. Equally important, the evaluation process must be effectively linked among functional support groups and with senior management in order to strategically align the project proposal and to unify the evaluation team and stakeholder community behind the mission objectives. All of this requires leadership and managerial skills in planning, organizing, and communicating. The paper suggests specific leadership actions, organizational conditions, and managerial processes for evaluating complex project proposals toward future value and success.

Keywords: Project evaluation and selection; Project management; Team leadership; Technology; Decision making; Rational; Judgmental

Challenges of determining potential for future success

Predicting project success has never been easy. The long list of prominent project failures, ranging from product developments to government social programs, from computers to pharmaceutical, and from public transit to supersonic transport, reminds us of this reality (Cicmil et al. 2006; Gulla 2012; Lemon et al. 2002; Standish Group 2013). Many projects do not live up to their expectations or outright fail even before their technical completion in spite of careful feasibility analysis during their proposal or selection stages (El Emam and Koru 2008; Shore 2008). Obviously, the ability to evaluate project proposals and assess future success and organizational value is critical to overall business performance. In fact, few decisions have more impact on enterprise performance than the resource allocations for new projects (Shenhar et al. 2007). Virtually, every organization evaluates, selects, and implements projects. Whether these projects are product developments,

organizational improvements, R&D undertakings or bid proposals, pursuing the 'wrong' project not only wastes company resources, but also causes the enterprise to (1) miss critical alternatives, (2) perform less agile in the market place, and (3) miss opportunities for leveraging core competencies. Yet, in spite of pressures to avoid these high-cost errors, predicting project success is difficult, and existing models are often unreliable. Especially in today's complex and changing business environment (see 'Today's Complex Project Environments' section), the process of evaluating and selecting the 'best' projects, most suitable and beneficial for the enterprise, has become both an art and a science, strongly influenced by human and organizational factors.

Objective and rationale for this study

While much has been published on specific methods for project evaluation, the literature lacks a summary and comparison of available tools and techniques. Moreover, few studies have examined the human side as a 'cross-functional tool' for collectively evaluating project proposals toward potential success, an important link that is

Correspondence: hthamhain@bentley.edu
Management Department Bentley University, 175 Forest Street Waltham, Massachusetts, USA

missing in the management literature. This field study contributes to both areas. *First*, the paper summarizes the benefits, challenges, and limitations of popular project evaluation methods, tools, and techniques. *Second*, the paper reports the lessons learned from a five-year field study of organizational conditions, leadership style, and decision making with regard to impact on project selection processes.

In today's increasingly complex and dynamic business environment ('Today's Complex Project Environments' section), predicting success is multifaceted. Typically, it includes not only technical, but also financial, marketing, social, legal, ethical, and technological dimensions, many of them *fuzzy or unknown* at the time of the proposed project evaluation (Thamhain 2011a,b). Hence, the DNA of success is highly complex, and outcomes are difficult to predict. Evaluating and selecting projects is not only an art and a science, but has to go beyond a simple cost-benefit analysis for most cases. To be effective, project opportunities must be analyzed relative to their potential value, strength, and importance to the enterprise. We have to understand and examine the whole spectrum of costs, risks, and benefits as part of the evaluation process, far beyond conventional project economics. This requires a comprehensive approach with sophisticated leadership, integrating organizational resources and facilitating a shared vision of risks across organizational borders, time, and space (Baker 2012; Bstieler 2005; Cicmil et al. 2006). Currently, *we are quite efficient on the analytical side of project evaluation/selection. However, we are still weak in dealing with the hidden, less obvious dimensions of predicting success that involve a broader spectrum of project performance variables, connecting to the enterprise and its socio-economic environment* (Thamhain and Skelton 2007). These are the issues explored in the study reported in this paper.

Today's complex project environments
Our complex, fast-changing business environment requires sophisticated decision making for selecting the right project with the most desirable outcome assuring success and enterprise performance. Companies are under pressures for quicker, cheaper, and smarter solutions. They are globally networked and can leverage their resources and accelerate their schedules by forming alliances, consortia, and partnerships with other firms, universities, and government agencies, which range from simple cooperative agreements to *open innovation*, a concept of scouting for new product and service ideas, anywhere in the world. Projects are complex in nature and imbedded in lots of technology. All of these issues contribute to the complexities and uncertainties of project selection which is a great challenge for managers in virtually every segment of industry and government.

They include computer, pharmaceutical, automotive, health care, transportation, and financial businesses, just to name a few. New technologies, especially computers and communications, have radically changed the workplace and transformed our global economy, with focus on effectiveness, value, and speed. These techniques offer more sophisticated capabilities for cross-functional integration, resources mobility, market responsiveness, and managerial decision support, but they also require more sophisticated skills both technically and socially, dealing effectively with higher levels of conflict, change, risks, and uncertainty. These challenges resulted in a shift in managerial focus from functional efficiency to effectiveness with attention to organizational interfaces, human factors, and overall enterprise performance.

What we know about project evaluation and selection
Managerial decision support for project selection has been known for a long time (Remer et al. 1993). Today, managers have available a large array of tools and techniques for project evaluation and selection (National Science Foundation NSF 2010) that can be grouped into three principle classes which are briefly summarized in Appendix 1 of this paper:

1. Primarily *quantitative* and *rational* approaches.
2. Primarily *qualitative* and *intuitive* approaches.
3. *Mixed approaches*, combining both quantitative and qualitative methods.

While in the past, decisions on project selection focused on quantitative approaches, such as *ROI*, *payback periods*, and *net present value* for determining desirability and potential success, today's managers take a more balanced approach between quantitative and qualitative methods. They cast a much wider net for capturing a broad spectrum of variable that go far beyond the scope and limitations of traditional financial evaluations, such as ROI or payback (Kumar 2006).

Most managers are keenly aware of the intricate connections of success variables among organizational systems and processes, which often limit the effectiveness of analytical methods. Especially for complex projects and business processes, managers argue that no single person or group within an enterprise has all the smarts and insight for assessing these multi-variable influences and their cascading effects (Shakhsi-Niaei et al. 2011). Further, no analytical model seems sophisticated enough to represent the complexities and dynamics of *all* factors that might affect success or failure of a major project (Kavadias and Loch 2004; Zhang et al. 2009; Loch et al. 2001). These managers realize that while analytical methods provide a critically important toolset for project evaluation and selection, these methods also take the collective thinking and collaboration of all the stakeholders and key personnel of the enterprise and its partners to identify and deal with the complexity of risks in today's business environment. As a result, an increasing number of organizations are complementing their analytical methods with managerial judgment and collective stakeholder experiences that include such broad measures as strategic desirability and projections from past performances (Hadad et al. 2012), hence moving beyond a narrow dependence on just analytical models. In addition, many companies have developed their own 'systems,' uniquely designed for dealing with uncertainties in their specific projects and enterprise environment (Henriksen and Traynor 2002; Kavadias and Loch 2004; Kumar 2006; Larson and Gray 2011). These systems emphasize the integration of various tools, often combining quantitative and qualitative methods to cast a wider net for capturing and assessing risk factors beyond the boundaries of conventional methods. Examples are well-known management tools, such as review meetings, Delphi processes, brainstorming, and focus groups, which have been skillfully integrated with analytical methods to leverage their effectiveness and improve their reliability. In addition, a broad spectrum of new and sophisticated tools and techniques, such as user-centered design (UCD), voice of the customer (VoC), and phase-gate processes, evolved which rely by and large on organizational collaboration and collective judgment processes to deal with the broad spectrum of risk variables that are dynamically distributed throughout the enterprise and its external environment.

The missing link

While project evaluation and selection methods have been studied extensively for several decades (Brenner 1994; Cook and Green 2000; Mantel et al. 2011), relatively little has been published on the role of collaboration across the total enterprise for evaluating potential project value (Oral et al. 1991). That is, we know little about management processes that involve the broader project community in a collective cross-functional way for dealing with project selection. The missing link is the people side as a cross-functional tool for collectively evaluating a project proposal toward potential success, *one of the areas that are being examined in this paper.*

Method

An exploratory field study format is used to investigate managerial approaches for evaluating and selecting project proposals. Specifically, this field study looks at both quantitative and qualitative models, and reports on their use and effectiveness for various project types and situations.

The work presented in this paper is a continuation of my ongoing research into project management alignment with enterprise strategy (Shenhar et al. 2007) and risk

management in complex project situations (Thamhain and Skelton 2007; Thamhain 2011a,b). The field study reported here focuses especially on data collected between 2008 and 2012.

Data

The unit of analysis used in this study is the project. Data were captured from 25 technology organizations between 2008 and 2012, as part of an ongoing research in the area of technology-oriented product development and team-based project management. The field study yielded data from 43 project teams with a total sample population of 425 professionals such as engineers, scientists, and technicians, plus their managers, including 10 supervisors, 53 project team leaders, 12 product managers, 5 directors of R&D, 4 directors of marketing, and 11 general management executives at the vice presidential level or higher. Together, the data covered more than 100 projects in 18 companies, as summarized in Table 1. The projects involved mostly high-technology product/service/process developments with budgets averaging US$28 million each. All project teams saw themselves working in a high-technology environment. The 18 host companies are large technology-based multinational companies of the FORTUNE 500 category. The data were obtained from three sources, *questionnaires, participant observation,* and *in-depth retrospective interviewing,* as discussed in the previous section. Specifically, in stage three, 138 interviews were held with team leaders, line managers, product managers, marketing directors, and general management executives. These discussions provided interesting and useful insight into the cross-functional issues and challenges involved in project evaluation and selection processes in complex business environment. Content analysis has been used in addition to standard statistical methods for evaluating the survey data.

Table 1 Summary of field sample statistics

Project environment	Metrics
Total sample population	425
Companies	18
Business units	25
Projects (product developments)	105
Project teams	43
Team members[a]	397
Product managers	12
R&D managers	5
Senior managers and directors	11
Average project budget	US$28 million
Average project life cycle	18 months

[a]Team = total sample minus product managers, R&D managers and senior managers.

Justification for the exploratory field study format

All components of this investigation, such as project management, product development, team work, decision making under uncertainty, technology, and business environment, involve highly complex sets of intricately related variables. Researchers have consistently pointed at the non-linear, often random nature of these processes that involve many facets of the organization, its members, and the environment (Bstieler 2005; Danneels and Kleinschmidt 2001; MacCormack, Nellore and Balachandra 2001; Thamhain 2008, 2009; Verganti and Buganza 2005). Investigating these organizational processes simultaneously is not a simple task, making it unlikely to find simple models appropriate for researching these environments. Because of these complexities and the still evolving nature of these components, their theories and constructs, an *exploratory field research format* was chosen for the investigation which uses questionnaires in addition to two qualitative methods: participant observation and in-depth retrospective interviewing.

Defining project value, performance, and success

All of these measures are multifaceted. Typically, they include not only technical, but also financial, marketing, social, legal, and ethical dimensions. For most projects, the DNA of success is highly complex, and outcomes are difficult to predict, especially long term. Special attention must be given to the assessment of these variables, with their many facets and reference points throughout the project lifecycle and beyond. Especially, *project success* is highly judgmental and difficult to determine at the outset, but often only years after project completion, as we can see in aerospace, pharmaceutical, automobile, infrastructure or government service projects. Yet, in spite of these challenges, senior managers, collectively, seem to have a good sense of the potential success and value of a proposed project. We validated this fact in previous studies by testing the agreement among senior managers from various parts and different levels of the enterprise regarding their judgment of project performance (Kruglianskas and Thamhain 2000; Thamhain 2005, 2006). These agreements were measured via Kruskal-Walles analysis of variance by ranks which provided evidence of the strong agreement among the managers on the degree of success expected for a given project under evaluation. Therefore, I used this finding as a basis for arguing the benefits of collective multifunctional judgment in selecting candidate projects, a discussion presented later in this paper. Yet, the topic of 'measuring success' and its underlying metrics is an area that needs additional study. Suggestions for future research include the relationship between short-term project performance and long-term project success, and underlying conditions and evaluation criteria that might

be useful for predicting future project performance and success.

Results

Based on the findings from this field study, an overview is first given on how companies evaluate and select complex projects, followed by specific recommendations for effectively managing the process. The first segment includes a discussion on the effectiveness of some of the most popular quantitative and qualitative methods available to managers, and an assessment of the situational value of quantitative versus qualitative methods.

How do companies evaluate and select projects

Based on the findings from this field study, supported by additional observations on hundreds of projects during my action research, a framework for effective project evaluation and selection is summarized first and then followed by specific recommendations for effectively managing the process. While any decision model needs to be fine tuned toward specific application, the basic framework and management philosophy of evaluating and selecting project proposals seem to be appropriate for most projects, in spite of their differences in complexities, technologies, organizational structure, and culture among companies.

Evaluation dimensions and decision phases

Project opportunities must be analyzed relative to their potential value, strength, and importance to the enterprise. *Four major dimensions* should be considered for project evaluation and compared to available alternatives as graphically shown in Figure 1: (1) added value of the new project, consistent with the organizational objectives; (2) resource requests such as cost, personnel, and facilities needed to complete the new project; (3) readiness and ability of the enterprise to execute the project; and (4) managerial belief and desire. A well-organized *project evaluation and selection process* provides the framework for systematic data gathering and informed decision making toward resource allocation. Typically, the decision process can be broken into phases which are often overlapping and executed concurrently, such as shown in Figure 1 and described below:

Phase I. *Deciding initial feasibility*: screening and filtering, quick decision on the viability of an emerging project for further evaluation;
Phase II. *Deciding strategic value to enterprise*: identifying alternatives and options to the proposed project;
Phase III. *Deciding detailed feasibility and value*: determining the specific value and chances of success for a proposed project;

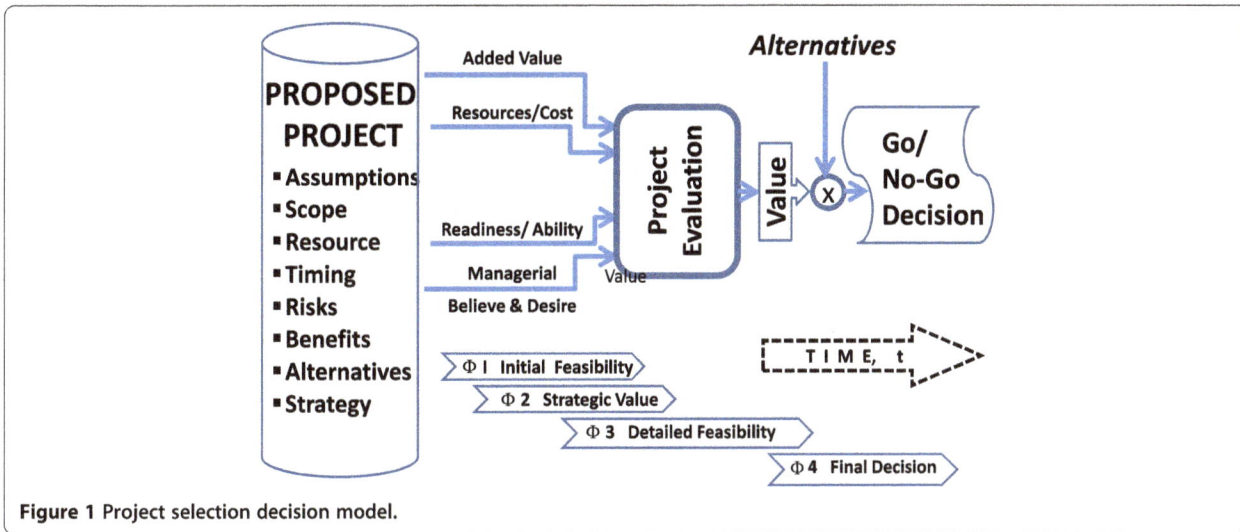

Figure 1 Project selection decision model.

Phase IV. *Deciding project go/no-go*: committing resources for a project implementation.

While the process seems simple, logical, and straightforward, developing meaningful support data is a complex undertaking. It is also expensive, time consuming, and often highly eclectic. Typically, decision making requires the following inputs: (1) specific resource requirements; (2) specific implementation risks; (3) specific benefits, i.e., economics, technology, markets, etc.; (4) benchmarking and comparative analysis of threats and opportunities; and (5) strategic perspective, including long- and short-term value assessment.

Estimating cost, schedules, risks, and benefits, such as those shown in Table 2, is always challenging. However, these estimates are relatively straightforward in comparison to *predicting project success*. The difficulty is in defining a meaningful *aggregate indicator for project value and success*. Methods for determining success range from purely intuitive to highly analytical. No method is seen as truly reliable in predicting success, especially for complex and technologically intensive projects. Yet, some companies have a better track record in selecting 'winning' projects than others. They seem to have the ability to create a more integrated picture of the potential benefits, costs, and risks for the proposed project relative to the company's strength and strategic objectives.

Methods of project evaluation and selection
Producing a comprehensive picture of future value that integrates the large array of potential costs and benefits is both a science and an art. During field interviews, managers confirm the changing paradigm of project evaluation that has been reported in the literature for

some time (Baker 2012; Kavadias and Loch 2004; Kumar 2006; Shakhsi-Niaei et al. 2011). That is, not too long ago, in the 1990s and before, *rational evaluation processes* prevailed by and large for supporting project selections. However, in today's more complex and dynamic business environment, managers point out that purely rational analytical processes apply only to a limited number of situations. Most of our technologically complex business scenarios require the integration of both analytical and judgmental techniques to capture the broad spectrum of variables affecting project success or failure, necessary for predicting success and making the *best choice*.

Table 2 Typical criteria considered for project evaluation and selection

The criteria relevant to the evaluation and selection of a particular project depend on the specific project type (i.e., product, service or process), business situation, industry, and market. Typically, evaluation procedures include the following criteria and measures:

• Cash flow, revenue, and profit	• Project cost
• Consistency with business plan	• Project duration (PLC)
• Cost-benefit	• Resource availability
• Impact on other business activities	• Return on investment
• Market share	• Risk
• Organizational readiness and strength	• Sales volume
• Product lifecycle (deliverables)	• Strategic value
• Project business follow on	• Technical complexity and feasibility

Each criterion is based on a complex set of parameters, variables, and assumptions.

Yet, in spite of the shift toward including judgmental methods more specifically into the selection process, systematic information gathering and standardized methods are at the heart of any project evaluation process and provide the best assurance for reliably predicting project outcome and repeatability of the decision process. In the sample of 105 projects of my field study, managers report the following breakdown of approaches used for project evaluation and selection:

- Primarily *quantitative* and *rational* approaches: used for 18% of all projects.
- Primarily *qualitative* and *intuitive* approaches: used for 24% of all projects.
- *Mixed approaches* (quantitative/qualitative combination): used for 58% of all projects.

Given the fact that projects of the field sample consisted of complex, largely high-technology projects, it might look surprising that a relatively large number of project evaluations are based on either primarily quantitative or qualitative methods. That is, one might expect more mixed approaches. Follow-up interviews provide additional insight. Of the sample's projects, 20% were very large undertakings of over US$100 million each, many of them defense contracts or alike. While these program evaluations included considerable qualitative inputs and judgment from focus groups, industry experts and management, the final go decision was strongly influenced by quantifiable data and highly rational methods of decision making. As a result, more than half of these 'very large' projects were judged as being selected by 'primarily quantitative and rational approaches'. Similarly, we found special situations in the 'primarily qualitative and intuitive' category. Approximately one half of the projects in this category were of R&D or exploratory nature with very limited quantifiable data, especially in the area of future value (i.e., future sales, revenue, market share, etc.). While all of these projects went through detailed technical feasibility, budgeting, and project planning processes, the final 'go decision' was based on primarily *qualitative* and *intuitive* approaches for approximately 60% of these projects.

However, regardless of the primary nature (quantitative, qualitative, or combined focus), all project evaluations included some quantitative and qualitative assessments. Equally important, we observed highly interactive interdisciplinary effort among the various resource groups of the enterprise and its partners during the evaluation and selection processes for all projects, regardless of their type, size, and classification. That is, while management may judge the process leading to the final go/no-go decision as 'primarily' one of the three approaches, all decisions included both quantitative and qualitative inputs to some degree. Often, many meetings were needed even before the formal project evaluation process started, just to gain a basic picture of potential benefits, costs, and risks involved, and to determine the type of data needed. All of these activities occurred regardless of primary evaluation/selection focus.

Yet, in all situations, there is the risk of relying too heavily on quantitative data. This was pointed out especially by senior managers. Most concerned were those managers and project leaders who lived through a disappointing project rollout or delivery, or outright project failure, after careful feasibility study and selection procedures at the front end. These managers observed in retrospect that many of their front-end decision processes relied too much on analytical data and models with assumptions that did not hold or forecasts that proved unreliable, while not casting a wide enough net for scanning the business environment. They felt that more experiential information had to be captured from senior management and technology, market and user communities, to complement and validate the analytical models. The issues and potential remedies to these challenges will be explored in the next section.

Going beyond simple formulas and quantitative methods

While quantitative methods of project evaluation have the benefit of producing relatively quickly a measure of merit for simple comparison and ranking, they also have many limitations, as summarized in Table 3. Yet, in spite of the limitations inherent to quantitative evaluation and the increased use of qualitative approaches, *virtually, every organization supports its project selections with some form of quantitative measures*; the most popular are ROI, cost-benefit, and payback period (see Appendix 1 and 2). However, driven by the growing complexity of the business environment, quantitative decision models are becoming less capable of capturing the full spectrum of variables associated with project success or failure. As pointed out by one senior manager in our field study, "Numbers like favorable ROI and projected revenue are great selling points toward project approval. However, often I'm not comfortable with the data. The numbers don't always seem to reflect market realities, customer reactions, changing technologies and dozens of other uncertainties and assumptions. Brainstorming the issues and defining critical success factors with experts from the areas critical, and getting inputs from user groups, often gives us a better sense of potential success or failure and how to plan for success that a cold number that is based on a complex financial model or projection that we don't really trust". These types of issues and inherent limitations of analytical methods concern managers across all industries, who started increasingly to augment quantitative methods and to explore alternatives. As observed in this field study and my own project work,

Table 3 Comparison of quantitative and qualitative approaches to project evaluation

Quantitative methods	Qualitative methods
Benefits	Benefits
Clear and simple comparison, ranking, and selection	Search for meaningful evaluation metrics
Repeatable process	Broad-based organizational involvement
Encourages data gathering and measurability	Understanding of problems, benefits, opportunities
Benchmarking opportunities	Problem solving as part of selection process
Programmable	Broadly distributed knowledge base
Useful input to sensitivity analysis and simulation	Multiple solutions and alternatives
Connectable to many analytical and statistical models	Multifunctional involvement leading to buy-in and risk sharing
Limitations	Limitations
Many success factors are not quantifiable	Complex, time-consuming process
Probabilities and weights may change	Biases introduced via organizational power and politics
True measures do not exist	Difficult to procedurize or repeat
Analyses and conclusions are often misleading	Conflict and disagreement over decision/outcome
Masking of hidden problems and opportunities	Does not fit conventional decision processes
Stifle innovative decision making	Intuition and emotion may obscure facts
Lack people involvement, buy-in, commitment	Used for justifying 'wants'
Ineffective in dealing with multifunctional issues, non-linearities and dynamic situations	Lead to more fact finding than decision making
May mask hidden costs and benefits	Temptation for unnecessary expansion of fact finding
Temptation for acting too quickly and prematurely	Process requires effective managerial leadership

and as a management consultant to hundreds of companies, there is an increasing trend for companies to supplement quantitative results with additional information for determining future cost-benefits and project success. Many of these contemporary decision making methods rely to a large degree on *qualitative, judgmental decision making* as summarized in Appendix 1. These data gathering methods cast a wider net and consider a broader spectrum of factors than those methods limited to quantitative measure. They can include fuzzy variables and decision parameters such as related to strategy and business ethics that may be difficult to describe or quantify, but are important in gaining overall perspective and a more comprehensive picture on potential benefits, risk, and challenges of the proposed project.

Discussion and recommendations

Effective evaluation and selection of project opportunities involve many variables of the organizational and technological environment, reaching often far beyond cost and revenue measures. While economic models provide an important dimension of the project selection process, most situations are too complex to use simple quantitative methods as the sole basis for decision making. Many of today's project evaluation procedures include a broad spectrum of variables and rely on a combination of rational and intuitive processes for defining the value of a new project venture to the enterprise.

The better organizations understand their business processes, markets, customers, and technologies, the better they will be able to evaluate the value, risks, and challenges of a new project venture. Further, manageability of the evaluation process is critical to its results, especially in complex situations. The process must have a basic structure, discipline, and measurability to be conducive to the intricate multivariable analysis. One method of achieving structure and manageability calls for grouping the evaluation variables into four categories: (1) consistency and strength of the project with the business mission, strategy, and plan; (2) multifunctional ability to produce the project deliverables and objectives, including technical, cost, and time factors; (3) success in the customer environment; and (4) economics, including profitability. Modern phase management, such as Stage-Gate® processes provide managers with the tools for organizing and conducting project evaluations in a systematic way. The following section summarizes suggestions that can help managers in effectively evaluating and selecting projects toward successful implementation.

Seek out relevant information

Meaningful project evaluations require relevant quality information. The four sets of variables related to the strategy, results, customer, and economics, as identified above, can provide a framework for establishing the proper metrics and detailed data gathering.

Ensure competence and relevancy

Ensure that the right people become involved in the data collection and judgmental processes.

Take top-down look first, detail comes later

Detail is less important than information relevancy and evaluator expertise. Do not get hung-up on missing data during the early phases of the project evaluation. Evaluation processes should be iterative. It does not make sense to spend a lot of time and resources on gathering perfect data to justify a 'no-go' decision.

Select and match the right people

Whether the project evaluation consists of a simple economic analysis or a complex multifunctional assessment, competent people from functions critical to the overall success of the project should be involved.

Define success criteria

Whether deciding on a single project or choosing among alternatives, evaluation criteria must be defined. They can be quantitative, such as ROI, or qualitative, such as the chances of winning a contract. In either case, these evaluation criteria should cover the true spectrum of factors affecting success and failure of the project(s). The success criteria should be identified by seasoned enterprise personnel. In addition, people from outside of the company, such as vendors, subcontractors, and customers, are often included in this expert group and critical to the development of meaningful success criteria.

Strictly quantitative criteria can be misleading

Be aware of evaluation procedures based on quantitative criteria only (ROI, cost, market share, MARR, etc.). The input data used to calculate these criteria are likely based on rough estimates and are often unreliable. Furthermore, a reliance on strictly quantitative data considers only a narrow spectrum of factors affecting project success or failure, thus ignoring many other important factors, especially those that influence project success in a dynamic or non-linear way, typical for many complex technologically sophisticated undertakings. Evaluations based on predominately quantitative criteria should at least be augmented with some expert judgment as a 'sanity check'.

Condense criteria list

Combine evaluation criteria, especially among the judgmental categories, to keep the list manageable. As a goal, try to stay within the 12 criteria for each category.

Gain broad perspective

The inputs to the project selection process should include the broadest possible spectrum of data from the business environment that affect success, failure, and limitations of the new project opportunity. Assumptions should be carefully examined.

Communicate across the enterprise

Facilitate communications among evaluators and functional support groups. Define the process for organizing the team and conducting the evaluation and selection process.

Ensure cross-functional representation and cooperation

People on the evaluation team must share a strategic vision across organizational lines. They also must have the desire to support the project if selected for implementation. The purpose, goals, objectives, and relationships of the project to the business mission should be clear to all parties involved in the evaluation/selection process.

Do not lose the big picture

As discussions go into detail during the evaluation, the team should maintain a broad perspective. Two global judgment factors can help focus on the big picture of project success: (1) overall cost-benefit perspective and (2) overall risk of failure assessment. These factors can be recorded on a ten-point scale, -5 to $+5$. This also leads to an effective two-dimensional graphic display for comparing competing project proposals.

Do your homework between iterations

Project evaluations are usually conducted progressively in iterative cycles. Therefore, the need for more information, clarification, and further analysis surfaces between each cycle. Necessary action items should be properly assigned and followed up to enhance the evaluation quality with each consecutive iteration.

Take a project-oriented approach

Plan, organize, and manage your project evaluation/ selection process as a *project*. Proposal evaluation and selection processes require valuable resources that must be justified and carefully managed.

Resource availability and timing

Do not forget to include in your selection criteria the availability and timing of resources. Many otherwise successful projects fail because they cannot be completed within a required time period.

Use red-team reviews

Set up a special review team of senior personnel. This is especially useful for large and complex projects with major impact on overall business performance. This review team examines the decision parameters, qualitative measures, and assumption used in the evaluation process.

Limitations, biases, and misinterpretations that may otherwise remain hidden can often be identified and dealt with.

Stimulate creativity and candor

Senior management should foster an innovative risk-shared ambience for the evaluation team. Especially, the evaluation of complex project situations involves intricate sets of variables. Criteria for success and failure are linked among many subsystems, such as organization, technology, and business, associated with a great deal of risks and uncertainty. Innovative approaches are required to evaluate the true potential of success for these projects. Risk sharing by senior management, recognition, visibility, and a favorable image in terms of high priority, interesting work, and importance of the project to the organization have been found strong drivers toward attracting and holding quality people on the evaluation team and toward gaining their active and innovative participation in the process.

Manage and lead

The evaluation team should be chaired by someone who has the trust, respect, and leadership credibility with the team members. Senior management can positively influence the work environment and the process by providing guidelines, charters, visibility, resources, and active support to the project evaluation team.

Conclusions

In summary, effective project evaluation and selection requires a broad scanning process across all segments of the enterprise and its environment to deal with the risks, uncertainties, ambiguities, and imperfections of data available for assessing the value of a new project venture relative to other opportunities. No single set of broad guidelines exists that guarantees the selection of successful projects. However, the process is not random! A better understanding of the organizational dynamics that affects project performance and the factors that drive cost, revenue, and other benefits can help in gaining a better, more meaningful insight into the future value of a prospective new project. Seeking out both quantitative and qualitative measures incorporated into a combined rational judgmental evaluation process often yields the most reliable predictor of future project value and desirability. Equally important, the process requires managerial leadership and skills in planning, organizing, and communicating. Above all, the leader of the project evaluation team must be a social architect who can unify the multifunctional process and its people. The leader must be able to foster an environment, professionally stimulating and conducive to risk sharing. It also must be effectively linked to the functional support groups needed for project implementation. Finally, organizational

strategy must be aligned and integrated with the evaluation/selection process, early and throughout its evaluation cycle. Senior management has an important role in unifying the evaluation team behind the mission objectives and in facilitating the linkages to the stakeholders and ultimate user community. Senior management should further help in providing overall leadership and in building mutual trust, respect, and credibility among the members of the proposal evaluation team, all critical drivers toward a strong partnership of all team members and the basis for an effective enterprise-wide decision-making system. Taken together, this is the environment conducive for cross-functional communication, cooperation, and integration of the intricate variables needed for effective evaluation and selection of project proposals in complex business environments.

Appendices
Appendix 1
Summary of project evaluation and selection techniques

Some of the popular project evaluation and selection tools, techniques, and approaches are summarized in this Appendix, grouped into three classes:

1. Primarily *quantitative* and *rational* approaches.
2. Primarily *qualitative* and *intuitive* approaches.
3. *Mixed approaches*, combining both quantitative and qualitative methods.

Quantitative approaches to project evaluation and selection

Quantitative approaches are often favored to support project evaluation and selections if the decisions require economic justification. They are also commonly used to *support* judgment-based project selections. One of the features of quantitative approaches is the generation of numeric measures for simple and effective comparison, ranking, and selection. These approaches also help establish quantifiable norms and standards and lead to repeatable processes. Yet, the ultimate usefulness of these methods depends on the assumption that the decision parameters, such as cash flow, risks, and the underlying economic, social, political, and market factors, can actually be quantified and reliably estimated over the project life cycle. Therefore, quantitative techniques are effective and powerful decision support tools, if meaningful estimates of cost-benefits, such as capital expenditures and future revenues, can be obtained and converted into net present values for comparison. Because of their importance, quantitative methods have been discussed in the literature extensively, ranging from simple return on investment (ROI) calculations to elaborate simulations of project scenarios. Many companies eventually developed their own project evaluation/selection models, customized to their specific needs. However, the backbone for most of these customized models is a set of economic/financial measures which tries to determine the cost-benefit of the proposed venture, usually for some point in the future. Specifically, five measures are especially popular:

1. Net present value (NPV)
2. ROI
3. Cost-benefit (CB)
4. Payback period (PBP)
5. Pacifico and Sobelman project ratings

The calculation and application of these measures to project evaluation/selection will be illustrated by case examples. Specifically, four project proposals (described in Table 4) will be evaluated in this chapter, using the above measures. The results are summarized in Table 5.

Net present value comparison

This method uses discounted cash flow as the basis for comparing the relative merit of alternative project opportunities. It assumes that all investment costs and revenues are known and that economic analysis is a valid basis for project selection. We can determine the NPV of a single revenue, or stream of future revenues, or costs expected in the future. Two types of presentations are common: (1) present worth (PW) and (2) net present value.

Present worth

This is the single revenue or cost (also called annuity A) which occurs at the end of a period n, subject to the *prevailing interest rate i*. Depending on the management philosophy and enterprise policies, this interest rate can be (1) the *internal rate of return (IRR)* realized by the

company on similar investments or (2) the *minimum attractive rate of return (MARR)* acceptable to company management, or the prevailing discount rate. The present worth is calculated as

$$\mathrm{PW}(A \mid i, n) = \mathrm{PW_n} = A \frac{1}{(1+i)^n}.$$

For the examples used in this chapter, we consider the IRR (defined as the average return realized on similar investments) to be the prevailing interest rate.

Net present value

The *net present value* is defined as a series of revenues or costs, A_n, over N periods of time at a prevailing interest rate i:

$$\mathrm{NPV}(A_n \mid i, N) = \sum_{n=1}^{N} A_n \frac{1}{(1+i)^n} = \sum_{n=1}^{N} \mathrm{PW}_n.$$

Three special cases exist for the net present value calculation: (1) *for a uniform series of revenues or costs* over N periods, $\mathrm{NPV}(A_n \mid i, N) = A[(1+i)^{N-1}]/i(1+i)^N$; (2) *for an annuity or interest rate i approaching zero*, NPV $= A \times N$; and (3) *for the revenue or cost series to continue forever*, NPV $= A/i$. Table 5 applies these formulas to the four project alternatives described in Table 4, showing the most favorable 5-year net present value of US$3,192 for project option P3.

Return on investment comparison

Perhaps one of the most popular measures for project evaluation is the ROI:

$$\mathrm{ROI} = \frac{\mathrm{Revenue\ (R) - Cost\ (C)}}{\mathrm{Investment\ (I)}}.$$

ROI calculates the ratio of net revenue over investment. In its simplest form, the stream of cash flow is *not* discounted. One can look at the revenue on a year-by-year basis, relative to the initial investment. For example, project option 1 in Table 3 would produce a 20% ROI each year, while project option 2 would produce a 75% ROI during the first year, 50% during the second year, and so on. In a somewhat more sophisticated way, we can calculate the *average ROI per year* over a given revenue cycle as shown in Table 3:

$$\mathrm{R\overline{O}I}(A_n, I_n \mid N) = \left[\sum_{n=1}^{N} \frac{(\mathrm{Revenue} R)_n - (\mathrm{Cost} C)_n}{(\mathrm{Investment} I)_n} \right] / [N].$$

We can then *compare the average ROI to the MARR*. Given a MARR of 10% for our project environment, all three project options P1, P2, and P3 compare favorable, with project P3 yielding the highest average return on investment of 54%. Although this is a popular measure,

Table 4 Description of four project proposals

Project proposal	Description
Project option P1	Management does not accept any new project proposal. Hence, neither investment capital is required nor is any revenue generated.
Project option P2	This opportunity requires a US$1,000 investment at the beginning of the first year and generates a US$200 revenue at the end of *each* of the following 5 years.
Project option P3	This opportunity requires a US$2,000 investment at the beginning of the first year and generates a variable stream of net revenues at the end of *each* of the next 5 years as follows: US$1,500, US$1,000, US$800, US$900, and US$1,200.
Project option P4	This opportunity requires a US$5,000 investment at the beginning of the first year and generates a variable stream of net revenues at the end of *each* of the next 5 years as follows: US$1,000, US$1,500, US$2,000, US$3,000, and US$4,000.

Table 5 Cash flow and net value calculations of four project options or proposals

	Given cash flow			
	Do-nothing option (P1)	Project option (P2)	Project option (P3)	Project option (P4)
End of year				
0	0	−1,000	−2,000	−5,000
1	0	200	1,500	1,000
2	0	200	1,000	1,500
3	0	200	800	2,000
4	0	200	900	3,000
5	0	200	1,200	4,000
Calculations				
Net cash flow ($\Sigma \cdot P$)	0	0	+3,400	+6,500
Net present value at the end of year 5 ($\text{NPV}|_{N\,=\,5}$)	0	−242	+2,153	+3,192
Net present value for revenue to continue ∞ ($\text{NPV}|_{N\,=\,\infty}$)	0	+1,000	+9,904	+28,030
Average annual return on investment ($\text{ROI}|_{N\,=\,5}$)	0	20%	54%	46%
Cost Benefit ($\text{CB} = \text{ROI}_{\text{NPV}|N\,=\,5}$)	0	76%	108%	164%
Payback period for MARR = 10% ($N_{\text{PBP}}|_{i\,=\,10}$)	0	8	1.8	3.8
Payback Period for MARR = 0% ($N_{\text{PBP}}|_{i\,=\,0}$)	0	5	1.5	3.3

MARR of $i = 10\%$ is assumed. Given for all four project proposals, (1) a single investment is being made at the beginning of the project life cycle (e.g., at the end of year 0), and (2) the internal rate of return (IRR) or the minimum attractive rate of return (MARR) is 10%.

it does not permit a meaningful comparative analysis of alternative projects with fluctuating costs and revenues. Furthermore, it does not consider the time value of money.

Cost-benefit

Alternatively, we can calculate the *net present value* of the total ROI over the project lifecycle. This measure, known as *cost-benefit* (CB) is calculated as the present value stream of net revenues divided by the present value stream of investments. It is an effective measure for comparing project alternatives with fluctuating cash flows:

$$\text{CB} = \text{ROI}_{\text{NPV}}(A_n, I_n | i, N) = \frac{\left[\sum_{n=1}^{N} NPV(A_n | i, N)\right]}{\left[\sum_{n=1}^{N} NPV(I_n | i, N)\right]}.$$

In our example of four project options (Table 3), project proposal P4 produces the highest cost-benefit of 164% under the given assumption of $i = \text{MARR} = 10\%$.

Payback period comparison

Another popular figure of merit for comparing project alternatives is the PBP. It indicates the time period of net revenues required to return the capital investment

made on the project. For simplicity, *undiscounted* cash flows are often used to calculate a quick figure for comparison, which is quite meaningful if we deal with an initial investment and a steady stream of net revenue. However, for fluctuating revenue and/or cost steams, the net present value must be *calculated for each period individually* and cumulatively added up to the 'break-even point' in time, N_{PBP}, when the net present value of revenue equals the investment. Mathematically,

$$N_{\text{PBP}} \text{ occurs when } \sum_{n=1}^{N} \text{NPV}(A_n | i) \geq \sum_{n=1}^{N} \text{NPV}(I_n | i).$$

In our example of four project options (Table 3), project proposal P3 produces the shortest, most favorable payback period of 1.8 years under the given assumption of $i = \text{MARR} = 10\%$.

Pacifico and Sobelman project ratings

The previously discussed methods of evaluating projects rely heavily on the assumption that technical and commercial success is assured, and all costs and revenues are predicable. Because these assumptions do not always hold, many companies have developed their own special procedures and formulas for comparing project alternatives. Two examples illustrate this special category of project evaluation metrics.

The project rating factor

This measure was originally developed by Carl Pacifico for assessing chemical products and predicting commercial success:

$$\mathrm{PR} = \frac{pT \times pC \times R}{TC}.$$

Pacifico's formula is in essence an ROI calculation adjusted for risk. It includes probability of technical success $(0.1 < pT < 1.0)$, probability of commercial success $(0.1 < pC < 1.0)$, total net revenue over project lifecycle (R), and total capital investment for product development, manufacturing set-up, marketing, and related overheads (TC).

Product development figure of merit

The formula developed by Sobelman

$$z = (P \times T_{\mathrm{LC}}) - (C \times T_{\mathrm{D}})$$

represents a modified cost-benefit measure which takes into account both the development time and commercial lifecycle of the product. It also includes average profit per year (P), estimated product lifecycle (T_{LC}), average development cost per year (C), and years of development (T_{D}).

Qualitative approaches to project evaluation and selection

While quantitative methods provide an important toolset for project proposal evaluation and selection, there is also a growing sense of frustration, especially among managers of complex and technologically advanced undertakings, that reliance on strictly quantitative methods does not always produce the most useful or reliable inputs for decision-making, nor are all methods equally suited for all situations. Therefore, it is not surprising that for project evaluations involving complex sets of business criteria, narrowly focused quantitative methods are often supplemented with broad scanning, intuitive processes, and collective, multifunctional decision making such as Delphi, nominal group technology, brainstorming, focus groups, sensitivity analysis, benchmarking, and UCD. Each of these techniques can either be used by itself to determine the *best, most successful, or most valuable* option, or these techniques are integrated into a comprehensive analytical framework for *collective multifunctional decision making*, which is being discussed next.

Collective, multifunctional evaluations

This process relies on subject experts from various functional areas for collectively defining and evaluating broad project success criteria, employing both quantitative and qualitative methods. *The first step* is to define the specific organizational areas critical to project success and to assign expert evaluators. For example, a product or service development project may typically include organizations such as R&D, engineering, testing, manufacturing, marketing, product assurance, and customer/field services. The function experts should be given the time and resources necessary for the evaluation. They also should have the commitment from senior management for full organizational support. Ideally, these evaluators should be members of the core team ultimately responsible for project implementation.

Evaluation factors

Early in the evaluation process, the team defines the factors which appear critical to the ultimate success of the projects under evaluation and arranges them into a list which includes both quantitative and qualitative factors. A mutually acceptable scale must be worked out for scoring the evaluation criteria. Studies of collective multifunctional assessment practices show that simple scales are most effective for leading to actionable team decisions. The four most popular and robust scales for judging situational outcomes are shown below:

1. *Ten-point judgment scale*: This scale ranges from +5 (most favorable) to – 5 (most unfavorable).
2. *Three-point judgment scale:* +1 (favorable), 0 (neutral or cannot judge), –1 (unfavorable).
3. *Five-point judgment scale:* A (highly favorable), B (favorable), C (marginally favorable), D (most likely unfavorable), F (definitely unfavorable).
4. *Five-point Likert scale:* 1 (strongly agree), 2 (agree), 3 (neutral), 4 (disagree), 5 (strongly disagree).

Weighing of criteria is *not* recommended for most applications as it complicates and often distorts the collective evaluation. Perspective and judgment are part of the strength and value of these qualitative methods, which can be lost by forcing too much of a quantitative framework on the qualitative evaluation.

The evaluation process

Evaluators first assess and then score all of the success factors they feel qualified to judge. Then, collective discussions follow. Initial discussions of project alternatives, their markets, business opportunities, and technologies involved are usually beneficial, but not necessary for the first round of the evaluation process. The objective of this first round of expert judgments is to get calibrated on the opportunities and challenges presented. Further, each evaluator has the opportunity to recommend (1) actions that could improve the quality and accuracy of the project evaluation, (2) additional data needed, and (3) suggestions for increasing project success. Before meeting at the next group session, agreed-on action items and activities for improving the decision process should be completed. The evaluation process is enhanced with each iteration by producing more accurate, refined, and comprehensive data. Typically, between 3 and 5 iterations are required before a go/no-go decision can be reached for a given project.

Mixed approaches, combining both quantitative and qualitative methods

Mixed approaches are the most common method of evaluating and selecting projects in today's complex business environment. Virtually all evaluations of project proposals include some form of quantitative *and* qualitative methods. However, to qualify as a mixed approach, the project evaluation and selection process has to contain a fairly balanced array of both classes of analytical and judgmental tools and techniques of the various types discussed above.

Appendix 2

The following is a summary description of terms, variables, and abbreviations used in this chapter:

- Cross-functional involves actions which span organizational boundaries.
- Phase management involves breaking of projects into natural implementation phases, such as development, production, and marketing, as a basis for project planning, integration, and control. Phase management also provides the framework for *concurrent engineering* and *Stage-Gate® processes*.
- Project success is a comprehensive measure, defined in both quantitative and qualitative terms which includes economic, market, and strategic objectives.
- Stage-Gate® process is a framework originally developed by R Cooper and S Edgett for executing projects within predefined stages with measurable deliverables (*at gates*) at the end of each stage. These gates also provide the review metrics for ensuring successful transition and integration of the project into the next stage.
- Weighing of criteria is a a multiplier associated with specific evaluation criteria.
- Annuity (*A*) is the present worth of a revenue or cost at the end of a period *n*.
- Cost benefit (CB) is the net present value of all ROIs in dollars.
- Prevailing interest rate (*i*).
- Investment (*I*).
- Internal rate of return (IRR) is the average return on investment realized by a firm on its investment capital.

- Minimum attractive rate of return (MARR) on new investments acceptable to an organization.
- Net present value (NPV) of a stream of future revenues or costs.
- Payback period (PBP) is the time period needed to recover the original investment.
- Project rating factor (PR) is a measure developed by Carlo Pacifico for predicting project success.
- Present worth (PW, also called annuity) is the present value of a revenue or cost at the end of a period *n*.
- Return on investment (ROI)
- Project rating factor (*z*) is a measure developed by Sobelman for predicting project success.

Competing interests
This author declares that he has no competing interests.

Authors' information
Dr. Hans Thamhain is a Professor of Management and Director of MOT and Project Management Programs at Bentley University, Boston/Waltham. Dr. Thamhain, held management positions with Verizon, General Electric and ITT, has written over seventy research papers and six professional reference books. He received the IEEE Engineering Manager Award in 2001, PMI's Distinguished Contribution Award in 1998 and PMI's Research Achievement Award in 2006. He is profiled in *Marquis Who's Who in America* and certified as NPDP and PMP.

References
Baker NR (2012) R&D project selection models: an assessment. R&D Manag 5(1):105–111

Brenner M (1994) Practical R&D project prioritization. Res Technol Manag 37(5):38–42

Bstieler L (2005) The moderating effects of environmental uncertainty on new product development and time efficiency. J Prod Innov Manag 23(3):267–284

Cicmil S, Williams T, Thomas J, Hodgson D (2006) Rethinking project management: researching the actuality of projects. Int J Proj Manag 24(8):675–686

Cook WD, Green RH (2000) Project prioritization: a resource-constrained data envelopment analysis approach. Socioecon Plann Sci 34(2):85–99

Danneels E, Kleinschmidt EJ (2001) Product innovativeness from the firm's perspective. J Prod Innov Manag 18(6):357–374

El Emam K, Koru A (2008) A replicated survey of it software project failures. Software (IEEE) 25(5):84–90

Gulla S (2012) Seven reasons why IT projects fail. IBM Systems Magazine. http://www.ibmsystemsmag.com/mainframe/tipstechniques/applicationdevelopment/project_pitfalls. Last accessed 22 Nov 2013

Hadad Y, Keren B, Laslo Z (2012) A decision-making support system module for project manager selection according to past performance. Int J Proj Manag 31(4):532–541

Henriksen AD, Traynor AJ (2002) A practical R&D project-selection scoring tool. IEEE Trans Eng Manag 46(2):158–170

Kavadias S, Loch CH (2004) Project selection under uncertainty: dynamically allocating resources to maximize value. Kluwer, Norwood, MA, USA

Kruglianskas I, Thamhain H (2000) Managing technology-based projects in multinational environments. IEEE Trans Eng Manag 47(1):55–64

Kumar PD (2006) Integrated project evaluation and selection using multiple-attribute decision-making technique. Int J Prod Econ 103(1):87

Larson E, Gray C (2011) "Organization strategy and project selection," Chapter 2 in *Project Management: The Management Process*. McGraw-Hill, New York, pp 22–63

Lemon WF, Bowitz J, Burn J, Hackney R (2002) Information systems project failure: a comparative study of two countries. J Glob Inf Manag 10(2):28–39

Loch CH, Pich MT, Terwiesch C, Urbschat M (2001) Selecting R&D projects at BMW: a case study of adopting mathematical programming models. IEEE Trans Eng Manag 48(1):70–80

Mantel S, Meredith J, Shafer S, Sutton M (2011) "Selecting projects to meet organizational objectives," Chapter Section 1.5 in *Project Management Practice*. Wiley, Hoboken, NJ, USA, pp 10–22

National Science Foundation (NSF) (2010) The 2010 user-friendly handbook for project evaluation. Division of Research and Learning in Formal and Informal Settings. National Science Foundation, Arlington, VA, USA

Oral M, Kettani O, Lang P (1991) A methodology for collective evaluation and selection of industrial R&D projects. Manag Sci 37(7):871–881

Remer DS, Re Stokdyk SB, Van Driel M (1993) Survey of project evaluation techniques currently used in industry. Int J Prod Econ 32(1):103–115

Shakhsi-Niaei M, Torabi S, Iranmanesh S (2011) A comprehensive framework for project selection problem under uncertainty and real-world constraints. Compt Ind Eng 61(1):226–237

Shenhar A, Milosevic D, Dvir D, Thamhain H (2007) Linking project management to business strategy. Project Management Institute (PMI) Press, Newtown, PA, USA

Shore B (2008) Systematic biases and culture in project failures. Proj Manag J 39(4):5–16

Standish Group (2013) Chaos Tuesday. http://blog.standishgroup.com/. Last accessed 22 Nov 2013

Thamhain H (2005) Team leadership effectiveness in technology-based project environments. IEEE Eng Manag Rev 33(2):11–25

Thamhain H (2006) Optimizing innovative performance of R&D teams in technology-based environments. Creativity Res J 18(4):435–436

Thamhain H (2008) Team leadership effectiveness in technology-based project environments. IEEE Eng Manag Rev 36(1):165–180

Thamhain H (2009) Leadership lessons from managing technology-intensive teams. Int J Innov Technol Manag 6(2):117–133

Thamhain H (2011a) Critical success factors for managing technology-intensive teams in the global enterprise. Eng Manag J 23(2):30–36

Thamhain H (2011b) Evaluating and selecting technology-based projects. In: Engineering measurements encyclopedia. Edited by Kutz M. Wiley, New York

Thamhain H, Skelton T (2007) Success factors for effective R&D risk management. Int J Technol Intell Plann (IJTIP) 3(4):376–386

Verganti R, Buganza T (2005) Design inertia: designing for life-cycle flexibility in internet-based services. J Prod Innov Manag 22(3):223–237

Zhang D, Zhang J, Lai KK, Lu Y (2009) An novel approach to supplier selection based on vague sets group decision. Expert Syst Appl 36(5):9557–9563

14

An empirical study of innovation-performance linkage in the paper industry

Parveen Farooquie[*], Abdul Gani, Arsalanullah K Zuberi and Imran Hashmi

Abstract

To enter new markets and remain competitive in the existing markets, companies need to shift their focus from traditional means and ways to some innovative approaches. Though the paper industry in India has improved remarkably on its technological and environmental issues, yet it shows a low rate of innovation. The present paper attempts to review the industry in the perspective of technological innovations and investigates empirically the role of innovations in performance improvement and pollution control. Multivariate analysis of variance and discriminant function analysis are applied for data processing. The findings reveal that the mean scores on the factors, such as sales, quality, and flexibility, are higher for the good innovators than those for the poor innovators. Conversely, the factors which are likely to be reduced as a result of innovations, such as time, cost, emissions, and disposal of waste, have shown higher means for the poor innovators.

Keywords: Discriminant function analysis, Paper industry, Performance, Multivariate analysis of variance, Technological innovation

Background

The basic philosophy behind any business organization is to produce the intended products and sell them to earn profit and satisfy customer requirements. Profitability and other such targets can be achieved, maintained, and excelled only when the performance is regularly measured and monitored. Organizations design their own systems of performance evaluation depending on the environment they work in and the nature of their operations. According to Fitzgerald et al. (1991), the framework for measuring the performance of any organization should integrate the measures that relate to results, such as competitiveness, and those that focus on the determinants of the results, such as quality and innovation. To enter new markets and remain competitive in the existing markets, companies need to shift their focus from traditional means and ways to some innovative approaches. Various studies have been reported in the literature to draw the attention of academicians and professionals towards the role of technological and non-technological innovations in performance improvement. According to the *Oslo Manual*, innovation means implementation of a new or significantly improved product, process, marketing method, or organizational method in business practices and organizations. Innovations in technical specifications, materials, and characteristics, of a product are said to be product innovation, whereas innovations pertaining to technique, process, and equipment are known as process innovation. The product and process types of innovations are together known as technological innovations and abbreviated as TPP. The non-technological category includes innovations related to marketing and organizational practices (OECD 2005).

With an annual output of over 6 million tons and an estimated turnover of US$3,400 million, the Indian paper industry is continuously progressing towards a projected demand of 8 million tons of paper in the year 2010 and 13 million tones by 2020 (Paper industry in India 2009). Though the industry has improved remarkably on its technological and environmental issues, yet it shows a low rate of innovation. Challenges such as pulp quality variation, high consumption of water and energy, raw material cleaning and storage, flexibility to reduce the multiplicity of paper grades, use of forest resources, water and air pollution, and production of solid waste are still alive.

* Correspondence: parveenfarooquie@yahoo.com
Department of Mechanical Engineering, Aligarh Muslim University (AMU), Aligarh, India

Paper looks simple compared to, for example, computers and mobile phones, but a lot of technology is involved in the pulp and paper process. Also, as technology develops, paper can become much more than what it is today (Karlsson 2009). Karlsson (2009) quotes a professor of Fibre and Cellulose Technology at Åbo Akademi University, 'It's difficult to say why the industry isn't that attractive any more. Perhaps young people find paper to be something old-fashioned and boring and think that its development is complete, but this is not the case at all. There is still a lot to explore, and we are in an exciting phase when innovative thinking is needed to take the industry forward. Now, more than ever, we need talented people.'

The present paper attempts to review the industry in the perspective of technological innovations. The study investigates empirically how such innovations contribute in performance improvement and pollution control.

Literature review

Neely et al. (1995) have defined performance measurement as quantification of effectiveness and efficiency. They have discussed important performance measures relating to cost, quality, time, and flexibility. Since there are numerous dimensions of each one of these performance determinants, researchers have applied them differently. Kaplan and Norton (1992) have a balanced scorecard approach of performance measurement which suggests four perspectives of any organization to be considered - financial (e.g., sales), internal business (e.g., flexibility), customer (e.g., quality, cost, and time), and innovation and learning (e.g., ability to innovate). Gomes et al. (2006) have identified 25 performance factors in manufacturing. They categorized them into six dimensions through factor analysis - operational responsiveness, market-related, costumer-driven, quality orientation, employees' involvement, working conditions, and innovation.

Innovation has emerged as a key to success for companies which want to remain competitive in the market or enter a new market. Researchers have been studying various issues pertaining to innovations in different industries such as pattern of innovation (Pavitt 1984; Freel 2005), determinants and measures of innovation (Wan et al. 2005; De Jong and Vermeulen 2006), process of innovation (Nieto 2004), manufacturing strategies and innovation performance (Prajogo et al. 2007), and impact of innovation on companies' performance (Lin and Chen 2007; Mansury and Love 2008). Manufacturing sector has particularly witnessed a positive relationship between firm innovation and its performance (Loof and Heshmati 2002). The study by Lin and Chen (2007) is about the role of innovation in performance gain in the context of small and medium enterprises (SMEs) in Taiwan. Their findings include that the majority of the sample companies have

done innovation in some form or the other. They also conclude that two types of innovation - technological and marketing - were more prominent than other types. Though the effect of innovation activities has been positive on the overall performance, yet they do not contribute strongly to the company sales. Mansury and Love (2008) have conducted a study in the service sector in the USA. Their findings also indicate a positive relationship between innovation and performance with productivity as an exception.

Results and discussion

The responding companies consist of manufacturers of paper, suppliers of raw material or equipments, and those which may be categorized as both supplier as well as manufacturer. The sample companies have been labeled as small or large based on the number of their full-time permanent employees. Companies with 100 or less employees are called small, and the ones with more than 100 are referred to as large.

Company innovativeness

Compatible with the *Oslo Manual* of OECD (2005) and based on the evidence from the literature (Wan et al. 2005; De Jong and Vermeulen 2006), a set of six questions was developed to measure the magnitude and novelty of innovation commercialized by the firms. The questions ask about the number of technologically new or improved processes or products (raw material in case of suppliers) implemented or introduced by a firm first time to itself or to the industry during the last three years (April 2006 to March 2009).

For computation convenience, an equal weightage has been assigned to each item, and hence, the simple average of the scores of a company on these six items represents its technological innovativeness (INNO). Such scores have ranged between 0.33 and 5.83 with a mean of 3.78. For further analysis, the companies with average score of 3.78 or less are designated as *poor innovators* (denoted by numeral 1); those with higher values, as *good innovators* (denoted by numeral 2). As shown in Figure 1, around 62% of the manufacturers have been assessed as good innovators, whereas there are more poor innovators (55%) in the *others* category. (Those who are suppliers-cum-manufacturers have been grouped with suppliers to form a single category, *others*). It could be understood that due to their nature of operations, the manufacturers might have scored higher than the suppliers on product and process innovation. This appears as one strong reason for them to outperform the *others* on the overall innovativeness.

Technological innovation and performance

Another set of questions measures the effect of innovations on company sales (SALE) and on other important determinants of company performance such as production time (TIME), production cost (COST), production flexibility (FLEX), and production quality (QUAL) (Neely et al. 1995; Kaplan and Norton 1992). Since the study deals with the paper industry, environmental factors such as emission of hazardous fumes (EMSN) and disposal of solid waste (DISP) are also included in the study as another two important (dependent) variables. This is evidenced from the literature that researchers have examined innovation as a dependent (Wan et al. 2005) as well as an independent variable (Lin and Chen 2007; Mansury and Love 2008). The present paper studies innovation as an independent variable.

Respondents were requested to report the effect of technological innovations, which they have introduced during the mentioned period of three years, on these performance parameters. A five-point Likert scale (Sawang 2006) ranging from improved significantly through improved moderately, no effect, worsened moderately to worsened significantly was proposed to them for this purpose. Bivariate correlation analysis is then run on the data obtained on company innovativeness.

Table 1 shows that whether performance is in the form of outputs (such as sales), production factors (such as production time, cost, flexibility, and quality) or environmental hazards (such as emissions and disposal of wastes), innovativeness is likely to have its impact on it. Except for production flexibility, all performance indicators have a significant correlation with the company's innovativeness. Sales, the most commonly used measure of performance, shows a highly significant positive relationship with innovative-ness. The ability of a company

to quickly incorporate the changes required in volume and deign (flexibility) and improve the quality has been found positively linked with how innovative the company is. However, the strength of relationship between innovativeness and flexibility is not significant. Since greater flexibility is more a factor of process innovation, this insignificant correlation may be explained by examining the companies' scores on process and product innovations separately. This argument is also supported by the literature, which reports that considering product and process innovation together or separately does influence the effect of innovation on performance (Michie and Sheehan 2003). The correlation coefficients also indicate that the greater the company innovativeness, the lower is the production time, cost, emission, and waste. This may be inferred that innovative companies not only perform better in competitive terms, but also are less harmful to the environment.

To further investigate the relationship between innovation and performance, MANOVA is carried out with company innovativeness as the predictor variable, whereas changes occurred due to innovation in the sales, time, cost, quality, flexibility, emissions, and waste production as dependent variables. The results are shown in Tables 2 and 3.

Descriptive statistics indicate that the mean scores on the factors such as sales, quality, and flexibility are higher for the good innovators than those for the poor innovators. Conversely, the factors which are likely to be reduced as a result of innovations, such as time, cost, emissions, and disposal of waste, have shown higher means for the poor innovators. This can also be observed from the standard deviation column that good innovators have been more consistent than their poor counterparts, except in cases of quality and flexibility, where wider dispersions are reported for the good innovators. The overall situation signals that innovation is likely to affect all variables positively.

The Box's test of equality of covariance matrices is conducted to validate the assumption of homogeneity before proceeding further. The result (Box's $M = 36.229$; $p = .530$) suggests that the assumption is valid, and hence, the multivariate tests (Table 3) are reliable (Field 2005). The *significance* column and *INNO* row of Table 3 indicates that innovations have a significant effect on performance. However, the results do not tell anything in detail. To investigate this effect with reference to the individual variables and their combinations, DFA is applied. Since there are only two groups (good innovators and poor innovators) involved in this analysis, there has to be a single discriminant function variate. The initial outcome of the DFA reveals that this variate is significant (Wilk's lamda = .361; $p = .000$). Finally, the structure

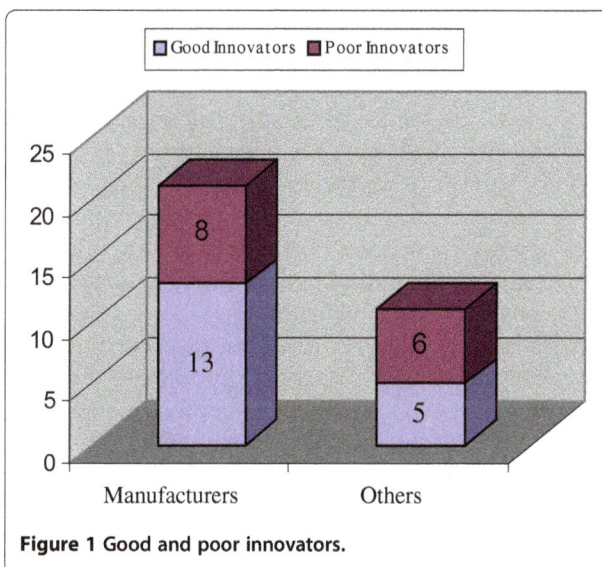

Figure 1 Good and poor innovators.

Table 1 Correlation matrix

Factors		INNO	TIME	COST	DISP	EMSN	SALE	FLEX	QUAL
INNO	Correlation	1	−.812[*]	−.779[*]	−.547[*]	−.670[*]	.837[*]	.111	.402[**]
	Significance		.000	.000	.001	.000	.000	.545	.023
	N	32	32	32	32	32	32	32	32
TIME	Correlation		1	.680[*]	.632[*]	.673[*]	−.716[*]	−.126	.296
	Significance			.000	.000	.000	.000	.492	.100
	N		32	32	32	32	32	32	32
COST	Correlation			1	.496[*]	.474[*]	−.668[*]	−.122	−.414[**]
	Significance			32	.004	.006	.000	.505	.018
	N				32	32	32	32	32
DISP	Correlation				1	.370[**]	−.456[*]	−.074	−.171
	Significance					.037	.009	.687	.351
	N				32	32	32	32	32
EMSN	Correlation					1	−.656[*]	−.047	−.192
	Significance						.000	.800	.292
	N					32	32	32	32
SALE	Correlation						1	.122	.403[**]
	Significance							.507	.022
	N						32	32	32
FLEX	Correlation							1	.196
	Significance								.283
	N							32	32
QUAL	Correlation								1
	Significance								
	N								32

*Correlation is significant at the 0.01 level (2-tailed); **Correlation is significant at the 0.05 level (2-tailed). *COST* production cost, *DISP* disposal of waste, *EMSN* hazardous emissions, *FLEX* production flexibility, *INNOV* technological innovativeness, *QUAL* production quality, *SALE* sales, *TIME* production time.

matrix (Table 4) is obtained, which explains the relationship between the dependent variables and the variate. The values (canonical variate correlation coefficients) in this matrix indicate the relative contribution of each variable (and its direction, positive or negative) in differentiating the two groups, poor innovators and good innovators, from each other.

Conclusions

In this study, a few selected manufacturers and suppliers belonging to the paper industry are examined for their status on innovativeness. According to the defined measure of technological innovativeness, manufacturers seem to be more innovative than suppliers in the industry. The authors have considered seven, most commonly used in literature, determinants of performance to investigate the effect of innovation on them. The MANOVA results in a significant difference between the overall performance of the good innovators and that of the poor innovators. The finding is well supported by the literature (Gomes et al. 2006; Loof and Heshmati 2002). The study conducted by

Gomes et al. (2006), for example, is indicative of two outcomes relevant to the present study. One, innovation aspect of performance measurement has been the least important among the Portugalis manufacturing industries out of the six dimensions identified by the authors. This justifies the need of the present study. The second outcome that the high performers have scored slightly better than the low performers on the use of innovation dimension of performance measurement, strengthens the finding of this study. Further, the DFA reveals that increased sales, reduced time and cost of production, better control over emissions and waste, and improved quality and flexibility discriminate good innovators from the poor ones. However, increase in sales, decrease in production time, and reduction in hazardous emissions emerge as more prominent outcomes of technological innovations.

Due to a small sample size, restricted to a particular region, the findings may not be generalized for other industries and regions. More valuable results are expected if process and product innovations are studied separately, added with more than two levels of innovativeness.

Table 2 Descriptive statistics (MANOVA)

INNO		Mean	Standard deviation	N
TIME	1	2.7857	.89258	14
	2	1.6667	.48507	18
	Total	2.1563	.88388	32
COST	1	3.0000	.87706	14
	2	1.9444	.63914	18
	Total	2.4063	.91084	32
DISP	1	2.7143	1.13873	14
	2	2.1667	.78591	18
	Total	2.4063	.97912	32
EMSN	1	3.0000	.87706	14
	2	1.8333	.61835	18
	Total	2.3438	.93703	32
SALE	1	3.0714	.82874	14
	2	4.5556	.51131	18
	Total	3.9063	.99545	32
FLEX	1	3.2143	.57893	14
	2	3.5000	.98518	18
	Total	3.3750	.83280	32
QUAL	1	3.5714	.64621	14
	2	4.0556	.80237	18
	Total	3.8438	.76662	32

Methods

Technological innovation is defined in this paper as the set of activities through which a new or significantly improved product or process has been launched by a company to itself or to the industry during the last 3 years (OECD 2005; Nieto 2004). A structured questionnaire (Additional file 1)

Table 3 Multivariate tests

Effect		Value	F	Hypo df	Error df	Significance
Intercept	Pillai's trace	.993	523.286	7.0000	24.000	.000
	Wilk's lambda	.007	523.286	7.0000	24.000	.000
	Hotelling's trace	152.625	523.286	7.0000	24.000	.000
	Roy's largest root	152.625	523.286	7.0000	24.000	.000
INNO	Pillai's trace	.639	6.066	7.0000	24.000	.000
	Wilk's lambda	.361	6.066	7.0000	24.000	.000
	Hotelling's trace	1.769	6.066	7.0000	24.000	.000
	Roy's largest root	1.769	6.066	7.0000	24.000	.000

Table 4 Structure matrix

Variate	Variable						
	SALE	TIME	EMSN	COST	QUAL	DISP	FLEX
1	.856	−.623	−.606	−.541	.252	−.221	.132

has been designed (Wan et al. 2005; Sawang 2006) and administered to a sample of SMEs operating under the paper industry in a few selected districts of northern India. The companies were identified from the business directories and through personal contacts. The questionnaire was pretested on a small sample of nine companies. Later on, 143 companies were contacted using convenience sampling method for actual data collection. With a low but sufficient response rate of 27%, 39 respondents filled the questionnaire. After scrutinizing and editing the received responses, 32 questionnaires were found usable for the analysis. *SPSS* version 13.0 of SPSS (IBM Corporation, Armonk, NY, USA) has been used to run multivariate analysis of variance (MANOVA), followed by discriminant function analysis (DFA) to obtain the results.

Additional file

Additional file 1: Structured questionnaire.

Authors' information

PF is an associate professor of Industrial Engineering at Aligarh Muslim University. AB, AKZ, and IH are of students of Industrial Engineering at Aligarh Muslim University.

References

De Jong JPJ, Vermeulen PAM (2006) Determinants of product innovation in small firms: a comparison across industries. International Small Business J 24(6):587–609

Field A (2005) Discovering Statistics using SPSS, 2nd edn. SAGE Publications, London

Fitzgerald L, Johnston R, Brignall S, Silvestro R, Voss C (1991) Performance measurement in service business. CIMA, London

Freel MS (2005) Patterns of innovation and skill in small firms. Technovation 25:123–134

Gomes CF, Yasin MM, Lisboa JV (2006) Key performance factors of manufacturing effective performance: the impact of customers and employees. The TQM Magazine 18(4):323–340

Kaplan RS, Norton DP (1992) The balanced scorecard – measures that drive performance. Harvard Business Review :71–79

Karlsson M (2009) Innovation is needed in the paper industry. http://web.abo.fi/meddelanden/english/2009_01/2009_01_paper_industry.sht. Accessed 20 June 2009

Lin CYY, Chen MYC (2007) Does innovation lead to performance? An empirical study of SMEs in Taiwan. Management Research News 30(2):115–132

Loof H, Heshmati A (2002) Knowledge capital and performance heterogeneity: a firm level innovation study. International Journal of Production Economics 76:61–85

Mansury MA, Love JH (2008) Innovation, productivity and growth in US business services: a firm level analysis. Technovation 28:52–62

Michie J, Sheehan M (2003) Labour market deregulation, flexibility and innovation. Cambridge J of Economics 27(1):123–143

Neely A, Gregory M, Platts K (1995) Performance measurement system design: a literature review and research agenda. International J of Operations & Production Management 15(4):80–116

Nieto M (2004) Basic propositions for the study of the technological innovation process in the firm. European J of innovation Management 7(4):314–324

OECD (2005) Oslo manual: proposed guidelines for collecting and interpreting technological innovation data, 3rd edn. OECD Publishing, Paris

Paper industry in India (2009) http://stationery.indiabizclub.com/info/ properties_of_paper/properties_of_paper_industry_in_india. Accessed 20 June 2009

Pavitt K (1984) Sectoral patterns of technical change: towards a taxonomy and a theory. Research Policy 13:343–373

Prajogo D, Laosirihongthong T, Sohal A, Boon-itt S (2007) Manufacturing strategies and innovation performance in newly industrialized countries. Industrial Management and Data Systems 107(1):52–68

Sawang S (2006) An empirical study: a role of financial and non-financial performance measurement and perceived innovation effectiveness. Proc IEEE Int Conf on Management of Innovation and Technology 2:1063–1065

Wan D, Ong CH, Lee F (2005) Determinants of firm innovation in Singapore. Technovation 25:261–268

15

Mahalanobis-Taguchi System-based criteria selection for strategy formulation: a case in a training institution

Seyed Ali Hadighi[1,2], Navid Sahebjamnia[3], Iraj Mahdavi[2*], Hadi Asadollahpour[2] and Hosna Shafieian[2]

Abstract

The increasing complexity of decision making in a severely dynamic competitive environment of the universe has urged the wise managers to have relevant strategic plans for their firms. Strategy is not formulated from one criterion but from multiple criteria in environmental scanning, and often, considering all of them is not possible. A list of criteria utilizing Delphi was selected by consultation with company experts. By reviewing the literature and strategy experts' proposals, the list is then classified into five categories, namely, human resource, equipment, market, supply chain, and rules. Since all the criteria may not be necessary for the decision process, as they are eliminated in the early stage traditionally, it is important to identify the prime set of criteria, which is a subset of the original criteria and affects decision making. Utilizing these criteria, a Mahalanobis-Taguchi System-based tool was developed to facilitate the selection of a prime set of criteria, which is a subset of the original criteria for ensuring that only ineffective subcriteria are eliminated and the conditions are prepared for relevant strategy formulation. Mahalanobis distance was used for making a measurement scale to distinguish ineffective subcriteria from significant criteria in the environmental scanning stage. The principles of the Taguchi method were used for screening the important criteria in the system and generate the prime set of criteria for each category. One can use these criteria within each category instead of all criteria for the identification of a suitable institution in training. To validate the proposed approach, a case study has been conducted for 38 educational institutions in Iran. The results demonstrated the usefulness of the proposed approach.

Keywords: Mahalanobis distance; Measurement scale; Service institution; Strategy formulation

Introduction

Strategy formulation is sometimes referred to as determining where you are now, where you plan to go, and finally how to get there. It consists of performing a situation analysis, self-evaluation, and competitor analysis in both inside and outside of the organization while setting the objectives concurrent with the assessment. Strategy formulation is the process of developing long-term goals for an effective management of environmental factors. Strategy formulation consists of two basic components: one is situation analysis which is the process of finding a strategic fit between external opportunities and internal strengths while working around external threats and internal weaknesses, and the other component is developing strategies based on goals. Many approaches and techniques can be used to analyze strategic cases in the process of strategic management (Dincer 2004).

Among several existing approaches, strengths, weaknesses, opportunities, and threats (SWOT) analysis, evaluating each of the indicated terms in an organization, is the most acclaimed (Hill and Westbrook 1997). SWOT analysis is the most significant part of strategic formulation. By identifying the strengths, weaknesses, opportunities, and threats, the organization can build strategies upon its strengths, eliminate its weaknesses, and exploit its opportunities or utilize them to encounter the threats. The strengths and weaknesses are considered as an internal organization environment appraisal, while the opportunities and threats are considered as an external organization environment appraisal (Dyson 2004).

* Correspondence: irajarash@rediffmail.com
[2]Department of Industrial Engineering, Mazandaran University of Science and Technology, Babol, Iran
Full list of author information is available at the end of the article

Hadighi et al. (2013) proposed a framework based on clustering algorithm for strategic formulation of corporate organization. They formed departmental clusters according to correlation among factors and goals in each department. Then, strategies are presented based on the generated organizational clusters. Goals, factors, and strategies are known as the three main elements in strategy formulation. The interrelationship among them should be considered as an integrated set, while in common methods, these relationships are vague. In another research, Hadighi et al. (2012) presented a strategy formulation framework for developing strategies on more accurate and objective bases by considering all the components. They identify whether the organization is intrinsically a production or a service company. Then, the environmental factors are explored including opportunities and threats, in the light of organization's strengths and weaknesses. At this stage, the factor-goal matrix is formed by considering the impact of factors on every individual goal. Finally, the goals with higher similarities are embedded within the same cluster, and strategies were generated for each of them.

Many approaches and techniques were developed for strategic management, such as the traditional SWOT analysis (Bellman and Zadeh 1970), analytical SWOT method (Chen et al. 1992), resource-based view (Paiva et al. 2008; Gordon et al. 2005), and quantitative SWOT methods (Chen and Hsieh 2000; David 2001). Fuzzy quantified SWOT (Kuo-liang and Shu-chen 2008), decision tree (Bunn and Thomas 1977), and quality function deployment (Killen et al. 2005) are used to support decision making in a competitive environment in a given organization.

In this paper, a new method of environmental factors' filtering has been developed. To this end, a list of criteria utilizing Delphi was selected by consultation with company experts. By reviewing the literature and strategy experts' proposals, the list is then classified into five categories, namely, human resource, equipment, market, supply chain, and rules. Since all the criteria may not be necessary for the decision process, as they are eliminated in the early stage traditionally, it is important to identify the prime set of criteria, which is a subset of the original criteria and affects decision making. Utilizing these criteria, a Mahalanobis-Taguchi System (MTS)-based tool was developed to facilitate the selection of a prime set of criteria, which is a subset of the original criteria for ensuring that only ineffective subcriteria are eliminated and the conditions are prepared for relevant strategy formulation. Mahalanobis distance (MD) was used for making a measurement scale to distinguish ineffective subcriteria from significant criteria in the environmental scanning stage. In this manner, the main contributions of this paper can be highlighted as follows:

- Consideration of all environmental factors
- Consideration of the interaction among factors
- Application of a MTS method for environmental scanning
- Utilization of a quantitative method for detecting abnormal factors

Organizations are considered to be of two general types: (1) service organization and (2) production organization. For each organization, according to its characteristics and nature of the problems of concern, the key indicators based on priorities would be significant.

Data gathering

Gathering data on factors was practiced at first in production companies. In emerging service companies, the significance of data gathering has spread to this ever increasing sector of the industry. So, data gathering is just as applicable to services as it is to production in general. Only live experiments with real customers and real transactions can provide the type of data needed for truly innovative services. However, live tests are difficult to control and risky to both customer relations and firm creditability, and therefore, most services are designed by brainstorming or trial and error, with limited success. Eventually, services are labor intensive, while manufacturing is more capital intensive (Russell and Taylor 2006).

The most important factors and subfactors as experts specified are shown in Table 1 and Figure 1. Since in most cases of SWOT analysis considering all factors is almost impossible, in general, a limited set of factors is being considered and some factors are eliminated according to an overall view of strategy makers. Since all the criteria may not be necessary for the decision process, it is important to identify the prime set of criteria, which is a subset of the original criteria. In order to identify the prime criteria, experimental design becomes complex and difficult to manage. In this study, an alternative new approach is identified to the experimental design, the MTS-based decision tool. The purpose is MTS-based criteria selection for suitable criteria for strategy formulation. In this paper, we are going to define a normal group which is called Mahalanobis space by utilizing MD. Mahalanobis space (MS) is a database for the normal group involving mean vector, standard deviation vector, and inverse of the correlation vector (Taguchi and Jugulum 2000). This space provides a reference point for the measurement scale. According to MTS theory, the average value of MDs is 1 for all the observations in MS (Taguchi et al. 2005).

Materials and methods

In this section the samples taken, decision criteria and the acceptable rating, normal and abnormal observations, and also MTS framework will be discussed.

Table 1 The steps to design and optimize the MTS

Subcriteria (before MTS)	Range of variations	Suitability ratings	Notation for implementation	Subcriteria (after MTS)
Human resource			(h)	
Appearance	0-100	≥70	h_1	-
Specialty	0-100	≥85	h_2	-
Courtesy	0-100	≥75	h_3	-
Experience	0-100	≥70	h_4	Experience
Attitude	0-100	≥75	h_5	Attitude
Public relation	0-100	≥80	h_6	Public relation
Motivation	0-100	≥80	h_7	Motivation
Education	0-100	≥85	h_8	Education
Performance	0-100	≥80	h_9	Performance
Timeliness	0-100	≥90	h_{10}	Timeliness
Equipment			(q)	
Functionality	0-4	≥3	q_1	Functionality
Aesthetic	0-4	≥2	q_2	-
Comfort ability	0-4	≥3	q_3	Comfort ability
Multifunction	0,1 (no, yes)	1	q_4	-
Market			(m)	
Demand	0%-100%	≥70%	m_1	Demand
Organization brand	1, 2, 3 (A, B, C)	≥B	m_2	Organization brand
Competitors	−5, +5	2-5	m_3	Competitors
Economic parameters	−5, +5	0-5	m_4	Economic parameters
Customer attitude	−5, +5	1-5	m_5	Customer attitude
Supply chain			(c)	
Reputation	1, 2, 3 (A, B, C)	≥B	c_1	Reputation
Financial ability	0, 1	1	c_2	Financial ability
Support services	0-9	≥6	c_3	Support services
Timeliness	0-9	≥7	c_4	Timeliness
Rules			(r)	
Social rules	−2, +2	0-2	r_1	-
Tax	−3, +3	1-3	r_2	-
Discipline	−3, +3	1-3	r_3	Discipline
Organizational rules	−5, +5	≥3	r_4	-

Samples

The samples are taken from 38 educational institutions all acting in the capital city of Tehran, Iran. The selected institutions are those which have been approved by the quality assurance department for utilization. Their main activity is in training and holding lectures for requesters.

Decision criteria and its appropriate ratings

The key factors and subfactors to be considered for selecting the suitable educational institutions according to experts' replies are the human resource factor with ten subfactors $(h_1,..., h_{10})$, namely, appearance, specialty, courtesy, experience, attitude, public relation, motivation,

education, performance, and timeliness. These attributes are purely qualitative, and their ranges of variations have been considered from 0 for the least to 100 for the maximum effect. Equipment has four subfactors $(q_1,..., q_4)$ such as functionality, aesthetic, comfort ability, with the range from 0 to 4 for the worst to the best condition, and multifunction (with the answer yes and no or 1 and 0, respectively). The factor market consists of five subfactors $(m_1,..., m_5)$ such as demand which is defined as the average percentage of having educated customers during the year (%); organization brand which is defined as A, B, or C, from the best to the worst (1, 2, or 3, respectively); competitors from −5 to the most +5;

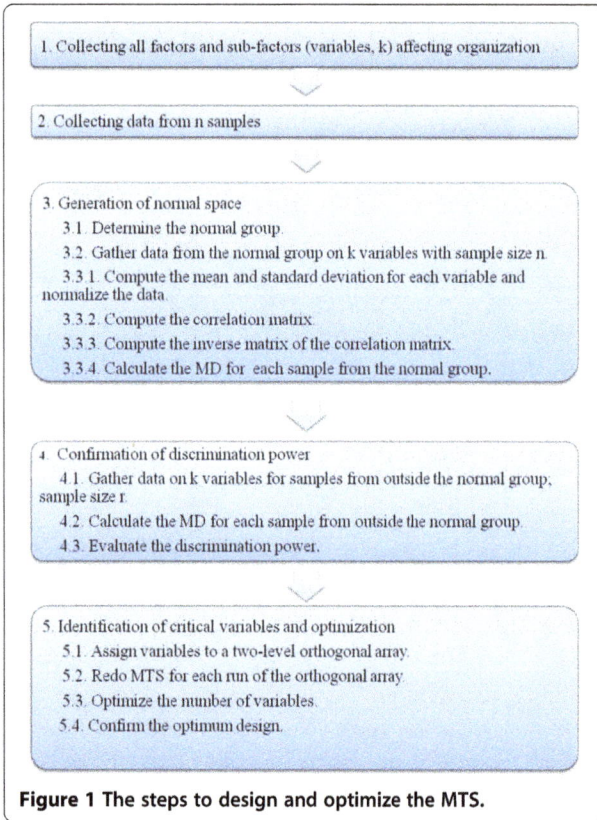

Figure 1 The steps to design and optimize the MTS.

economic parameters also expressed as −5 to the most +5; and finally, customer attitude from −5 to +5, too. The factor supply chain (c_1,..., c_4) also includes reputation defined as A, B, or C, from the most famous to the least (or 1, 2, or 3, respectively); financial ability expressed as the average current ratio per month, less than 1 and equal or greater than 1 (0 and 1, respectively); support service from 0 for the least to 9 for the greatest; and timeliness also from 0 for the minimum to 9 for the maximum. Finally, the factor rules (r_1,..., r_4) includes social rules from the rules by worst effect −2 to the best effect +2, tax with the worst effect −3 to the best effect +3, discipline also from −3 to +3, and organizational rules stated as −5 to +5 from worst to the best, respectively. The whole factors and subfactors with their range of variability, limits of acceptability, and notations for implementation before and after MTS are shown in Table 1.

Normal and abnormal observations

The data used in this study were collected from 38 educational institutions including normal and abnormal observations. The normal group which is called Mahalanobis space was formed based on observations on 29 sample institutions, those which do not have any value out of the acceptable limit. The abnormal group was built

according to nine institutions, which have one or more criteria values beyond the acceptable range.

Mahalanobis-Taguchi System

The selection of a prime set of subcriteria for an individual main criterion was calculated iteratively utilizing the following steps (Taguchi and Jugulum 2000, 2002; Taguchi et al. 2001). The algorithm of processing the article is shown in Table 1.

Making a measurement scale with MS as the reference scale

The first step for the construction of measurement scale was the collection of normal observations and then the normalization of these observations by using mean and standard deviation obtained from the normal observations. MDs corresponding to all these observations were computed using the inverse of the correlation matrix method (Taguchi et al. 2001).

$$
\mathrm{MD}_i = D_i^2 = \frac{1}{k} Z_{ij}^T A^{-1} Z_{ij}
$$

$$
= \frac{1}{k} (z_{i1}, z_{i2}, ..., z_{ik}) A^{-1} \begin{pmatrix} z_{i1} \\ \vdots \\ z_{ik} \end{pmatrix}, \tag{1}
$$

where Z_i is the normalized vector obtained by normalizing the values of X_j (j = 1, 2, 3..., k)

$$
z_{ij} = \frac{X_{ij} - \bar{X}_j}{S_j}, \tag{2}
$$

$$
\bar{X}_j = \frac{\sum_{i=1}^{n} X_{ij}}{n_i}, \tag{3}
$$

where X_{ij} is the value of jth subcriteria in ith observation

$$
S_j = \sqrt{\frac{\sum_{i=1}^{n} \left(X_{ij} - \bar{X}_j\right)^2}{n-1}}, \tag{4}
$$

where S_j is the standard deviation of jth subcriteria, k is the number of subcriteria, n is the number of observations, T is the transpose of the normalized vector, and A^{-1} is the inverse of the correlation matrix.

It can be considered that MD in Equation 1 was obtained by scaling (that is by dividing with k) the original MD. Since MDs were used to define the normal group, we called this group as Mahalanobis space. Mahalanobis space is a database for the normal group involving mean vector, standard deviation vector, and inverse of the correlation vector (Taguchi and Jugulum 2000). This space provides a reference point for the measurement scale. According to MTS theory, the average value of MDs is 1 for all the observations in MS (Taguchi et al. 2001).

Table 2 L_{32} (2^{27}) OA and average response for the larger-the-better S/N ratio

Run	h_1	h_2	h_3	h_4	h_5	h_6	h_7	h_8	h_9	h_{10}	q_1	q_2	q_3	q_4	m_1	m_2	m_3	m_4	m_5	c_1	c_2	c_3	c_4	r_1	r_2	r_3	r_4	28	29	30	31
1	1	1	1	1	1	1	1	1	1	1	1	1	1	1	1	1	1	1	1	1	1	1	1	1	1	1	1	1	1	1	1
2	1	1	1	1	1	1	1	1	1	1	1	1	1	1	1	2	2	2	2	2	2	2	2	2	2	2	2	2	2	2	2
3	1	1	1	1	1	1	1	2	2	2	2	2	2	2	2	1	1	1	1	1	1	1	1	2	2	2	2	2	2	2	2
4	1	1	1	1	1	1	1	2	2	2	2	2	2	2	2	2	2	2	2	2	2	2	2	1	1	1	1	1	1	1	1
5	1	1	1	2	2	2	2	1	1	1	1	2	2	2	2	1	1	1	1	2	2	2	2	1	1	1	1	2	2	2	2
6	1	1	1	2	2	2	2	1	1	1	1	2	2	2	2	2	2	2	2	1	1	1	1	2	2	2	2	1	1	1	1
7	1	1	1	2	2	2	2	2	2	2	2	1	1	1	1	1	1	1	1	2	2	2	2	2	2	2	2	1	1	1	1
8	1	1	1	2	2	2	2	2	2	2	2	1	1	1	1	2	2	2	2	1	1	1	1	1	1	1	1	2	2	2	2
9	1	2	2	1	1	2	2	1	1	2	2	1	1	2	2	1	1	2	2	1	1	2	2	1	1	2	2	1	1	2	2
10	1	2	2	1	1	2	2	1	1	2	2	1	1	2	2	2	2	1	1	2	2	1	1	2	2	1	1	2	2	1	1
11	1	2	2	1	1	2	2	2	2	1	1	2	2	1	1	1	1	2	2	1	1	2	2	2	2	1	1	2	2	1	1
12	1	2	2	1	1	2	2	2	2	1	1	2	2	1	1	2	2	1	1	2	2	1	1	1	1	2	2	1	1	2	2
13	1	2	2	2	2	1	1	1	1	2	2	2	2	1	1	1	1	2	2	2	2	1	1	1	1	2	2	2	2	1	1
14	1	2	2	2	2	1	1	1	1	2	2	2	2	1	1	2	2	1	1	1	1	2	2	2	2	1	1	1	1	2	2
15	1	2	2	2	2	1	1	2	2	1	1	1	1	2	2	1	1	2	2	2	2	1	1	2	2	1	1	1	1	2	2
16	1	2	2	2	2	1	1	2	2	1	1	1	1	2	2	2	2	1	1	1	1	2	2	1	1	2	2	2	2	1	1
17	2	1	2	1	2	1	2	1	2	1	2	1	2	1	2	1	2	1	2	1	2	1	2	1	2	1	2	1	2	1	2
18	2	1	2	1	2	1	2	1	2	1	2	1	2	1	2	2	1	2	1	2	1	2	1	2	1	2	1	2	1	2	1
19	2	1	2	1	2	1	2	2	1	2	1	2	1	2	1	1	2	1	2	1	2	1	2	2	1	2	1	2	1	2	1
20	2	1	2	1	2	1	2	2	1	2	1	2	1	2	1	2	1	2	1	2	1	2	1	1	2	1	2	1	2	1	2
21	2	1	2	2	1	2	1	1	2	1	2	2	1	2	1	1	2	1	2	2	1	2	1	1	2	1	2	2	1	2	1
22	2	1	2	2	1	2	1	1	2	1	2	2	1	2	1	2	1	2	1	1	2	1	2	2	1	2	1	1	2	1	2
23	2	1	2	2	1	2	1	2	1	2	1	1	2	1	2	1	2	1	2	2	1	2	1	2	1	2	1	1	2	1	2
24	2	1	2	2	1	2	1	2	1	2	1	1	2	1	2	2	1	2	1	1	2	1	2	1	2	1	2	2	1	2	1
25	2	2	1	1	2	2	1	1	2	2	1	1	2	2	1	1	2	2	1	1	2	2	1	1	2	2	1	1	2	2	1
26	2	2	1	1	2	2	1	1	2	2	1	1	2	2	1	2	1	1	2	2	1	1	2	2	1	1	2	2	1	1	2
27	2	2	1	1	2	2	1	2	1	1	2	2	1	1	2	1	2	2	1	1	2	2	1	2	1	1	2	2	1	1	2
28	2	2	1	1	2	2	1	2	1	1	2	2	1	1	2	2	1	1	2	2	1	1	2	1	2	2	1	1	2	2	1
29	2	2	1	2	1	1	2	1	2	2	1	2	1	1	2	1	2	2	1	2	1	1	2	1	2	2	1	2	1	1	2
30	2	2	1	2	1	1	2	1	2	2	1	2	1	1	2	2	1	1	2	1	2	2	1	2	1	1	2	1	2	2	1
31	2	2	1	2	1	1	2	2	1	1	2	1	2	2	1	1	2	2	1	2	1	1	2	2	1	1	2	1	2	2	1
32	2	2	1	2	1	1	2	2	1	1	2	1	2	2	1	2	1	1	2	1	2	2	1	1	2	2	1	2	1	1	2

Validation of the measurement scale

The accuracy of the scale was justified by measuring the MDs of the known abnormal observations. The data from abnormal observations were normalized using mean and standard deviation obtained from the normal observations. The MDs were obtained from the abnormal observations using the correlation matrix of normal observations in Equation 1. After calculating MDs, the average MD for the normal observations was compared with that for abnormal observations. The higher values of MDs for the abnormal group validate the accuracy of the measurement scale.

Table 3 Average response for the larger-the-better S/N ratio (human resource)

	h_1	h_2	h_3	h_4	h_5	h_6	h_7	h_8	h_9	h_{10}
Level 1	−31.57	−33.31	−31.29	−29.38	−28.41	−29.82	−28.39	−28.46	−26.79	−27.64
Level 2	−29.19	−27.46	−29.47	−31.38	−32.35	−30.94	−32.37	−32.30	−33.97	−33.12
Gain	−2.38	−5.85	−1.82	2.00	3.94	1.12	3.99	3.84	7.18	5.48

Table 4 Average response for the larger-the-better S/N ratio (equipment)

	q_1	q_2	q_3	q_4
Level 1	−28.38	−31.01	−28.39	−30.86
Level 2	−32.39	−29.76	−32.37	−29.90
Gain	4.01	−1.25	3.98	−0.96

Based on the MTS theory, the MD of abnormal observations will be larger than the MD of normal observations if this is a good scale. Otherwise, one has to resample or find new subcriteria (if applicable) to build the MS again.

Identification of the prime set of subcriteria
At this stage, the prime set of subcriteria was identified by applying orthogonal arrays (OAs) and signal-to-noise (S/N) ratios. The orthogonal arrays are used so that the interactions between control factors are almost evenly distributed to other columns of the orthogonal arrays and confounded to various main effects (Taguchi et al. 2005). Then, the appropriate OA was selected depending on the total degrees of freedom required to study the individual subcriteria. The number of degrees of freedom is one less than the number of levels associated with the subcriterion (Antony and Antony 2001). The individual 'k' subcriteria under study were assigned to the first k columns of the identified OA (Taguchi 1987). Level 1 in the identified OA column represents the presence of a subcriterion, and level 2 represents the absence of that subcriterion. Inside an OA, each row represents the experimental combination of a run (Table 2).

Using the subcriteria combinations in the identified OA, MDs for the known abnormal observations were obtained using Equation 1. From the MDs, the larger-the-better S/N ratio was obtained for the qth run using the formula (Taguchi et al. 2001)

$$\frac{S}{N}\text{ratio} = -10 \log_{10}\left[\left(\frac{1}{t}\right) \sum_{i=1}^{t} \frac{1}{\mathrm{MD}_i^2}\right] \quad j$$
$$= 1, 2, ..., t, \tag{5}$$

where t is the number of subcriteria presented for a given combination of the experimental run.

An average S/N ratio was calculated for each subcriterion at levels 1 and 2. Subsequently, gain in S/N ratio values was calculated as

Table 6 Average response for the larger-the-better S/N ratio (supply chain)

	c_1	c_2	c_3	c_4
Level 1	−29.55	−29.29	−29.68	−30.16
Level 2	−31.22	−31.47	−31.08	−30.60
Gain	1.67	2.18	1.40	0.44

$$\text{Gain} = \left(\text{average of } \frac{S}{N}\text{ratio}\right)_{\text{level 1}} - \left(\text{average of } \frac{S}{N}\text{ratio}\right)_{\text{level 2}}. \tag{6}$$

If the gain is positive, we keep the subcriterion; if not, then we exclude it for the next step.

Confirmation run
A confirmation run was performed with the prime set of subcriteria identified from the step before. The reduced measurement scale was constructed by utilizing the prime set of subcriteria identified, and then MDs corresponding to the abnormal observations were obtained using Equation 1. In the next step, the average MD based on the abnormal observations obtained with the prime set of subcriteria identified was compared to that gained from all the subcriteria originally used. If the average MD of abnormal group with the prime set of subcriteria identified was lower than that of all the subcriteria, then retain the excluded subcriteria also in the prime set of subcriteria identified; if not, then consider only the subcriteria which were identified as the prime set from the previous step, for the anticipation of further observations.

Results and discussion

The Mahmoudabad Training Center is an educational center for training personnel of petroleum industries (about 100,000 persons) and also staffs from other organizations needing special on-the-job and recruitment trainings. The center utilizes different educational organizations as supplier. We conducted the proposed method with the aid of 29 normal observations; the MDs corresponding to all these observations were calculated. Mahalanobis space was defined for all main criteria with the help of the MDs obtained for 29 observations. The

Table 5 Average response for the larger-the-better S/N ratio (market)

	m_1	m_2	m_3	m_4	m_5
Level 1	−29.78	−27.13	−26.69	−26.60	−28.02
Level 2	−30.99	−33.64	−34.07	−34.17	−32.74
Gain	1.21	6.51	7.38	7.57	4.72

Table 7 Average response for the larger-the-better S/N ratio (rules)

	r_1	r_2	r_3	r_4
Level	−31.80	−31.28	−26.10	−29.03
Level2	−28.96	−29.49	−34.66	−31.74
Gain	−2.84	−1.79	8.56	−2.71

data from the 9 abnormal observations were normalized utilizing mean and standard deviation gained from the 29 normal observations. The MDs corresponding to all these observations were estimated for the main criteria. Since the average MD of the abnormal group with the prime set of subcriteria identified was lower than that of all the subcriteria, it would be valid.

There are many orthogonal arrays in Taguchi's method (David 2001). Commonly, for surveying of significant subcriteria from the main criteria, the orthogonal arrays of L_{12} (2^{11}), L_s (2^7), and L_4 (2^3) are used for the individual main criterion appropriately. The important set of subcriteria will be assigned to each main criterion accordingly. The weakness of such a method is that we only compare subcriteria which belong to one type of main criterion isolately. However, in this method, the L_{32} (2^{27}) orthogonal array was utilized for the identification of the prime set of subcriteria with minimum number of subcriteria combinations. Hereby, we have considered the interactions among all criteria as whole. The strength of such a method is considering the whole combinations of subfactors from the main criteria concurrently.

In Tables 3, 4, 5, 6, and 7, '1' (or level 1) represents inclusion of the subcriterion and '2' (or level 2) the exclusion of the subcriterion. An average S/N ratio was obtained for each subcriterion at levels 1 and 2. Subsequently, gain in S/N ratio was calculated by the difference between the average S/N ratio value at level 1 and level 2 (Tables 3, 4, 5, 6, and 7).

From Table 3, it is clear that in the main criteria human resource, the seven subcriteria such as h_4, h_5, h_6, h_7, h_8, h_9, and h_{10} have positive gains. That means these subcriteria have higher average responses when they are part of the system (level 1). Hence, these subcriteria were considered to be useful for the confirmation run. A similar interpretation was true for the rest of the main criteria in Tables 4 and 7.

The obtained results from Tables 3, 4, 5, 6, and 7 show the subcriteria which are significant and more effective in formulating strategy as mentioned in Table 1. Based on the samples taken, it can be stated that the subcriteria relating to human resource, supply chain, and market as main criteria have more influences and should be considered more closely in strategy formulation.

Conclusion

A simple-to-use MTS-based decision tool that assists in the selection of significant criteria which will be useful for the identification of suitable educational institutions has been developed. One can use the MTS-based tool in the education system without much knowledge of statistics. If the discrimination is performed alone by an individual criterion, it may produce a misleading result. The proposed MTS decision tool combines all criteria

into an MD index with consideration of the correlation among all criteria. The advantage of this method compared to other methods such as SWOT is that it considers not only the whole factors in environmental scanning but also the interaction among factors by using orthogonal array analysis. The proposed decision tool can be easily adapted to other closely related industries such as tourism, agriculture, and environmental engineering.

Competing interests
The authors declare that they have no competing interests.

Authors' contributions
SAH, NS, and IM carried out environmental scanning, explored and labeled factors, designed the case study, and prepared the manuscript. HA and HS participated in the case study and performed the statistical analysis. All authors read and approved the final manuscript.

Acknowledgments
The authors wish to express their appreciation to the National Iranian Oil Company (NIOC) for sponsorship, data preparation, and cooperation. Also, we would like strongly to appreciate the valuable comments received from four referees through this research process.

Author details
[1]National Iranian Oil Company - Mahmoudabad Training Center, Mahmoudabad, Iran. [2]Department of Industrial Engineering, Mazandaran University of Science and Technology, Babol, Iran. [3]School of Industrial Engineering, College of Engineering, University of Tehran, Tehran, Iran.

References
Antony J, Antony FJ (2001) Teaching the Taguchi method to industrial engineers. Work Study 50:141–149

Bellman RE, Zadeh LA (1970) Decision-making in a fuzzy environment. Manag Sci 17(4):141–164

Bunn D, Thomas H (1977) Decision analysis and strategic policy formulation. Long Range Plann 10(6):23–30

Chen SH, Hsieh CH (2000) Representation, ranking, distance, and similarity of L-R type fuzzy number and application. Aust J Intell Inform Process Syst 6(4):217–229

Chen SJ, Huang CL, Huang FP (1992) Fuzzy multiple attribute decision making methods and applications. Springer, Berlin

David FR (2001) Strategic management, concepts and cases, 8th edn. Prentice Hall, New Jersey

Dincer O (2004) Strategy management and organization policy. Beta Publication, Istanbul

Dyson RG (2004) Strategic development and SWOT analysis at the University of Warwick. Eur J Oper Res 152(1):631–640

Gordon JRM, Lee PM, Lucas HC (2005) A resource-based view of competitive advantage at the Port of Singapore. J Strateg Inf Syst 14(1):69–86

Hadighi SA, Mahdavi I, Sahebjamnia N, Mahdavi-Amiri N (2012) A new approach in strategy formulation using clustering algorithm: an instance in a service company. Int J Ind Eng 23(2):125–142

Hadighi SA, Sahebjamnia N, Mahdavi I, Shirazi MA (2013) A framework for strategy formulation based on clustering approach: a case study in a corporate Organization. Knowl-Based Syst 49:37–49

Hill T, Westbrook R (1997) SWOT analysis: it's time for a product recall. Long Range Plann 30(1):46–52

Killen CP, Walker M, Hunt RA (2005) Strategic planning using QFD. Int J Qual Reliab Manag 22(1):17–29

Kuo-liang L, Shu-chen L (2008) A fuzzy quantified SWOT procedure for environmental evaluation of an international distribution center. Inf Sci 178:531–549

Paiva EL, Roth AV, Fensterseifer JE (2008) Organizational knowledge and the manufacturing strategy process: a resource-based view analysis. J Oper Manag 26(1):115–132

Russell RS, Taylor BW III (2006) Operations management. Wiley, New York,
    pp 200–211
Taguchi G (1987) System of experimental design. Dearborn, Michigan, and White
    Plain. ASI Press and UNIPUB-Kraus International Publications, New York
Taguchi G, Jugulum R (2000) New trends in multivariate diagnosis. Indian J Stat
    62(1):233–248
Taguchi G, Jugulum R (2002) The Mahalanobis-Taguchi Strategy. Wiley, New York
Taguchi G, Chowdhury S, Wu Y (2001) The Mahalanobis-Taguchi System.
    McGraw-Hill, New York
Taguchi G, Chowdhury S, Wu Y (2005) Taguchi's quality engineering handbook.
    Wiley, New York, p 595

16

Multi-objective design of fuzzy logic controller in supply chain

Mahdi Ghane[1] and Mohammad Jafar Tarokh[2*]

Abstract

Unlike commonly used methods, in this paper, we have introduced a new approach for designing fuzzy controllers. In this approach, we have simultaneously optimized both objective functions of a supply chain over a two-dimensional space. Then, we have obtained a spectrum of optimized points, each of which represents a set of optimal parameters which can be chosen by the manager according to the importance of objective functions. Our used supply chain model is a member of inventory and order-based production control system family, a generalization of the periodic review which is termed 'Order-Up-To policy.' An auto rule maker, based on non-dominated sorting genetic algorithm-II, has been applied to the experimental initial fuzzy rules. According to performance measurement, our results indicate the efficiency of the proposed approach.

Keywords: Supply chain, Fuzzy logic controller, Multi-objective optimization

Background

A supply chain is a system of organizations, people, technology, activities, information, and resources involved in performing the functions of procurement of raw materials, transformation of raw materials into intermediate and finished product that is delivered to the end customer. Recently, enterprises have exposed a growing interest in efficient supply chain management. This is due to the rising cost of manufacturing and transportation, the globalization of market economies, and the customer demand for diverse products of short life cycles. A properly designed supply chain system is essential for competitive performance. Control theory advocates a wide range of attributes and standard measures for proper design (Towill 1982). There are different methods to evaluate the performance of a supply chain especially in different case studies. Wang et al. (2007) evaluated the performance of a supply chain for mass customization.

Fundamentally, there are two common objective functions in supply chain systems (Wang et al. 2007): (1) inventory-level recovery and (2) attenuation of demand rate fluctuations on the ordering rate. Proper inventory-level recovery results in lower inventory costs and better

customer services, although in order to optimize the system performance, the designer has to select fixed or variable stock values (agile production systems; Towill and McCullen 1999). The second objective function aims at the reduction of the 'bullwhip' effect. According to Lee et al. (1997), a small variation in the demands of the downstream end-customer may cause remarkable variation in the upstream supplier's side which is known as the bullwhip effect. The term bullwhip is not a new concept (Forrester 1961; Burbidge 1991; Figure 1).

Many researchers have worked on the bullwhip effect and its attenuation. Christer and Robert (2002) studied the complexities of bullwhip by using fuzzy numbers in the bullwhip models. Peter and Dennis (2002) verified how proven material flow control principles considerably reduce the bullwhip in a supply chain. Some researchers are dedicated to forecasting policy. The bullwhip problem is studied by exponential smoothing algorithms in both 'stand-alone' passing-on-orders mode and within inventory-controlled feedback systems in Dejonckheere et al. (2003). Also, Xiaolong (2004) derived and measured a forecasting procedure that minimizes the mean squared forecasting error for the specified demand process. Disney and Towill (2003) proposed a good analytical expression in the inventory position and pipeline position to quantify the bullwhip effect. Jiuh-Biing (2005)

* Correspondence: mjtarokh@kntu.ac.ir
[2]Industrial Engineering Department, K.N.Toosi University of Technology, Tehran 19697 64499, Iran
Full list of author information is available at the end of the article

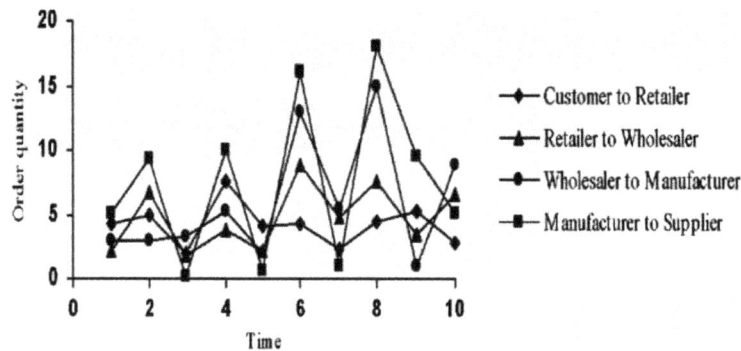

Figure 1 Increasing variability of orders up the supply chain. Adapted from Yu and Zhang (2010).

presented a multi-layer demand-responsive logistics control strategy for alleviating the bullwhip effect.

In the case of inventory-level recovery, the work of Disney and Towill (2003) can be mentioned which uses control theory method to evaluate this objective function in the general supply chain to cover the dynamic behavior of the chain. However, for a simplified system, we can make an analytical measure. Disney and Towill (2003) tried to show how Ti, Ta, and Tw change as the balance between inventory-carrying costs (first objective) and production on-costs (second objective) alters. The sum of objectives in a simplified supply chain model is defined in Equation 1. Their solution only returns a single set of parameters and cannot be extended to a general model; in addition, it is not able to add more constraints to the model. Moreover, it does not cover the dynamic effect of different demand signals.

$$\text{Score} = (K \times VR_{\text{ORATE}}) + VR_{\text{AINV}} \quad (1)$$

Achieving a well-designed supply chain (SC) is very difficult since various sources of uncertainty and complex interactions among various entities (suppliers, manufacturers, distributors, retailers) exist in the SC. Structured uncertainty and unstructured uncertainty are the two major uncertainties in each model. One of the best solutions to deal with uncertainty is fuzzy logic. In Wang and Shu (2005), fuzzy set theory was used to minimize SC holding cost.

A good attempt to optimize and design a supply chain using a multi-objective viewpoint is presented by Mahnadm et al. (2009). In their work, fuzzy logic is used to handle the uncertainty of different suppliers. Banerjee and Roy (2009) considered the application of the intuitionistic fuzzy optimization in the constrained multi-objective stochastic inventory model. Larbani (2009) surveyed most of the approaches for solving non-cooperative fuzzy games in normal form. In addition, applications of these games were also discussed. Yu and Zhang (2010) proposed a generalized form of fuzzy game and extended as a cooperative fuzzy game. They also

provided a practical application for production problem. Chen et al. (2010) formulated a game framework for the strategic behavior of supply chain partners based on fuzzy multi-objective programming, but fuzzy rules mostly were fixed and defined by expert knowledge. Due to the lack of an intelligent mechanism to derive fuzzy rules, fuzzy membership functions based on different weights between objective functions were clear in these attempts.

In this study, a multi-objective approach is used to optimize fuzzy controllers and forecasting policy simultaneously. Structured uncertainty is one of the main challenges in a well-designed supply chain. In this regard, we use a fuzzy logic controller in inventory policy and work in process policy.

The paper is organized as follows: simulation results are presented in the 'Results and discussion' section; concluding remarks and suggestions for future research directions, in the 'Conclusions' section. This paper ends with the 'Methods' section which offers a review on basic theory of the suggested method.

IOBPCS model description

The first attempts for modeling and controlling decentralized systems were done by Forrester (1961). The objective of this work was to perform a dynamic analysis, simulation of industrial systems using discrete dynamic mass balances, and linear and non-linear delays in the distribution channels and manufacturing sites. Various alternative methods have been proposed for modeling supply chains which are classified into four groups (Beamon 1998; Balan et al. 2007).

A simple decentralized supply chain is shown in Figure 2.

For modeling each echelon of this supply chain, a powerful generic order-based production control system (IOBPCS) model was presented by Towill (1982) in a block diagram form (Figure 3).

The IOBPCS family has been considered in both continuous and discrete time using the Laplace and z-transform (Chen et al. 2010; Deb 2001). Via specific parameter settings, a range of well-known replenishment algorithms

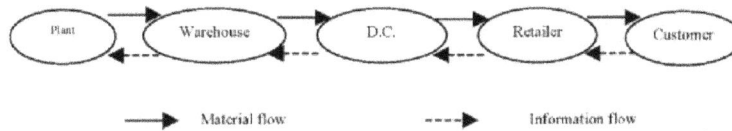

Figure 2 A simple supply chain model. Adapted from Yu and Zhang (2010).

could be implemented. The IOBPCS family consists of a range of PIC systems with five main components:

- Demand policy
- Lead-time
- Inventory policy (inventory feedback loop)
- Pipeline policy (work in progress (WIP) feedback loop)
- Target stock setting.

The demand policy is a feed-forward loop, which, in essence, is a forecasting mechanism that averages the current market demand to reach smoother orders placed on a supplier. The more accurate this forecast, the fewer inventories will be required in the supply chain (Hosoda and Disney 2005). The lead time simply represents the time between placing an order and receiving the goods into inventory. The inventory policy, which is a feedback loop, is an error-compensating mechanism based on the inventory or net stock levels. The pipeline policy, which is a feedback loop, determines the rate at which work in process (WIP) deficit between desired WIP level and actual WIP level is recovered.

As is common practice in the design of engineering systems, and assuming that lead times are not too long, we have incorporated a proportional controller which, in this paper, is triggered by multi-objective optimization into the inventory feedback loop to shape its dynamic response. The target stock setting can be either fixed or

a multiple of current average sale rates. Standard nomenclature used in industrial dynamics is adopted to stand for input, output, and intermediate signals in the block diagram (Haralambos et al. 2008):

- AINV: actual inventory holding
- AVCON: average consumption
- AWIP: actual WIP holding
- COMRATE: completion rate
- CONS: consumption or market demand
- DINV: desired inventory level
- DWIP: desired WIP
- EINV: error in inventory holding
- EWIP: error in WIP
- ORATE: order rate

Recently, researchers have studied different properties of the so-called inventory order-based production control system (IOBPCS) model. As a case in point, the stability of the discrete-time IOBPCS model has been investigated by Disney and Towill (2003, 2005): they presented a general methodology to derive the critical stability boundary using a transfer function.

Fuzzy rule base optimization

We used a fuzzy logic controller (FLC) in inventory policy and work in process policy. The FLC has two inputs (premises): error, e(t), and error derivative, de(t), and

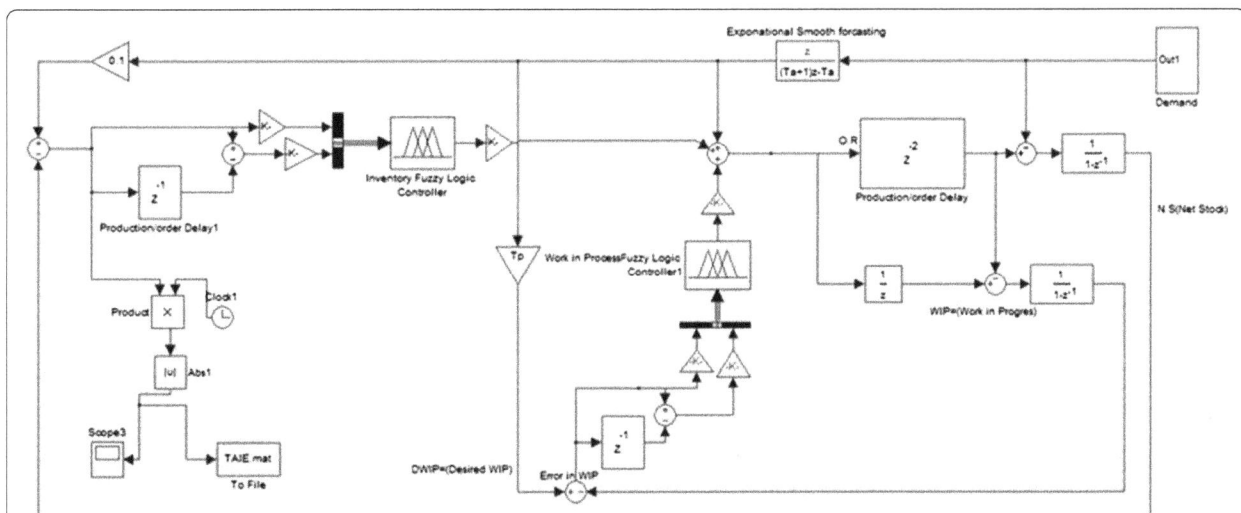

Figure 3 Block diagram of APVIOBPCS with fuzzy controllers.

one output (consequent): control action, u(t). Each of these three controller variables is evaluated over a normalized space range of $[-1,1]$, using five membership functions (NB, NS, Z, PS, PB). The Mamdani interference method and the AND connective are utilized with equal rule weighting. Then, a rule base consisting of 25 rules is produced.

One approach could be to optimize the rule base using all possible combinations of premise/consequent possibilities. We, therefore, have to test 450 different rules to achieve the best set of rules, which is time-consuming. Another alternative method is to diminish the rules. In this respect, there are different rule decreasing methods (Foran 2002). Before applying the method described in Foran (2002), some assumptions should be considered which are essentially based on experience.

The magnitude of the output control action is consistent with the magnitude of the input value; in other words, an extreme input value (premise) results in an extreme output value (consequent), a mid-range input value causes a mid-range output value, and a small/zero input value is synonymous with a small/zero output value.

If a large negative (positive) input generates a large negative (positive) response, then it is likely that smaller negative (positive) inputs will demand a response with the same polarity, but it would be with a smaller magnitude, and so on until a zero-crossing point is reached at the point whose response polarity changes.

The above-mentioned approach is a variation of the method used in Ross (1995). The consequent rule space is then 'overlaid' upon the premise coordinate system and is partitioned into regions where each region represents a consequent fuzzy set. The rule base is then extracted by determining the consequent region in which each premise combination point lies. Different possible consequent space partitions are defined using two parameters: consequent-line angle, θ, and consequent-region spacing, Cs.

The consequent-line angle defines the slope of the consequent line, which is used to create the space partitions. Cs is a proportion of the fixed distance between premises on the coordinate system and is used to define the distance between consequent points along the consequent line defined by angle θ.

This method is illustrated in Figure 4, considering $\theta = 50°$ and Cs = 1 with fixed premise spacing which is equal to consequent spacing. It should be noted that this may not be a real rule base in our study and it is just used to show the method.

Results and discussion
Structure description
In this part, we are going to present the results of our simulation to reflect the efficiency and the flexibility of our proposed method in finding optimal fuzzy controller

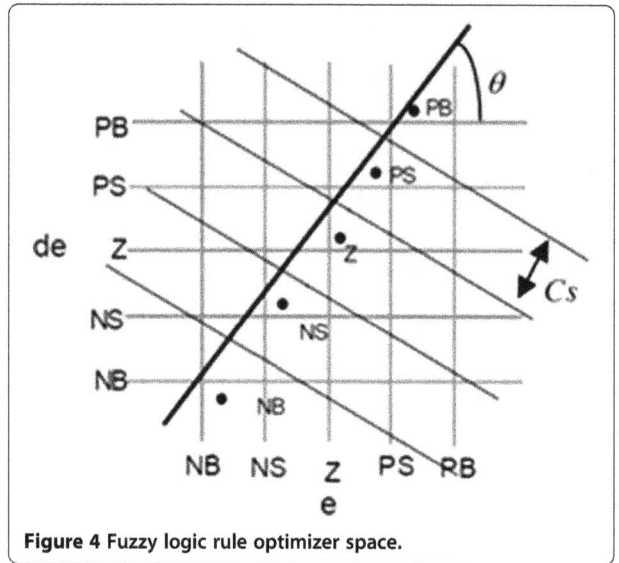

Figure 4 Fuzzy logic rule optimizer space.

parameters. We will use a discrete model of the most general member of the IOBPCS family, the automatic pipeline inventory and the order-based production control system (APVIOBPCS).

Figure 5 illustrates a block diagram of our proposed structure. Inputs of the multi-objective optimizer block are system objective functions. Notwithstanding the fact that this is a general structure, in our study, objective functions are inventory recovery response and attenuating bullwhip. According to these objective functions, and using the non-dominated sorting genetic algorithm-II (NSGA-II) method as an algorithm for multi-objective optimization, two sets of optimal parameters are produced. The first set directly affects the system dynamics, and the other set is used as the fuzzy tuning block input to optimize rule base and scaling gains. Outputs of multi-objective optimizer block and boundaries are shown in Table 1.

For evaluating objective functions, bullwhip is quantified as the ratio of output variance (ORATE) to the variance of input (Demand) while white noise represents

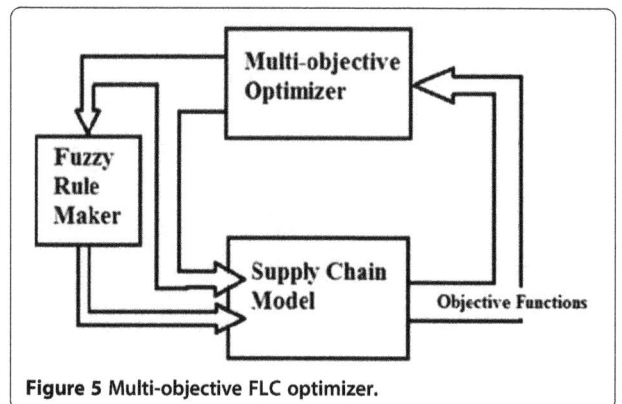

Figure 5 Multi-objective FLC optimizer.

Table 1 Fuzzy controller parameters and boundaries

		Parameters	Boundaries	Sample
Inventory policy	FLC	θ_inv	[5,170]	79.53
		Cs_inv	[0.5,2]	1.346
	Scaling gain	G1_e	[0.5,3]	0.9282
		G1_de	[0.5,3]	0.7175
		G1_out	[−2,2]	1.0013
Work in process policy	FLC	θ_wip	[5,170]	17.496
		Cs_wip	[0.5,2]	0.9529
	Scaling gain	G2_e	[0.5,3]	1.1328
		G2_de	[0.5,3]	0.7797
		G2_out	[−2,2]	0.9529
Forecasting		Ta	[1,10]	2.578

random demand in the market place, and the metric for quantifying inventory recovery responsiveness is the integral of time multiplied by absolute error (ITAE) which is defined by Equation 2 (Disney and Towill 2003):

$$\text{ITAE}_{\text{AINV}} = \sum_{n=0}^{\infty} n|E|. \qquad (2)$$

This error is the difference between desired inventory and actual inventory. Also, the desired inventory level may be fixed or be a multiple of average order.

Results and comparisons
As we mentioned, NSGA-II is used inside the multi-objective optimizer block. In this study, we fix the population of NSGA-II to 165 individuals. The optimal Pareto front, obtained through the multi-objective optimization of the parameters, is presented in Figure 6.

Figure 6 compares the fuzzy controller with the p-action controller. In some area, the fuzzy controller acts better than the p-action controller, and in some area, it does not. The major feature of fuzzy controllers is their ability in dealing with uncertainty, which is discussed in the subsequent part; nevertheless, in this study, we decrease the number of rules and use only 165 individuals in our simulation. Probably, it could be more optimal if a different setting is used for multi-objective block.

Uncertainty analysis
P-action controller is compared with fuzzy controller in the case of structured uncertainty. These may happen in inventory policy or forecasting policy. Exponential smoothing forecasting is used with the parameter Ta, which is shown in Figure 3. Lead time (Tp) is the parameter of forecasting policy which plays a crucial role because it defines the order of the system. A sample is selected from the Pareto front. Optimized parameters and their boundaries are shown in Table 1 for fuzzy controller, and p-action controller parameters are shown in Table 2.

Rule bases which resulted from our method for inventory loop and work in process loop are presented in Figures 7 and 8, respectively.

The order quantity of the system is presented for these controllers in normal situation in Figure 9. Normal parameters are shown in Tables 1 and 2. Now, we assume that there is some structured uncertainty in both forecasting mechanism and lead time. Hence, we assume Tp = 1.5 and Ta = 3 instead of Tp = 1 and Ta = 1.7213. The response is shown in Figure 10.

Although Ta is an important parameter in forecasting mechanism, uncertainty in Tp as lead time is more important because it defines the order of the system. It is clear that fuzzy controller acts much better than p-action controller.

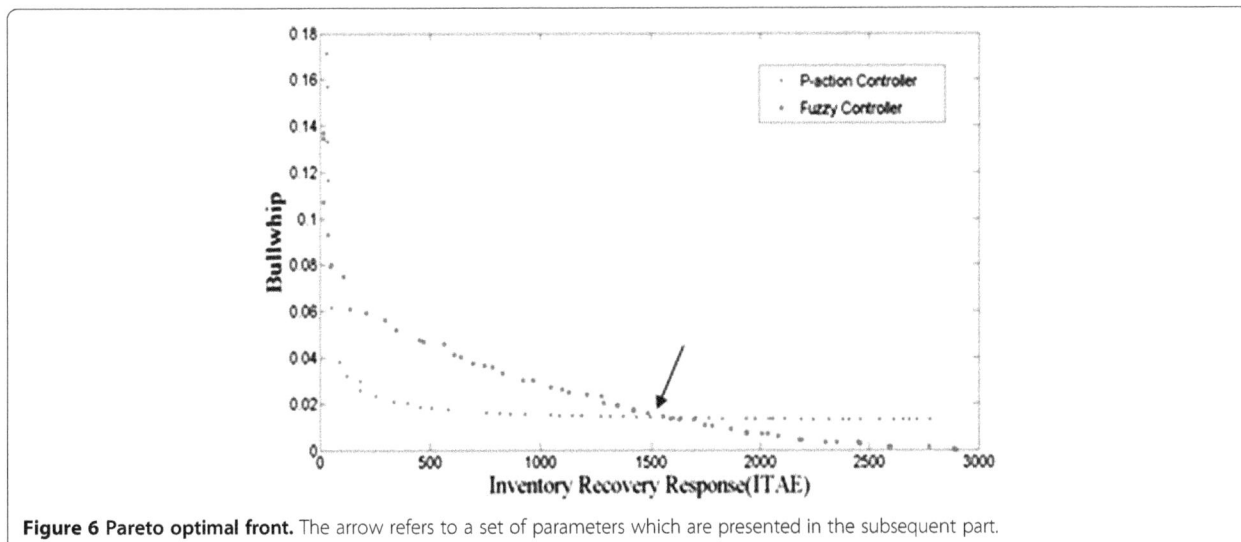

Figure 6 Pareto optimal front. The arrow refers to a set of parameters which are presented in the subsequent part.

Table 2 P-action controller parameters and boundaries

	Parameters	Sample
P-action	Ti	7.9353
	Ta	1.7213
	Tw	2.2507

de/e	NB	NS	Z	PS	PB
NB	NB	NB	NS	Z	PS
NS	NB	NS	Z	PS	PB
Z	NB	NS	Z	PS	PB
PS	NB	NS	Z	PS	PB
PB	NS	Z	PS	PB	PB

Figure 8 Rule base for work in process loop.

Conclusions

In supply chain, control parameter settings for different policies and different cases are very important. It is more critical when we face uncertainty. In this study, based on a fuzzy controller, the bi-objective supply chain model has been analyzed. In this manner, a rule base-maker algorithm has been used to make new fuzzy rule bases to achieve the optimum rule base. Using a multi-objective manner based on NSGA-II, the optimum parameters for rule maker's input and scaling and forecasting gains have been obtained, and then they have been used for an optimum design of fuzzy controller. In case of the presence of uncertainty, both p-action and fuzzy controllers have been compared, and the obtained results indicate excellent robustness in designed fuzzy controller compared to p-action controller. This robustness becomes more important when we face uncertain lead time because it changes system dynamic (the order of the system). This new bi-objective approach for supply chain design has been proposed which, in the case of uncertainty, is more robust than formerly used p-action controllers.

The results also show that combination of fuzzy logic and other soft computing methods is a good candidate for future development work because of its non-linear approximation capability and adaptability. Further perspective and attractive challenges for future research are the modification of different membership functions and the establishment of a non-linear-based modeling approach for real supply chain, e.g., the improvement of forecasting mechanism due to its fundamental effect on bullwhip.

Methods

NSGA-II implementation for multi-objective optimization

Multi-objective optimization (MOO) methods are utilized when two or more objective functions are necessary to be optimized simultaneously (Deb 2001). In the case of MOO methods, relative importance of objective functions is not generally known until the system's best capabilities are determined and trade-off between the objective functions is fully understood. This feature is the main advantage of MOO methods in comparison with simply weighted cost functions. The definition of an MOO problem requires substantial acquaintance with some preliminary definitions which are described as follows:

Objective functions: some functions of decision variables and criteria for estimating the appropriateness of a response. There are numbers of $k \geq 2$ objective functions in MOO which are shown as the vector $f(\bar{x}) = [f_1(\bar{x}), \ldots, f_k(\bar{x})]$.

Decision variables: a set of variables whose values suggest the response the response and can be right or wrong. These variables are presented in a form of $\bar{x} = [x_1, x_2, \ldots, x_r]^T$ where r is the number of variables.

- Constraints: defined in a form of some functions of decision variables, such as equalities or inequalities:

$$\text{Equality} : g_i(\bar{x}) = 0; i = 1, \ldots, p, \qquad (3)$$

$$\text{Inequality} : h_i(\bar{x}) \leq 0; i = 1, \ldots, m. \qquad (4)$$

- Feasible region: it is defined as the set of whole decision variables satisfying all constraints.
- Dominancy: NSGA-II is used in non-dominated sorting for fitness assignments. All individuals not dominated by any other individuals are assigned front number 1. All individuals only dominated by individuals in front number 1 are assigned front number 2, and so on. Selection is made using tournament between two individuals. The individual with the lowest front number is selected if the two individuals are from different fronts. The individual with the highest crowding distance is selected if they are from the same front.

de/e	NB	NS	Z	PS	PB
NB	NB	NB	NS	NS	Z
NS	NS	NS	NS	Z	Z
Z	NS	Z	Z	Z	PS
PS	Z	Z	PB	PS	PS
PB	Z	PS	PB	PB	PB

Figure 7 Rule base for inventory loop.

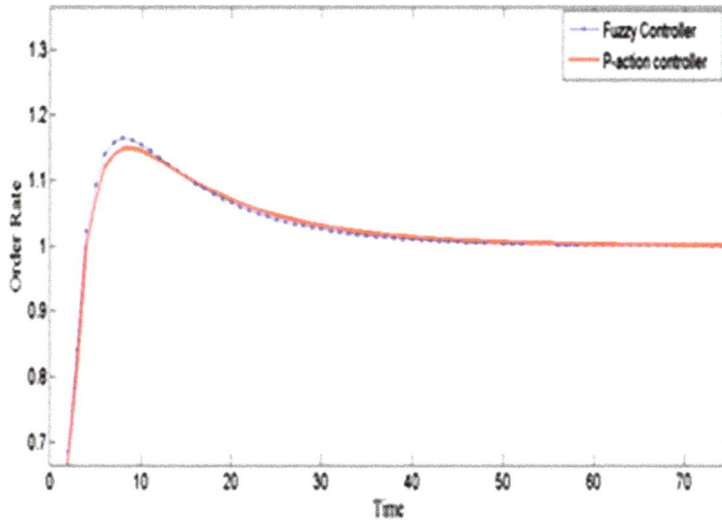

Figure 9 Order rate placed on a supplier.

In other words, the decision vector $\bar{x}_2 \in F$ is dominated by the decision vector $\bar{x}_1 \in F$, if and only if the decision vector $\bar{x}_1$ is better than or equal to $\bar{x}_2$ in all objectives:

$$\bar{x}_1 \succ \bar{x}_2 \leftrightarrow f(\bar{x}_1) \leq f(\bar{x}_2). \tag{5}$$

Pareto: when the cost functions have no confliction with each other, in MOO problems, it is possible to come across a unit optimal response which optimizes the whole cost functions. Otherwise, it is not feasible to find a response which optimizes all cost functions. In fact, since $F(x)$ is a vector, any other components of $F(x)$ are competing with each other and there is no unique solution for this problem. If $\bar{x}_2 \in F$ does not exist to dominate $\bar{x}_1 \in F$, the decision vector $\bar{x}_1$ is called an optimal Pareto response. The set of optimal Pareto responses is also known as a Pareto optimal front. All solutions on the Pareto optimal front are optimal.

In fact, according to Equation 5, the aim of MOO is to find the optimal Pareto front in which every response minimizes at least one of the two or more objective functions. In other cases, we have to revise Equation 1. For example, if we want both objective functions be maximized, Equation 5 must be rewritten as

$$\bar{x}_1 \succ \bar{x}_2 \leftrightarrow f(\bar{x}_1) \geq f(\bar{x}_2). \tag{6}$$

Now, MOO is defined as follows: Finding vector $\bar{x}^* = [x_1{}^*, x_2{}^*, \ldots, x_r{}^*]^T$ in a way that p equality constraints (Equation 3) and m inequality constraints (Equation 4)

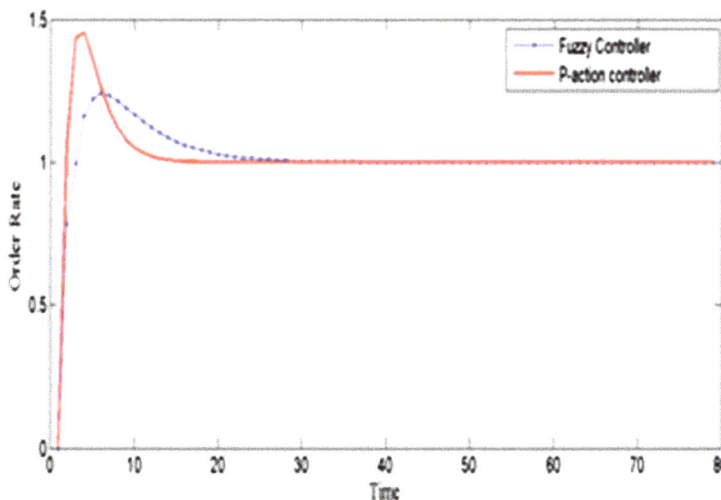

Figure 10 Order rate placed on a supplier with uncertainty.

would be satisfied; likewise, the function vector $f(\bar{x}) = [f_1(\bar{x}), \ldots, f_k(\bar{x})]$ is optimized. In other words, we estimate a specific $\bar{x}^*$ in a feasible region which leads to an optimal amount for the whole amount of k in the cost function.

In this article, in order to solve the available MOO problem, MATLAB software is utilized.

Competing interests
Both authors declare that they have no competing interest.

Authors' contributions
The work presented here was carried out as a collaboration between MJT and MG. Both authors read and approved the final manuscript.

Authors' information
Dr. MJT is an assistant professor in the Industrial Engineering Department at K.N.Toosi University of Technology, Tehran, Iran. Mr. MG is a faculty at Eqbal Lahouri Institute for Higher Education, Mashad, Iran.

Acknowledgments
The authors would like to thank Dr. Mahdi Aliyari, Dr. Hamid Khaloozadeh, and Dr. Mehrdad Kazerooni for their insightful comments and constructive suggestions that have improved the paper.

Author details
[1]Mechatronics Department, K.N.Toosi University of Technology, Tehran 19697 64499, Iran. [2]Industrial Engineering Department, K.N.Toosi University of Technology, Tehran 19697 64499, Iran.

References
Balan S, Prem V, Pradeep K (2007) Reducing the Bullwhip effect in a supply chain with fuzzy logic approach. Int J Integrated Supply Management 3:3
Banerjee S, Roy TK (2009) Application of the intuitionistic fuzzy optimization in the constrained multi-objective stochastic inventory model. J Tech 41:83–98
Beamon BM (1998) Supply chain design and analysis: models and methods. International Journal of Production Economics 55:281–294
Burbidge JL (1991) Period Batch Control (PBC) with GT—the way forward from MRP. In: Paper presented at the 26th BPICS annual conference, Birmingham, 14–16 November 1991
Chen Y-W, Larbani M, Liu C-H (2010) Simulation of a supply chain game with multiple fuzzy goals. Fuzzy Set Syst 161:1489–1510
Christer C, Robert F (2002) A position paper on the agenda for soft decision analysis. Fuzzy Set Syst 131:3–11
Deb K (2001) Multi-objective optimization using evolutionary algorithms. Wiley, Chichester
Dejonckheere J, Disney SM, Towill DR (2003) Measuring and avoiding the Bullwhip effect: a control theoretic approach. European Journal of Operational Research 147:567–590
Disney SM, Towill DR (2003) On the Bullwhip and inventory variance produced by an ordering policy. Omega 31:157–167
Disney SM, Towill DR (2005) Eliminating inventory drift in supply chains. International Journal of Production Economics 331–344:93–94
Foran J (2002) Fuzzy controller optimisation using genetic algorithms. M. Eng Dissertation, School of Electronic Engineering, DCU
Forrester JW (1961) Industrial dynamics. MIT, Cambridge
Haralambos S, Panagiotis P, Tarantilis CD, Kiranoudis CT (2008) Dynamic modeling and control of supply chain systems: a review. Computers & Operations Research 35:3530–3561
Hosoda T, Disney SM (2005) On variance amplification in a three echelon supply chain with minimum mean squared error forecasting. Omega: The International Journal of Management Science 34:344–358
Jiuh-Biing S (2005) A multi-layer demand-responsive logistics control methodology for alleviating the Bullwhip effect of supply chains. European Journal of Operations Research 161:797–811
Larbani M (2009) Non cooperative fuzzy games in normal form: a survey. Fuzzy Set Syst 160:3184–3210

Lee HL, Padmanabhan V, Whang S (1997) The bullwhip effect in supply chains. Sloan Management Review Spring 38:93–102
Mahnadm M, Yadollahpour MR, Famil-Dardashti V, Reza Hejazi S (2009) Supply chain modeling in uncertain environment with bi-objective approach. Computers & Industrial Engineering 56:1535–1544
Peter MC, Denis T (2002) Diagnosis and reduction of Bullwhip in supply chains. Supply Chain Management: An International Journal 7:64–179
Ross TM (1995) Fuzzy logic with engineering applications. McGraw-Hill, New York
Towill DR (1982) Dynamic analysis of an inventory and order based production control system. International Journal of Production Research 20:671–687
Towill DR, McCullen P (1999) The impact of agile manufacturing program on supply chain dynamics. International Journal of Logistics Management 10 (1):83–96
Wang J, Shu YF (2005) Fuzzy decision modeling for supply chain management. Fuzzy Set Syst 150:107–127
Wang Z, Qi G, Gu X, Pan X (2007) Supply chain performance evaluation system for mass customization. International Journal of Zhejiang University, Eng Sci 41(9):1567–1571
Xiaolong Z (2004) The impact of forecasting methods on the Bullwhip effect. Int J Production Economics 88:15–27
Yu X, Zhang Q (2010) An extension of cooperative fuzzy games. Fuzzy Set Syst 161:1614–1634

17

An inventory model for deteriorating items with time-dependent demand and time-varying holding cost under partial backlogging

Vinod Kumar Mishra[1*], Lal Sahab Singh[2] and Rakesh Kumar[2]

Abstract

In this paper, we considered a deterministic inventory model with time-dependent demand and time-varying holding cost where deterioration is time proportional. The model considered here allows for shortages, and the demand is partially backlogged. The model is solved analytically by minimizing the total inventory cost. The result is illustrated with numerical example for the model. The model can be applied to optimize the total inventory cost for the business enterprises where both the holding cost and deterioration rate are time dependent.

Keywords: Inventory model, Deteriorating items, Shortage, Time-dependent demand, Time-varying holding cost

Background

In the traditional inventory models, one of the assumptions was that the items preserved their physical characteristics while they were kept stored in the inventory. This assumption is evidently true for most items, but not for all. However, the deteriorating items are subject to a continuous loss in their masses or utility throughout their lifetime due to decay, damage, spoilage, and penalty of other reasons. Owing to this fact, controlling and maintaining the inventory of deteriorating items becomes a challenging problem for decision makers.

Harris (1915) developed the first inventory model, Economic Order Quantity, which was generalized by Wilson (1934) who gave a formula to obtain economic order quantity. Whitin (1957) considered the deterioration of the fashion goods at the end of the prescribed shortage period. Ghare and Schrader (1963) developed a model for an exponentially decaying inventory. Dave and Patel (1981) were the first to study a deteriorating inventory with linear increasing demand when shortages are not allowed. Some of the recent work in this field has been done by Chung and Ting (1993); Wee (1995) studied an inventory model with deteriorating items. Chang and Dye (1999) developed an inventory model with time-varying

demand and partial backlogging. Goyal and Giri (2001) gave recent trends of modeling in deteriorating item inventory. They classified inventory models on the basis of demand variations and various other conditions or constraints. Ouyang and Cheng (2005) developed an inventory model for deteriorating items with exponential declining demand and partial backlogging. Alamri and Balkhi (2007) studied the effects of learning and forgetting on the optimal production lot size for deteriorating items with time-varying demand and deterioration rates. Dye et al. (2007) find an optimal selling price and lot size with a varying rate of deterioration and exponential partial backlogging. They assume that a fraction of customers who backlog their orders increases exponentially as the waiting time for the next replenishment decreases.

In 2008, Roy developed a deterministic inventory model when the deterioration rate is time proportional. Demand rate is a function of selling price, and holding cost is time dependent. Liao (2008) gave an economic order quantity (EOQ) model with non instantaneous receipt and exponential deteriorating item under two level trade credits

Pareek et al. (2009) developed a deterministic inventory model for deteriorating items with salvage value and shortages. Skouri et al. (2009) developed an inventory model with ramp-type demand rate, partial backlogging, and Weibull's deterioration rate. Mishra and Singh (2010) developed a deteriorating inventory model for waiting

* Correspondence: vkmishra2005@gmail.com
[1]Department of Computer Science & Engineering, B. T. Kumaon Institute of Technology, Dwarahat, Almora, Uttarakhand 263653, India
Full list of author information is available at the end of the article

time partial backlogging when demand and deterioration rate is constant. They made the work of Abad (1996, 2001) more realistic and applicable in practice.

Mandal (2010) gave an EOQ inventory model for Weibull-distributed deteriorating items under ramp-type demand and shortages. Mishra and Singh (2011a, b) gave an inventory model for ramp-type demand, time-dependent deteriorating items with salvage value and shortages and deteriorating inventory model for time-dependent demand and holding cost with partial backlogging. Hung (2011) gave an inventory model with generalized-type demand, deterioration, and backorder rates.

In classical inventory models, the demand rate and holding cost is assumed to be constant. In reality, the demand and holding cost for physical goods may be time dependent. Time also plays an important role in the inventory system; therefore, in this inventory system, we consider that demand and holding cost are time dependent.

In this paper, we made the work of Roy (2008) more realistic by considering demand rate and holding cost as linear functions of time and developed an inventory model for deteriorating items where deterioration rate is expressed as a linearly increasing function of time. Shortages are allowed and partially backlogged. The assumptions and notations of the model are introduced in the next section. The mathematical model and solution procedure are derived in the 'Mathematical formulation and solution' section, and the numerical and graphical analysis is presented in the 'Results and discussion' section. The article ends with some concluding remarks and scope of future research.

Methods
Assumption and notations
This inventory model is developed on the basis of the following assumption and notations:

   i. Deterioration rate is time proportional.
   ii. $\theta(t) = \theta t$, where θ is the rate of deterioration; $0 < \theta < 1$.
   iii. Demand rate is time dependent and linear, i.e., $D(t) = a + bt$; $a, b > 0$ and are constant.
   iv. Shortage is allowed and partially backlogged.
   v. C_2 is the shortage cost per unit per unit time.
   vi. β is the backlogging rate; $0 \leq \beta \leq 1$.
   vii. During time t_1, the inventory is depleted due to the deterioration and demand of item. At time t_1, the inventory becomes zero and shortage starts occurring.
   viii. There is no repair or replenishment of deteriorating item during the period under consideration.
   ix. Replenishment is instantaneous; lead time is zero.
   x. T is the length of the cycle.
   xi. The order quantity of 1 cycle is q.

   xii. Holding cost $h(t)$ per unit time is time dependent and is assumed $h(t) = h + \alpha t$, where $\alpha > 0$; $h > 0$.
   xiii. C is the unit cost of an item.
   xiv. IB is the maximum inventory level during the shortage period.
   xv. I_0 is the maximum inventory level during $(0, T)$.
   xvi. S is the lost sale cost per unit.

Mathematical formulation and solution
The rate of change of the inventory during the positive stock period $(0, t_1)$ and shortage period (t_1, T) is governed by the following differential equations:

$$\frac{dI_1(t)}{dt} = -D(t) - \theta(t)I_1(t), \qquad 0 \leq t \leq t_1, \qquad (1)$$

$$\frac{dI_2(t)}{dt} = -\beta D(t), \qquad t_1 \leq t \leq T. \qquad (2)$$

The initial inventory level is I_0 unit at time $t = 0$; from $t = 0$ to $t = t_1$, the inventory level reduces, owing to both demand and deterioration, until it reaches zero level at time $t = t_1$. At this time, shortage is accumulated which is partially backlogged at the rate β. At the end of the cycle, the inventory reaches a maximum shortage level so as to clear the backlogged and again raises the inventory level to I_0 (Figure 1).

Thus, boundary conditions are as follows:

$$I_1(0) = I_0, \qquad I_1(t_1) = 0, \qquad I_2(t_1) = 0.$$

The solutions of Equations 1 and 2 with boundary conditions are as follows:

$$I_1(t) = -e^{-\frac{\theta t^2}{2}} \int_t^{t_1} e^{\frac{\theta t^2}{2}} (a + bt)\,dt \qquad (3)$$
$$0 \leq t \leq t_1,$$

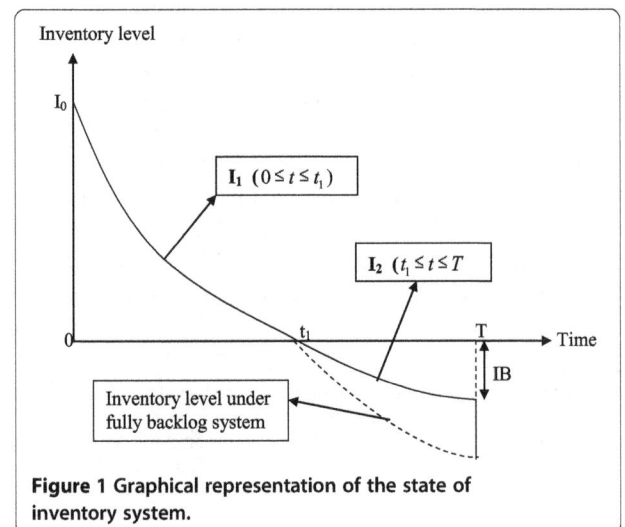

Figure 1 Graphical representation of the state of inventory system.

$$I_2(t) = -\beta\left[a(T - t) + \frac{b}{2}\left(T^2 - t_1^2\right)\right] \qquad (4)$$

$$t_1 \leq t \leq T.$$

Using Equation 3, we get the following:

$$I_0 = \int_0^{t_1} (a + bt)e^{\frac{\theta t^2}{2}}dt. \qquad (5)$$

Inventory is available in the system during the time interval $(0, t_1)$. Hence, the cost for holding inventory in stock is computed for time period $(0, t_1)$ only.

Holding cost is as follows:

$$
\begin{aligned}
\mathrm{HC} &= \int_0^{t_1} h(t)I_1(t) \\
\mathrm{HC} &= \int_0^{t_1} (h + \alpha t)I_1(t) \\
&= \int_0^{t_1} (h + \alpha t)e^{-\frac{\theta t^2}{2}}\int_t^{t_1}(a + bu)e^{\frac{\theta u^2}{2}}\,du\,dt \\
&= \begin{bmatrix} ah\left(\frac{1}{2}t_1^2 + \frac{1}{12}\theta t_1^4 + \frac{1}{90}\theta t_1^6\right) \\ + bh\left(\frac{1}{3}t_1^3 + \frac{1}{15}\theta t_1^5 + \frac{1}{105}\theta t_1^7\right) \\ + a\alpha\left(\frac{1}{6}t_1^3 + \frac{1}{40}\theta t_1^5 + \frac{1}{136}\theta t_1^7\right) \\ + b\alpha\left(\frac{1}{8}t_1^4 + \frac{1}{48}\theta t_1^6 + \frac{1}{384}\theta t_1^8\right) \end{bmatrix}.
\end{aligned}
\qquad (6)
$$

Shortage due to stock out is accumulated in the system during the interval (t_1, T).

The optimum level of shortage is present at $t = T$; therefore, the total shortage cost during this time period is as follows:

$$
\begin{aligned}
\mathrm{SC} &= C_2\int_{t_1}^{T} - I_2(t)dt \\
&= \beta a C_2(T - t_1)^2 + \frac{1}{2}\beta b C_2(T - t_1)^2(T + t_1).
\end{aligned}
\qquad (7)
$$

Due to stock out during (t_1, T), shortage is accumulated, but not all customers are willing to wait for the next lot size to arrive. Hence, this results in some loss of sale which accounts to loss in profit.

Lost sale cost is calculated as follows:

$$
\begin{aligned}
\mathrm{LSC} &= S\int_{t_1}^{T}(1 - \beta)D(t)dt \\
\mathrm{LSC} &= S(1 - \beta)\left[a(T - t_1) + \frac{1}{2}b\left(T^2 - t_1^2\right)\right].
\end{aligned}
\qquad (8)
$$

Purchase cost is as follows:

$$
\begin{aligned}
\mathrm{PC} &= C\left(I_0 + \int_{t_1}^{T}\beta D(t)dt\right) \\
\mathrm{PC} &= CI_0 + C\beta a(T - t_1) + \frac{1}{2}C\beta a\left(T^2 - t_1^2\right).
\end{aligned}
\qquad (9)
$$

Total cost is as follows:

$$\mathrm{TC} = \mathrm{OC} + \mathrm{PC} + \mathrm{HC} + \mathrm{SC} + \mathrm{LSC}$$

$$
\mathrm{TC} = \begin{bmatrix} A + CI_0 + \beta C + \left[a(T - t_1) + \frac{b}{2}\left(T^2 - t_1^2\right)\right] \\ + ah\left[\frac{1}{2}t_1^2 + \frac{1}{12}\theta t_1^4 + \frac{1}{90}\theta t_1^6\right] \\ + bh\left[\frac{1}{3}t_1^3 + \frac{1}{15}\theta t_1^5 + \frac{1}{105}\theta t_1^7\right] \\ + \alpha a\left[\frac{1}{6}t_1^3 + \frac{1}{40}\theta t_1^5 + \frac{1}{136}\theta t_1^7\right] \\ + \alpha b\left[\frac{1}{8}t_1^4 + \frac{1}{48}\theta t_1^6 + \frac{1}{136}\theta t_1^8\right] \\ + \beta a C_2(T - t_1)^2 + \frac{1}{2}\beta b C_2(T - t_1)^2(T + t_1) \\ + S(1 - \beta)\left[a(T - t_1) + \frac{1}{2}b\left(T^2 - t_1^2\right)\right] \end{bmatrix}.
\qquad (10)
$$

Differentiating Equation 10 with respect to t_1 and T, we then get the following:

$$\frac{\partial \mathrm{TC}}{\partial t_1} \quad \text{and} \quad \frac{\partial \mathrm{TC}}{\partial T}.$$

To minimize the total cost $\mathrm{TC}(t_1, T)$ per unit time, the optimal value of T and t_1 can be obtained by solving the following equations:

$$\frac{\partial \mathrm{TC}}{\partial t_1} = 0 \quad \text{and} \quad \frac{\partial \mathrm{TC}}{\partial T} = 0, \qquad (11)$$

providing that Equation 10 satisfies the following conditions:

$$\left(\frac{\partial^2 \mathrm{TC}}{\partial t_1^2}\right)\left(\frac{\partial^2 \mathrm{TC}}{\partial T^2}\right) - \left(\frac{\partial^2 \mathrm{TC}}{\partial t_1\partial T}\right)^2 > 0 \quad \text{and} \quad \left(\frac{\partial^2 \mathrm{TC}}{\partial t_1^2}\right) > 0. \qquad (12)$$

By solving (11), the value of T and t_1 can be obtained, and with the use of this optimal value, Equation 10 provides the minimum total inventory cost per unit time of the inventory system. Since the nature of the cost function is highly nonlinear, thus the convexity of the function is shown graphically in the next section.

Results and discussion

The following numerical values of the parameter in proper unit were considered as input for numerical and graphical analysis of the model, $A = 2,500$, $a = 10$, $b = 50$, $C = 10$, $C_2 = 4$, $h = 0.5$, $\theta = 0.8$, $\alpha = 20$, $\beta = 0.8$, and $S = 8$. The output of the model by using maple mathematical software (the optimal value of the total cost, the time when the inventory level reaches zero, and the time when the maximum shortages occur) is TC = 2,463.65, $t_1 = 1.127$, and $T = 1.562$.

If we plot the total cost function (10) with some values of t_1 and T such that fixed T at 1.562 and t_1 varies from 0.8 to 1.2, fixed t_1 at 1.127 and T varies from 1.2 to 1.8, and $t_1 = 0.08$ to 1.2 with equal interval $T = 1.2$ to 1.8, then we get the strictly convex graph of total cost function (TC) given by Figures 2, 3, 4, respectively.

The observation from Figures 2, 3, 4 is that the total cost function is a strictly convex function. Thus, the optimum value of T and t_1 can be obtained with the help of the total cost function of the model provided that the total inventory cost per unit time of the inventory system is minimum.

Conclusion

This paper presents an inventory model of direct application to the business enterprises that consider the fact that the storage item is deteriorated during storage periods and in which the demand, deterioration, and holding cost depend upon the time. In this paper, we developed a deterministic inventory model with time-dependent demand and time-varying holding cost where deterioration is time proportional. The model allows for shortages, and the demand is partially backlogged. The model is solved analytically by minimizing the total inventory cost. Finally, the proposed model has been verified by the numerical and graphical analysis. The obtained results indicate the validity and stability of the model. The proposed

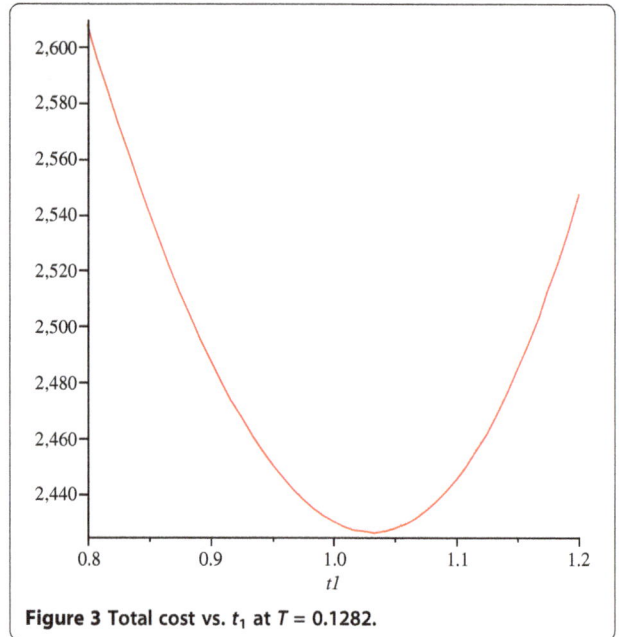

Figure 3 Total cost vs. t_1 at $T = 0.1282$.

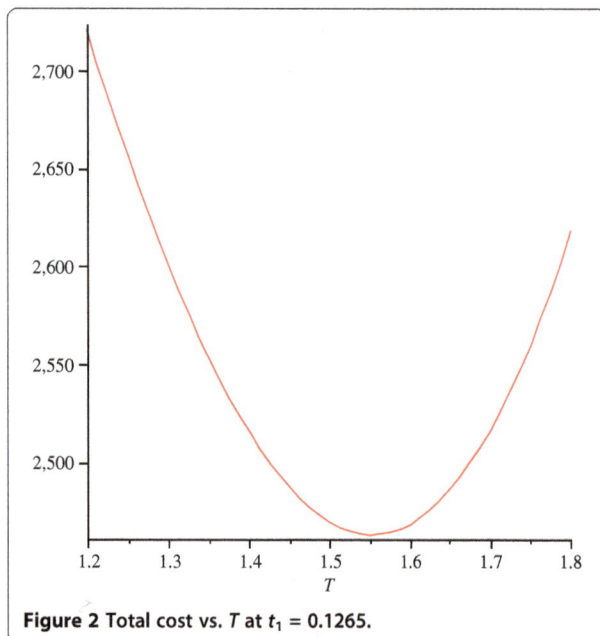

Figure 2 Total cost vs. T at $t_1 = 0.1265$.

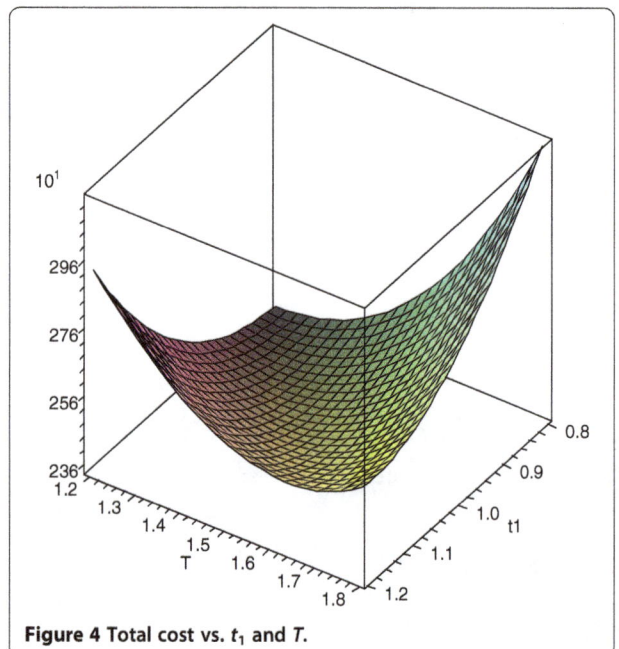

Figure 4 Total cost vs. t_1 and T.

model can further be enriched by taking more realistic assumptions such as finite replenishment rate, probabilistic demand rate, etc.

Competing interests
The authors declare that they have no competing interests.

Authors' contributions
VKM, LSS, and RK formulated the problem of deterministic inventory model for deteriorating items for time-dependent demand and time-varying holding cost under partial backlogging. VKM and RK performed the literature review and solved the formulated problem. VKM and LSS carried out the numerical and graphical analysis. All authors read and approved the final manuscript.

Acknowledgments
The authors would like to thank the editor and anonymous reviewers for their valuable and constructive comments, which have led to a significant improvement in the manuscript.

Author details
[1]Department of Computer Science & Engineering, B. T. Kumaon Institute of Technology, Dwarahat, Almora, Uttarakhand 263653, India. [2]Department of Mathematics & Statistics, Dr. Ram Manohar Lohia Avadh University, Faizabad, Uttar Pradesh 224001, India.

References

Abad PL (1996) Optimal pricing and lot-sizing under conditions of perishability and partial backordering. Manage Sci 42:1093–1104

Abad PL (2001) Optimal price and order-size for a reseller under partial backlogging. Comp Oper Res 28:53–65

Alamri AA, Balkhi ZT (2007) The effects of learning and forgetting on the optimal production lot size for deteriorating items with time varying demand and deterioration rates. Int J Prod Econ 107:125–138

Chang HJ, Dye CY (1999) An EOQ model for deteriorating items with time varying demand and partial backlogging. J Oper Res Soc 50:1176–1182

Chung KJ, Ting PS (1993) A heuristic for replenishment for deteriorating items with a linear trend in demand. J Oper Res Soc 44:1235–1241

Dave U, Patel LK (1981) (T, Si) policy inventory model for deteriorating items with time proportional demand. J Oper Res Soc 32:137–142

Dye CY, Ouyang LY, Hsieh TP (2007) Deterministic inventory model for deteriorating items with capacity constraint and time-proportional backlogging rate. Eur J Oper Res 178(3):789–807

Ghare PM, Schrader GF (1963) A model for an exponentially decaying inventory. J Ind Engineering 14:238–243

Goyal SK, Giri BC (2001) Recent trends in modeling of deteriorating inventory. Eur J Oper Res 134:1–16

Harris FW (1915) Operations and cost. A. W, Shaw Company, Chicago

Hung K-C (2011) An inventory model with generalized type demand, deterioration and backorder rates. Eur J Oper Res 208(3):239–242

Liao JJ (2008) An EOQ model with noninstantaneous receipt and exponential deteriorating item under two-level trade credit. Int J Prod Econ 113:852–861

Mandal B (2010) An EOQ inventory model for Weibull distributed deteriorating items under ramp type demand and shortages. Opsearch 47(2):158–165

Mishra VK, Singh LS (2010) Deteriorating inventory model with time dependent demand and partial backlogging. Appl Math Sci 4(72):3611–3619

Mishra VK, Singh LS (2011a) Inventory model for ramp type demand, time dependent deteriorating items with salvage value and shortages. Int J Appl Math Stat 23(D11):84–91

Mishra VK, Singh LS (2011b) Deteriorating inventory model for time dependent demand and holding cost with partial backlogging. Int J Manage Sci Eng Manage 6(4):267–271

Ouyang W, Cheng X (2005) An inventory model for deteriorating items with exponential declining demand and partial backlogging. Yugoslav J Oper Res 15(2):277–288

Pareek S, Mishra VK, Rani S (2009) An inventory model for time dependent deteriorating item with salvage value and shortages. Math Today 25:31–39

Roy A (2008) An inventory model for deteriorating items with price dependent demand and time varying holding cost. Adv Modeling Opt 10:25–37

Skouri K, Konstantaras I, Papachristos S, Ganas I (2009) Inventory models with ramp type demand rate, partial backlogging and Weibull deterioration rate. Eur J Oper Res 192:79–92

Wee HM (1995) A deterministic lot-size inventory model for deteriorating items with shortages and a declining market. Comput Oper 22:345–356

Whitin TM (1957) The theory of inventory management, 2nd edition. Princeton University Press, Princeton

Wilson RH (1934) A scientific routine for stock control. Harv Bus Rev 13:116–128

18

A simple approach to the two-dimensional guillotine cutting stock problem

Mir-Bahador Aryanezhad[1^], Nima Fakhim Hashemi[1*], Ahmad Makui[1] and Hasan Javanshir[2]

Abstract

Cutting stock problems are within knapsack optimization problems and are considered as a non-deterministic polynomial-time (NP)-hard problem. In this paper, two-dimensional cutting stock problems were presented in which items and stocks were rectangular and cuttings were guillotine. First, a new, practical, rapid, and heuristic method was proposed for such problems. Then, the software implementation and architecture specifications were explained in order to solve guillotine cutting stock problems. This software was implemented by C++ language in a way that, while running the program, the operation report of all the functions was recorded and, at the end, the user had access to all the information related to cutting which included order, dimension and number of cutting pieces, dimension and number of waste pieces, and waste percentage. Finally, the proposed method was evaluated using examples and methods available in the literature. The results showed that the calculation speed of the proposed method was better than that of the other methods and, in some cases, it was much faster. Moreover, it was observed that increasing the size of problems did not cause a considerable increase in calculation time.

In another section of the paper, the matter of selecting the appropriate size of sheets was investigated; this subject has been less considered by far. In the solved example, it was observed that incorrect selection from among the available options increased the amount of waste by more than four times. Therefore, it can be concluded that correct selection of stocks for a set of received orders plays a significant role in reducing waste.

Keywords: Cutting stock problem, Trim loss, Two-dimensional cutting, Guillotine cutting

Background

Cutting problems are derived from various industrial processes, for example, textile production systems, glass industries, steel, adhesive tape, wood, paper, etc. (Gonçalves 2007; Ben Messaoud et al. 2008; Leung et al. 2001; Erjavec et al. 2009).

The first research on cutting and packing problems was probably performed by Kantorovich in 1939 and Brooks in 1940, but extensive scientific work in this field began in 1960 as cited in Faina (1999), Dyckhoff (1990), and Javanshir and Shadalooee (2007). Among the most famous studies from that time are those of Gilmore and Gomory (Gilmore and Gomory 1961, 1963, 1965; Alvarez-Valdes et al. 2001; Eshghi and Javanshir 2005). Since the 1960s, many other cutting and packing

problems with the same logic and structure emerged but were given different names in the literature, such as cutting stock and trim-loss problems, bin packing, strip packing, knapsack problems, loading pallets, and loading containers (Faina 1999; Eshghi and Javanshir 2005).

In 1990, Dyckhoff presented a systematic and consistent approach to the integration of different types of problems into a comprehensive typology based on the logical structure of cutting and packing problems. His goal was to assimilate the notations for various uses in the literature and to focus future research on specific types of problems (Dyckhoff 1990).

There are several types of cutting problems based on different characteristics, which Dyckhoff divided into four groups based on their salient characteristics (dimension, type assignment, assortment of large objects, and small items) (Dyckhoff 1990; Dyckhoff and Finke 1992).

Cutting is categorized according to different conditions, such as the cutting style, which can be either

* Correspondence: n_hashemi@ind.iust.ac.ir
^Deceased
[1]Department of Industrial Engineering, Iran University of Science and Technology, Tehran 16846-13114, Iran
Full list of author information is available at the end of the article

guillotine or non-guillotine cutting. Guillotine cutting, which is also called edge-to-edge cutting, is a common method for cutting rectangular forms, in which the cutting edge starts on one side and continues to the opposite side (Suliman 2006; Ben Messaoud et al. 2008; Belov and Scheithauer 2006; Mellouli and Dammak 2008).

Another significant condition is whether the items have rotation. Fixed orientation is a state in which there is no possible rotation of the items. Cutting wallpaper or packing articles in a newspaper are examples of this condition (Ben Messaoud et al. 2008; Dyckhoff 1990). This condition should also be considered for the cardboard used for making cartons because of their inter-leaved layers, which play a significant role in the stability of the resulting carton.

In Suliman (2006), a sequential heuristic procedure for a two-dimensional rectangular guillotine cutting stock problem is introduced. The paper provides a three-step solution.

In Gonçalves (2007), the main topic is a two-dimensional orthogonal packing problem, where a fixed group of small rectangles must be fitted into a large rectangle so that most of the material is used, and the unused area of the large rectangle is minimized. The algorithm combines a replacement method with a genetic algorithm based on random numbers; the new algorithm, called 'a hybrid genetic algorithm-heuristic,' includes a fitness function and a new heuristic replacement policy.

In Ben Messaoud et al. (2008), the authors focus on guillotine constraints. First, for a cutting problem to be in a guillotine pattern, they provide a necessary condition and then a polynomial algorithm for checking this condition. These methods are useful when the raw materials for cutting are expensive. They also claimed that their study is the first to formulate the guillotine constraints in terms of linear inequalities. In fact, an integer programming formulation is provided that considers the guillotine constraints explicitly. However, an adaptation of their algorithm enables it to convert an initial non-guillotine pattern to a guillotine pattern and to recognize all non-guillotine regions in a given pattern.

In Alvarez-Valdés et al. (2002), a number of heuristic algorithms for two-dimensional cutting problems (on large scales) are developed. In this study, there is a large primary stock that has to be cut into smaller pieces so as to maximize the value of the pieces. They developed a greedy randomized adaptive search procedure (GRASP), which is a fast method that provides good-quality answers. They also developed a Tabu search algorithm, which was more complex but gave better results in reasonable computational time. They also stated that the GRASP method should be used if quick solutions are needed, whereas the Tabu search algorithm should be selected if quality is more important.

In Faina (1999), the two-dimensional cutting stock problem is solved using a general optimization algorithm and simulated annealing. The algorithm includes the guillotine and non-guillotine constraints. The author also noted that for large numbers of items, the NON-GA is far superior to the GA because it obtains better cutting patterns in almost the same time.

Assembly line balancing problems have attracted attention for years. These problems were considered in around 1950s. The studies of Bryton in 1954 and Salveson in 1955 as cited in Malakooti (1991, 1994), Fathi et al. (2011), and Roy and Khan (2010, 2011a, 2011b) can be referred to as examples. In general, assembly line balancing seeks to assign a number of tasks to workstations considering precedence (priority) relations so that a product is ready in a given cycle time and efficiency is maximized (Malakooti 1991, 1994; McMullen and Tarasewich 2003; Capacho and Pastor 2008; Fathi et al. 2011; Roy and Khan 2010, 2011a, 2011b).

Most of the methods developed for two-dimensional cutting stock problems are difficult, time-consuming, and costly. In addition, most are suitable only for special kinds of problems, so most factories are not interested in using such methods but rely on traditional procedures (based on the author's personal experience in the carton industry). The main goal of this paper is to obtain a simple and practical approach by finding and exploiting a meaningful relation between cutting stock problems, assembly line balancing problems, and number of workstation problems.

The other subject that we will discuss, which has been neglected to some extent in the literature thus far, is the problem of selecting the best stock size from the available stocks in the market and the effect of size selection on the level of waste.

The contents below are organized as follows: the 'Problem description' section describes the problem; the 'Suggested cutting approach' section suggests the cutting approach and describes the algorithm; the 'Minimum number of stocks' section determines the minimum number of stocks required; the 'Stock selection' section focuses on the stock selection; the 'Results and discussion' section reports the experimental results; and the 'Conclusions' section gives the conclusions and suggests subjects for future research.

Methods

Problem description

Two-dimensional regular cutting stock problem

Packing problems or two-dimensional cutting problems are given in the following form: given a series of large rectangles of a standard size (stocks) and a series of small rectangles (items or pieces), the aim is to cut or pack all items while minimizing the waste (Ben Messaoud et al. 2008).

Assumptions

In this research, we make the following assumptions:

a) Cuts are guillotine cuts.
b) The stocks and items are rectangular.
c) The orientation is fixed, and the items are not rotated.
d) The assignment is of the third kind (all items, selection of objects) (Dyckhoff and Finke 1992; Wäscher et al. 2007) because all items must be cut to serve all demands.
e) The model is proposed under deterministic conditions (inputs are deterministic).
f) The remaining sheets in each size, when all demands have been met, are considered as waste, although they may be used for further orders. (Because the stocks are not assumed to be in rolls and are separated, if there are no orders left, the remaining parts of any size are considered temporarily as waste). This assumption causes the waste percentage to be higher, but possibly only temporarily; as new demands are received and new parts are cut from the waste, the waste percentage decreases.

Suggested cutting approach

As observed in Neapolitan and Naimipour (2004), packing and cutting stock problems are considered to be in the field of knapsack problems, which are non-deterministic polynomial-time (NP)-hard problems. One of the methods discussed in this book is approximation algorithm design. This algorithm is used for a NP-hard optimization problem. There is no proof that it offers an optimal solution, but it offers solutions that are logically close to the optimal solution. Neapolitan and Naimipour (2004) present an approximation algorithm for the bin-packing problem, which, in fact, is equivalent to a one-dimensional cutting stock problem; this simple strategy is called 'first-fit', and the limit of the closeness of the approximate solution to the optimal solution is well defined.

The theorem is as follows:

Let 'opt' be the optimum number of bins for a packing problem and 'approx' be the number of bins used by the non-descending first-fit algorithm; then,

$$\text{approx} \leq 1.5 \times \text{opt}. \tag{1}$$

The above theory was proved in the ninth chapter of the book by Neapolitan and Naimipour.

In Dyckhoff (1990), the author states that the first-fit-decreasing algorithm (FFD) consumes up to 22.3% more stock than under-optimized conditions and that the unnecessary waste is no more than 18.2%. He then considers a condition under which the waste is reduced to 1.9% and

finally observes that if the number of items tends to infinity, the average amount of waste tends to zero.

In Dósa (2007), the author first reviews the previous bounds, then observes that minimizing the constant in the equation has been an open question over the years, and then states,

$$\text{FFD}(I) \leq 11/9\text{OPT}(I) + 6/9. \tag{2}$$

He also notes that this is a tight bound.

As described in Neapolitan and Naimipour (2004), these methods are not subject to specific limitations and complexities, and normally they provide a convenient solution. The main advantage of cutting stock problems is that to evaluate the quality of a solution, there is no need to know the exact optimum solution, as the waste percentage itself describes the quality of the solution.

In this part, we suggest an easy way to decrease trim loss and reduce stock sheet usage. We also examine the guillotine form of two-dimensional cutting stock problems. According to the definition of guillotine cutting, it seems that the waste in this form of problem is greater than or equal to that in the non-guillotine form, as the guillotine cut imposes a limitation and, as we know, as the limitations increase, if the solution space does not shrink, at best, it will be constant and not grow.

The first principle that comes to mind in the guillotine form is that the width of the pieces in each strip should be equal or as close as possible to the width of the strip, so as to reduce the waste at the edges of the strips.

The cutting problems in this research are assumed to be almost the same as the balancing problems. Therefore, we have tried to take advantage of their methods as much as possible and adapt them as necessary. In this part, we use a method similar to the ranked positional weight (RPW) proposed by Helgeson and Bernie in 1961 for balancing problems as cited in Capacho Betancourt (2007).

The RPW method is also very similar conceptually to the FFD method, which has been used in one-dimensional cutting stock problems (Javanshir and Shadalooee 2007). We have therefore tried to suggest a suitable method for two-dimensional guillotine cutting problems using the above concepts. As mentioned before, we have tried to suggest a simple, applicable, quick, and easy calculation method that is user-friendly and easily applicable for industry.

Similarity of the RPW and FFD methods

The RPW method is a method for the assignment of tasks to workstations in line balancing problems. The stages of the RPW method are as follows:

1. First step: Calculation of weights for each task according to their positions.

2. Second step: Descending arrangement and ranking of the tasks according to their positional weights.
3. Third step: Initial assignment of tasks to stations according to the cycle time and precedence from the beginning of the list.

The FFD method is conceptually similar to the RPW method. The approach suggested in this paper for the two-dimensional guillotine cutting problem will make use of these concepts and offer some suggestions. It is necessary to mention that the precedence mentioned in the third step of the PRW method also exists in the two-dimensional guillotine cutting problem as a different form to be considered. In fact, when creating a strip in two-dimensional guillotine cutting, the width of each piece cut from the strip should be less than or equal to the width of the strip.

Suggested approach for two-dimensional guillotine cutting
The assumed problem is two-dimensional guillotine cutting, which we can describe in terms of the three steps of the RPW method as follows:

1. First step: Assigning priorities to the pieces to be cut (weights).

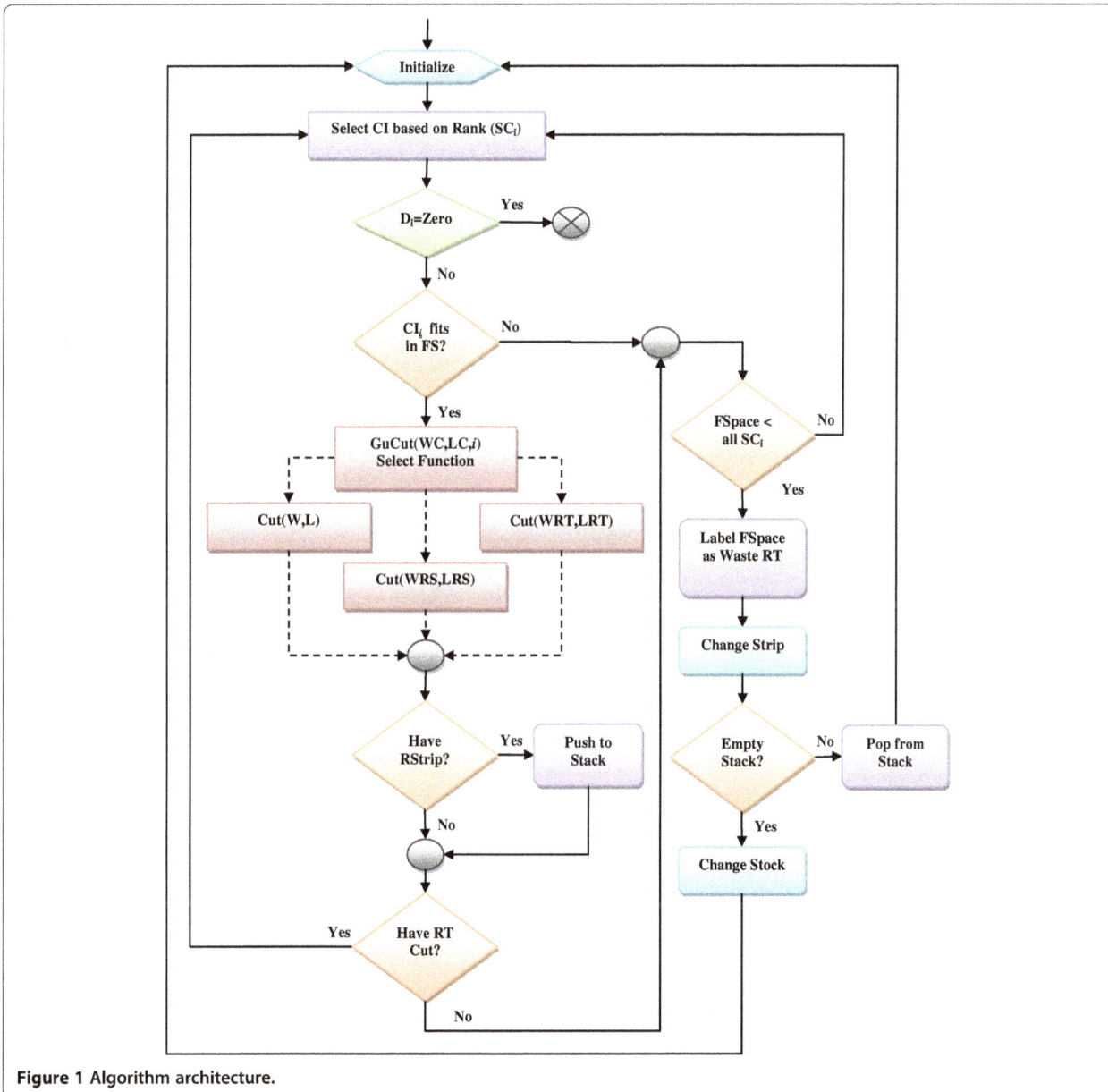

Figure 1 Algorithm architecture.

Table 1 Example specifications

Reference	Example	Stock size	Items (quantity)	Types
Burke, Kandall, Whitwell (chapter 3)(Whitwell 2004)	N1	40×40	10	10
	N2	30×50	20	20
	N3	30×50	30	30
	N4	80×80	40	40
	N5	100×100	50	50
Burke, Kandall, Whitwell (chapter 4)(Whitwell 2004)	MT1	400×600	100	12
	MT2	300×900	150	10
	MT3	400×900	200	23
	MT4	$600 \times 1,300$	300	24
	MT5	$800 \times 1,400$	500	47

2. Second step: Descending arrangement of the pieces by priority (weights).
3. Third step: Cutting the pieces from the beginning of the list produced in the previous step, considering the length and width of the stock.

Suggested algorithm implementation

In this algorithm, W is the width of stock; L, length of stock; N, quantity of needed stocks; w_i, width of the pieces ($w_i \leq W \quad \forall \quad i = 1,2,\ldots, n$); l_i, length of the pieces ($l_i \leq L \quad \forall \quad i = 1,2,\ldots, n$); and d_i, number of demands for each kind of pieces.

Stages of the algorithm. The following are the stages of the algorithm:

1. Initializing the primary quantities:
   In this step, the dimensions of the stock, the requested pieces, and the demand for each item are input: W, L, w_i, l_i, d_i
   ($i = 1,2,\ldots, n$)
2. Primary calculations:

   (a) Calculating the areas of the pieces.
   $$a_i = w_i \times l_i, (i = 1, 2, \ldots, n) \qquad (3)$$

(b) Multiply the demand by the area.
$$a'_i = a_i \times d_i, (i = 1, 2, \ldots, n) \qquad (4)$$

(c) Calculating the total area of the demand.
$$A_t = \sum_{i=1}^{n} a'_i \qquad (5)$$

3. Assigning priorities to the pieces to be cut:
   Here, all of the demands will be arranged in descending order by the width of the pieces. If the widths of two pieces are the same, then the longer piece will be chosen. In this step, the priorities can also be assigned by the sum of the widths and lengths or by the area. While the cuts in this paper are assumed to be guillotine, it makes sense to assign priorities by the widths of the demand pieces because most of the waste in the guillotine cuts is created in the strips and while trimming, and in this suggested approach, the widths of the pieces in each strip are equal or as close as possible. It is possible that assigning priorities based on the sum of widths, lengths, or areas might produce better results in some non-guillotine cases. It is even possible to use a

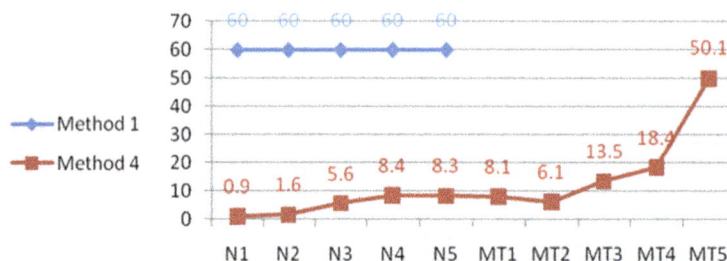

Figure 2 Comparing calculation time of Methods 1 and 4.

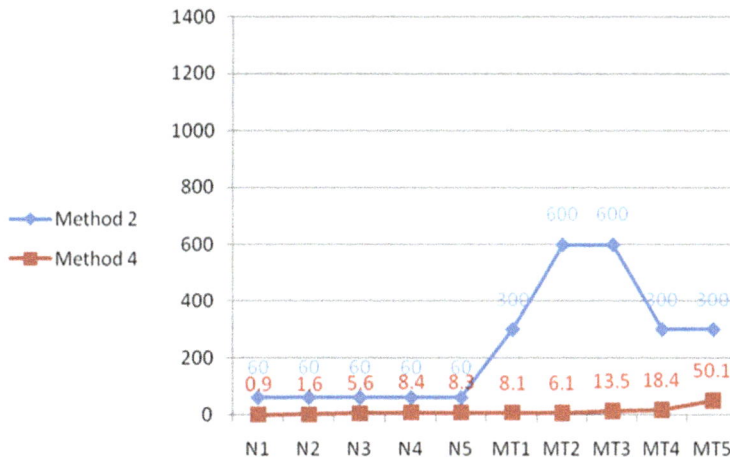

Figure 3 Comparing calculation time of Methods 2 and 4.

combination of one of these methods and the bottom left method.

In two-dimensional guillotine cutting, the widths of the pieces should also be controlled in each cut because the widths of the pieces should be less than or equal to the width of the strip created in the guillotine cuts. For this reason, assigning priority by the width of the pieces also has the additional advantage of reducing the controls in each cut. In this case, there is no need to compare the widths of the pieces with the width of the strip because surely they are less than or equal to the width of the strip, and this increases the speed of the algorithm. Finally, a list of cutting priorities is created.

4. Loading the stock with width W and length L.
5. The first undone item in the cutting priority list (CPL) is loaded, and its width is cut from the remaining stock if possible. Then, the item's lengths are cut out of the cut piece. If the width or the length of the demanded pieces was greater than the width or the length of the remaining stock, proceed to the next step.
6. Repeat step 5 for the other demands in order of CPL. If the width or the length of the remaining stock is less than all remaining demands, then go to step 4; if not, go to the next step.
7. If all of the demands have been completed, the algorithm stops, performs final calculations, and reports the result.

Algorithm performance criteria. The algorithm performance is defined as the ratio of total area of the demanded pieces to the total area of the stocks consumed in order to fulfil the demand. This ratio is the

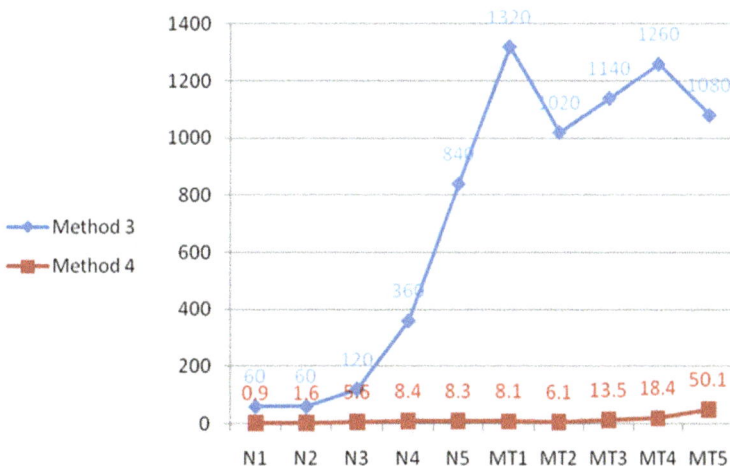

Figure 4 Comparing calculation time of Methods 3 and 4.

same as the efficiency in balancing problems, which is derived as follows:

$$\text{Efficiency} = \frac{\sum_{i=1}^{n} t_i}{m \times t_c}, \tag{6}$$

where t_i is the processing time of task, t_c is the cycle time, and m is the number of workstations. The efficiency for the proposed approach is calculated as follows using the same method:

$$E = \frac{\sum_{i=1}^{n} a_i \times d_i}{N \times W \times L} \times 100. \tag{7}$$

We could also define the percentage of waste as follows:

$$\%\text{Wastage} = \frac{(N \times W \times L) - \sum_{i=1}^{n} a_i \times d_i}{N \times W \times L} \times 100. \tag{8}$$

Proposed algorithm architecture

The first step in the algorithm (Figure 1) is the Initialization phase, in which all the variables are set according to the current iteration and the initial parameters. Then, an item is selected based on the priority table for cutting. This step is represented in the Select Current Item part of the flowchart. Following this step, the Termination Condition will be checked to determine whether all the demands have been satisfied. The algorithm stops when there are no more demands left. In cases where there are still unsatisfied demands, the process continues to compare the dimensions of the Current Item to the Free Space left on the strip. If the cutting is feasible, then the Cut Function will be called.

There are three different scenarios in which the main cutting function can be called. In every run, one of these cases occurs. The first case is when the free space is equal to the raw stock (W,L) size. From the second iteration and after the cutting procedure, there will be two remainder strips from the stock. The width and length of the upper strip are represented by WRS and LRS, respectively. In every cycle, newly generated strips are placed on the top of the stack, which forms the second scenario. Conversely, the lower strips are used to produce new items in the third scenario. They will be used until there are no more feasible cuts, which lead to waste items. Then, the control flow will change to the stack items and the pop function will be called. The cutting algorithm continues to run until the control flow reaches the waste items. In a situation where there are no strips and the stack is empty, a new raw stock will be used. The above algorithm has been implemented using the C++ language.

The design and implementation of the applied algorithm for solving such problems has a direct influence on the reduction of calculation time. The software architecture presented in this paper can be extended to different cutting approaches.

Minimum number of stocks

In this section, we will determine the minimum number of stocks required. When our organization receives a list of demands, the first step is to provide the stock sheets. This step should be done as soon as possible so that the process can start without delay. Additionally, in this case, the cutting problem is treated as a balancing problem (number of workstations).

The first kind of balancing problem, the simple assembly line balancing problem (SALBP-1), assumes that the cycle time is given and the target is to minimize the number of workstations (Capacho Betancourt 2007).

We now assume that the cutting problem is a SALBP-1, in which the area of the stock corresponds to the given cycle time in the SALBP-1. Balancing problems are problems of calculating a limit for the number of workstations.

The lower bound on the number of workstations is calculated using the following formula:

$$m_{\min} = \left\lceil \sum_{i=1}^{n} t_i / t_c \right\rceil, \tag{9}$$

where $m_{\min}$ is the lower bound for the minimum number of workstations, t_i is the required time for each task, and t_c is the cycle time.

Using the above formula, we can calculate the lower bound on the number of stocks:

$$S_{\min} = \left\lceil \sum_{i=1}^{n} a_i \times d_i / W \times L \right\rceil, \tag{10}$$

where $S_{\min}$ is the lower bound on the number of stocks, a_i is the area of each item, d_i is the number of demand for each item, and W and L are the width and length of the stock, respectively.

We can thus calculate the minimum number of required stocks. This lower bound is valid for every cutting method.

The upper bound for the number of workstations in balancing problems is

$$m_{\max} = n, \tag{11}$$

where 'n' refers to the number of tasks, but this upper bound is not applicable for the required number of stocks. (In further studies, the applicable upper bound based on this approach could be examined).

Stock selection

In this section, the main target is to select the proper stock. Stock producers and suppliers generally produce stocks in various sizes, and companies have several choices available. Therefore, we have exploited this fact and assumed that depending on demands received, the companies could choose the stock size that ultimately results in the least waste.

Suppose the organization has received an order. To satisfy this order, the stocks satisfying the following condition could be chosen: $\{W \geq w_i, L \geq l_i \ \forall \ i = 1,2,\ldots, n\}$.

The waste is calculated for each of the available stocks in the market for which this condition applies, and finally, the stock with the lowest waste is selected.

$$\text{Best stock} = \text{Min} \left\{ \%\text{wastage}_1, \%\text{wastage}_2, \ldots, \%\text{wastage}_k \right\},$$

where wastage$_k$ is the waste of the kth-type stock.

The important point is that this criterion is used under the assumption that the price is calculated based on the area or weight of the sheets. (In weight mode, it is assumed that the weight is proportional to the area). Otherwise, the conditions will be different. For example, if the price is not calculated based on area or weight, but each type of stock has a unique price (not related to area or weight directly), then a different selection criterion must be defined, which will be described later on.

Here, the selection criterion is

$$\text{Best stock} = \text{Min}\{(N_1 \times C_1), (N_2 \times C_2), \ldots, (N_k \times C_k)\},$$

where Ck is the kth-type stock cost and Nk is the number of the kth-type stocks used.

Results and discussion

In this section, computational results for a number of examples are given to evaluate the algorithm's efficiency. The results were derived by running the software on a Core 2 Duo 2.53 GHz PC with 4 GB of RAM.

Approach evaluation

In this section, we evaluate the numerical results obtained by the approach discussed in the 'Suggested cutting approach' section of this article and by some pre-existing methods in the literature on rectangular sheets and parts (Whitwell 2004). Ten examples including N1, N2,..., N5 and MT1, MT2,..., MT5 were considered. The specifications of these examples including sheet dimensions and the number and type of pieces which indicated the size of problems are given in Table 1.

All the examples were solved using the proposed approach, and they were compared with the results of

three other methods in Figures 2, 3, and 4 (Method 1: metaheuristic bottom-left-fill (MBLF); Method 2: best-fit (BF) + MBLF; Method 3: interactive approach (IA); Method 4: the proposed method of this paper).

The comparison criterion was the calculation speed. Each of the examples was placed on the horizontal axis, and their solving time (in seconds) was specified on the vertical axis.

As can be observed in Figure 2, the calculation time of the proposed method (Method 4) is much lower than that of Method 1 in five initial examples. For Method 1 in the following five examples, no results have been reported in the literature.

Also, it can be seen in Figures 3 and 4 that the proposed method (Method 4) conducts and ends the calculations faster than Methods 2 and 3. Moreover, the calculation time of the proposed method (Method 4) shows no noticeable increase by increasing the size of problems. It should be mentioned that the time declared for the proposed method (Method 4) includes the time spent for generating the report in addition to the solving time.

Influence of stock selection on amount of waste

We present an example of the influence of stock selection on the amount of waste produced. Suppose demands are received as in Table 2.

Suppose there are 20 different kinds of stock available on the market, naturally, the best choice is the stock size that leads to the least waste. The waste percentages for these 20 stocks are listed in Table 3.

As seen here, the choice of stock for a series of demands has a noticeable influence on the waste; the

Table 2 List of items and demands

i	w_i	l_i	d_i
1	25	40	16
2	60	120	11
3	10	10	18
4	25	50	14
5	35	50	10
6	15	20	25
7	70	95	11
8	15	25	24
9	58	70	15
10	5	10	6
11	42	60	6
12	15	15	16
13	40	60	10
14	30	65	15
15	20	35	4

Table 3 Waste percentage for each kind of stock

N_i	Stock size	S_{min}	S_{used}	%Waste	Best choice
1	70 × 120	43	50	14.8524	
2	80 × 120	38	49	23.9753	
3	100 × 120	30	37	19.455	
4	100 × 150	24	30	20.5289	
5	100 × 180	20	24	17.2176	
6	100 × 200	18	19	5.88947	√
7	100 × 220	17	19	14.445	
8	100 × 230	16	18	13.6184	
9	100 × 250	15	16	10.595	
10	100 × 260	14	15	8.30256	
11	100 × 320	12	12	6.86979	
12	120 × 120	25	28	11.3046	
13	140 × 220	12	14	17.064	
14	140 × 260	10	11	10.6843	
15	140 × 400	7	7	8.77041	
16	200 × 310	6	7	17.5991	
17	200 × 330	6	6	9.69192	
18	200 × 340	6	6	12.348	
19	220 × 320	6	6	15.3362	
20	300 × 400	3	4	25.4958	

example shows that an improper stock selection could increase the waste by more than four times. Therefore, it seems logical that factories should choose the stocks that lead to the least waste. This means that it is also important to consider the raw material selection. It is important to remember that the waste amounts reported in the table could be temporary, as it is possible to decrease the waste by processing further orders.

Conclusions

Considering that cutting stock problems are NP-hard, heuristic methods, like the one presented in this paper, can be efficiently applied for such problems. As was observed before, the proposed method did not have specific complexities of these types of problems. Also, it was observed that it had better speed in solving the problems in comparison with other methods.

The proposed algorithm was designed and implemented using C++ language in a way that it recorded the report of all the functions and, at the end, the user accessed all the information related to the cutting. At the same time, it can be easily used by operators. Most industries and workshops are not interested in using scientific methods since they are time-consuming, costly, complex, and limited. As described above, the proposed approach is a simple and quick method, which makes it easier to use.

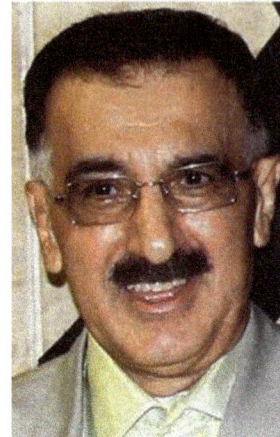

Figure 5 Professor Mir-Bahador Aryanezhad.

Furthermore, better selection of sheets from among the available options (options on the stock) has a considerable effect on decreasing the amount of waste. It was observed in the sample solved above that, by correct selection of the stock size, the amount of sheet waste decreased by more than one-fourth.

Presenting simple and combinatorial approaches (hybrid methods) for non-guillotine problems can be considered in future studies. Also, meta-heuristic methods can be applied in this regard.

Competing interests
The authors declare that they have no competing interests.

Authors' contributions
MBA proposed, designed, and managed the study. NFH worked on the study details (proposed the algorithms and approaches and carried them out) and drafted the manuscript. AM and HJ helped and participated in its design and coordination. All authors read and approved the final manuscript.

Authors' information
MBA was a professor, NFH is an M.Sc. graduate, and AM is an associate professor in the Department of Industrial Engineering, Iran University of Science and Technology, Tehran, Iran. HJ is an assistant professor in the Department of Industrial Engineering, Islamic Azad University, South Tehran Branch, Tehran, Iran.

Acknowledgements
Here, Nima Fakhim Hashemi would like to thank all the honorable teachers and professors during his study years, especially Professor Aryanezhad who recently passed away. He was a kind, well-educated, and morally distinctive professor who was always guiding and encouraging his students. His memories will always be with us. May God bless his soul (Figure 5).

Author details
[1]Department of Industrial Engineering, Iran University of Science and Technology, Tehran 16846-13114, Iran. [2]Department of Industrial Engineering, Islamic Azad University, South Tehran Branch, Tehran 11518-63411, Iran.

References

Alvarez-Valdes R, Parajon A, Tamarit JM (2001) A computational study of heuristic algorithms for two-dimensional cutting stock problems. 4th metaheuristics international conference (MIC'2001) 16–20, Porto

Alvarez-Valdés R, Parajón A, Tamarit JM (2002) A tabu search algorithm for large-scale guillotine (un)constrained two-dimensional cutting problems. Comput Oper Res 29(7):925–947

Belov G, Scheithauer G (2006) A branch-and-cut-and-price algorithm for one-dimensional stock cutting and two-dimensional two-stage cutting. Eur J Oper Res 171(1):85–106

Ben Messaoud S, Chu C, Espinouse M (2008) Discrete optimization characterization and modelling of guillotine constraints. Eur J Oper Res 191(1):112–126

Capacho Betancourt L (2007) ASALBP: the alternative subgraphs assembly line balancing problem. Technical University of Catalonia, Thesis (Ph.D.)

Capacho L, Pastor R (2008) ASALBP: the alternative subgraphs assembly line balancing problem. Int J Prod Res 46(13):3503–3516

Dósa G (2007) The tight bound of first fit decreasing bin-packing algorithm is FFD (I) ≤ 11/9 OPT (I) + 6/9. Lect Notes Comput Sci 4614:1–11

Dyckhoff H (1990) A typology of cutting and packing problems. Eur J Oper Res 44(2):145–159

Dyckhoff H, Finke U (1992) Cutting and packing in production and distribution: a typology and bibliography. Physical-Verlag, Heidelberg

Erjavec J, Gradišar M, Trkman P (2009) Renovation of the cutting stock process. Int J Prod Res 47(14):3979–3996

Eshghi K, Javanshir H (2005) An ACO algorithm for one-dimensional cutting stock problem. J Ind Eng Int 1(1):10–19

Faina L (1999) An application of simulated annealing to the cutting stock problem. Eur J Oper Res 114(3):542–556

Fathi M, Jahan A, Ariffin MKA, Ismail N (2011) A new heuristic method based on CPM in SALBP. J Ind Eng Int 7(13):1–11

Gilmore PC, Gomory RE (1961) A linear programming approach to the cutting-stock problem. Oper Res 9(6):849–859

Gilmore PC, Gomory RE (1963) A linear programming approach to the cutting-stock problem–part II. Oper Res 11(6):863–888

Gilmore PC, Gomory RE (1965) Multistage cutting stock problems of two and more dimensions. Oper Res 13(1):94–120

Gonçalves JF (2007) A hybrid genetic algorithm-heuristic for a two-dimensional orthogonal packing problem. Eur J Oper Res 183(3):1212–1229

Javanshir H, Shadalooee M (2007) The trim loss concentration in one-dimensional cutting stock problem (1D-CSP) by defining a virtual cost. J Ind Eng Int 3(4):51–58

Leung TW, Yung CH, Troutt MD (2001) Applications of genetic search and simulated annealing to the two-dimensional non-guillotine cutting stock problem. Comput Ind Eng 40(3):201–214

Malakooti B (1991) A multiple criteria decision making approach for the assembly line balancing problem. Int J Prod Res 29(10):1979–2001

Malakooti B (1994) Assembly line balancing with buffers by multiple criteria optimization. Int J Prod Res 32(9):2159–2178

McMullen PR, Tarasewich P (2003) Using ant techniques to solve the assembly line balancing problem. IIE Transaction 35:605–617

Mellouli A, Dammak M (2008) An algorithm for the two-dimensional cutting-stock problem based on a pattern generation procedure. Inf Manag Sci 19(2):201–218

Neapolitan R, Naimipour K (2004) Foundations of algorithms using C++ pseudocode, 3rd edn. Jones and Bartlett Publishers, Sudbury

Roy D, Khan D (2010) Assembly line balancing to minimize balancing loss and system loss. J Ind Eng Int 6(11):1–5

Roy D, Khan D (2011a) A new type of problem to stabilize an assembly setup. Manag Sci Lett 1(3):271–278

Roy D, Khan D (2011b) Optimum assembly line balancing by minimizing balancing loss and a range based measure for system loss. Manag Sci Lett 1(1):13–22

Suliman SMA (2006) A sequential heuristic procedure for the two-dimensional cutting-stock problem. Int J Prod Econ 99(1–2):177–185

Wäscher G, Haußner H, Schumann H (2007) An improved typology of cutting and packing problems. Eur J Oper Res Research 183(3):1109–1130

Whitwell G (2004) Novel heuristic and metaheuristic approaches to cutting and packing. University of Nottingham, Thesis (Ph.D.)

19

An investigation of model selection criteria for technical analysis of moving average

Milad Jasemi[1*] and Ali M Kimiagari[2]

Abstract

Moving averages are one of the most popular and easy-to-use tools available to a technical analyst, and they also form the building blocks for many other technical indicators and overlays. Building a moving average (MA) model needs determining four factors of (1) approach of issuing signals, (2) technique of calculating MA, (3) length of MA, and (4) band. After a literature review of technical analysis (TA) from the perspective of MA and some discussions about MA as a TA, this paper is structured to highlight the effects that each of the first three factors has on performance of MA as a TA. The results that based on some experiments with real data support the fact that deciding about the first and second factors is not much critical, and more attention should be paid to other factors.

Keywords: Moving average, Technical analysis, Trend forecasting, Investment decision, Portfolio selection

Findings

The site of www.finance.yahoo.com has contributed the most to the market scholars and practitioners by providing the very details of the most notable stock markets of the world freely. So in this study the mentioned database was the foundation. There is no limitation in access to the site, while in this study the extracted data were applied to investigate the technique of Moving Average structurally on a practical and real base. Because of the fact that the data are real everyone has a better sense to the results.

Availability and requirements

- Project name: My simple technical analysis.
- Project home page: -
- Operating system(s): Windows XP (Professional)
- Programming language: Matlab
- Other requirements: Microsoft Excel
- License: -
- Any restrictions to use by non-academics: No

Introduction

Developing a model for predicting returns in order to make investment decisions is an important goal for academics and practitioners. Typically, the financial services industry relies on three main approaches to make investment decisions: the fundamental approach that uses fundamental economic principles to form portfolios, the TA approach, and the mathematical approach that is based on mathematical models (Leigh et al. 2002). The first two approaches dominate practice because of their applicability; however, our paper focuses on TA approach. Blanchet-Scalliet et al. (2007), from the perspective of mathematical models, justify this very well and say it is impossible to specify and calibrate mathematical models that can capture all the sources of parameter instability during a long time interval if one considers a non-stationary economy.

TA remains very popular despite a lack of theoretical foundation and has been used by professional investors for more than a century (Blanchet-Scalliet et al. 2007), and there is a little dispute that it is very common among practitioners (Roberts 2005). Brorsen and Irwin (1987) report that only 2 of 21 large commodity fund managers surveyed used no objective TA. According to Cesari and Cremonini (2003), TA is perhaps the oldest device designed to beat the market. It has a secular history given that its origins can be traced to the seminal articles published by Charles H. Dow in the *Wall Street Journal* between 1900 and 1902, and its basic concepts became popular after the contributions of Hamilton (1922) and Rhea (1932).

* Correspondence: miladj@aut.ac.ir
[1]Department of Industrial Engineering, Islamic Azad University, Masjed Soleyman Branch, Masjed Soleyman, 61649, Iran
Full list of author information is available at the end of the article

The definitions of TA that have been presented in the literature by different scholars are almost the same. Tian et al. (2002) know TA as a search for recurrent and predictable patterns in stock prices. Dourra and Siy (2002) define it as an attempt to predict future stock price movements by analyzing the past sequence of stock prices because of the fact that forces of supply and demand affect those prices. They believe that it dismisses such factors as the fiscal policy of the government, economic environment, industry trends, and political events as being irrelevant in attempting to predict future stock prices. Roberts (2005) knows it as a broad collection of methods and strategies which attempts to forecast future prices on the basis of past prices or other observable market statistics, such as volume or open interest. According to Wang and Chan (2007), TA studies records or charts of past stock prices, hoping to identify patterns that can be exploited to achieve excess profits and so many other definitions that imply the same meanings and implications.

TA literature can be divided into two periods. The first period encompasses decades of 60s and 70s, while the second period cover 80s and after. Some results obtained during the first period like Alexander (1964) and Fama and Blume (1966) supported the impracticability of applying TA for prediction of the future. Van Horne and Parker (1967) and James (1968) examine MA rules and indicated that this trading strategy did not yield returns that were superior to a buy-and-hold strategy even before transaction costs were taken into account. Over the succeeding years, many researchers like Jensen and Benington (1970) reached similar conclusions, especially when transactions costs were included in the analysis. Although these decades are known as years of skeptical attitude of academic community toward TA, there were also some supporting studies like Levy (1967) that employed relative strength.

During the last 25 years, TA has been enjoying a rebirth in the academic world, and a considerable amount of theoretical and empirical works has been developed supporting the TA. Thus, theoretical models have been proposed by Hellwig (1982), Treynor and Ferguson (1985), Brown and Jennings (1989), and Blume et al. (1994). Also, many empirical papers provide evidence of the profitability of technical trading rules, outstanding among others are Brock et al. (1992), Levich and Thomas (1993), Knez and Ready (1996), Gençay (1996), Neely et al. (1997), and Chang and Osler (1999). It is to be noted that many researchers believe that the study of Brock et al. (1992) has played the role of a turning point in the history of TA. They demonstrated the profitability of simple trading rules, MA and trading range breakout. They, after applying 26 trading rules on the basis of MA and trading range breakout rules to the Dow Jones Industrial Average, found that they significantly outperform a

benchmark of holding cash. They document that buy signals generate higher returns than sell signals, and the returns following buy signals are less volatile than the returns on sell signals.

There is a fairly comprehensive literature related to TA in various financial domains, but the remainder of this part addresses the ones that are more related to MA. For example, Pruitt and White (1988) developed the CRISMA trading system, which combined trading rules of on balance volume, relative strength, and MA and confirmed the profitability of technical trading rules. Sweeney (1988), Allen and Taylor (1990), and Taylor and Allen (1992) find that trading rules can outperform statistical models in predicting exchange rates and stock prices. Bessembinder and Chan (1995) and Ratner and Leal (1999) following Brock et al. (1992) also demonstrated the profitability of simple trading rules, MA, and trading range breakout. Hudson et al. (1996) examine prices for the Financial Times Industrial Ordinary Index from 1935 to 1994 and showed that MA trading rules can be utilized for USA and UK markets. As a matter of fact, this study provides novel evidence on the predictive ability of technical trading rules in developed markets with long series of price histories (Gunasekarage and Power 2001). Mills (1997) analyzes daily data on the London Stock Exchange FT30 index for the period 1935 to 1994. It is found that the trading rules worked, in the sense of producing a return greater than a buy-and-hold strategy, for most of the sample period, at least up to the early 1980s. Gencay and Stengos (1998) use the daily Dow Jones Industrial Average Index from 1963 to 1988 to examine the linear and non-linear predictability of stock market returns with some simple technical trading rules. Some evidence of non-linear predictability in stock market returns is found using the past buy-and-sell signals of the moving average rules. Sullivan et al. (1999) examine close to 8,000 technical trading rules and repeat Brock et al. (1992) study while correcting it for data snooping problems. Gunasekarage and Power (2001) showed that technical trading rules have predictive ability in South Asian stock markets and reject the null hypothesis that the returns to be earned from studying MA values are equal to those achieved from a naive buy-and-hold strategy. Tian et al. (2002) focus in markets with different efficiency level. They found that these simple trading rules are quite successful in Chinese markets in 1990s, while do not beat the US index during the same period. Ausloos and Ivanova (2002) present a generalization of the classical TA concepts taking into account the volume of transactions. Lastly, the purpose of Andrada-Felix and Fernandez-Rodriguez (2008) is to improve moving average trading rules with boosting and statistical learning methods. In fact, MA is seen in most of the academic studies of TA that has mainly adopted quantitative indicators as prediction variables, and the literature is full of such studies.

MA as a TA

One common component of many technical rules is MA (Gencay and Stengos 1998). MAs are one of the most popular and easy-to-use tools available to technical analysts. According to the categorization of Reilly and Brown (1994) on different technical trading rules which are practiced by US technicians, MA falls into the fourth group of stock price and volume techniques. For the application of MA as a TA, first of all the approach of calculating MA, i.e., simple or exponential, and then the mechanism of issuing signals, i.e., direction or crossover, should be determined.

Equations 1 and 2 show how the MA is calculated simply and exponentially, respectively:

$$\text{SMA}_{N,T} = \frac{\sum_{i=0}^{N-1} P_{T-i}}{N}, \tag{1}$$

$$\text{EMA}_{N,T} = \frac{2}{N+1} \times \left(P_T - \text{EMA}_{N,T-1}\right) + \text{EMA}_{N,T-1}, \tag{2}$$

(StockCharts.com Inc. 1999) where $\text{SMA}_{N,T}$ is the simple moving average of length N on day T; $\text{EMA}_{N,T}$ the exponential moving average of length N on day T; P_T, the stock price on day T; and N; the length of moving average.

Independent of the method that is used to calculate MA, there are two mechanisms of direction and crossover to make investment decision. The first mechanism uses the direction of MA to determine the trend. If the MA is rising/declining by an amount larger than the band, the trend is considered up/down, and a buy/sell signal is issued. The second mechanism is based on the location of the shorter MA relative to the longer MA. If the shorter MA is above/below the longer MA by an amount larger than the band, the trend is considered up/down. Two variations of crossover technique are variable-length MA (VLMA) and fixed-length MA (FLMA). The difference between them is their band, i.e., with a band of zero we have FLMA otherwise VLMA. Band is an amount of difference that is needed to generate a signal. It is to be noted that the approach of price location is exactly the same as crossover with shorter MA of length 1. If $band = \alpha$, direction and crossover issue signals according to Equations 3 and 4, respectively:

$$\begin{cases} \text{Buy} & \textit{if } \text{MA}_{N,T} > (1+\alpha)\text{MA}_{N,T-1} \\ \text{Sell} & \textit{if } \text{MA}_{N,T} < (1-\alpha)\text{MA}_{N,T-1} , \\ \text{Hold} & \text{Otherwise} \end{cases} \tag{3}$$

$$\begin{cases} \text{Buy} & \textit{if } \text{MA}_{N-n,T} > (1+\alpha)\text{MA}_{N,T} \\ \text{Sell} & \textit{if } \text{MA}_{N-n,T} < (1-\alpha)\text{MA}_{N,T} , \\ \text{Hold} & \text{Otherwise} \end{cases} \tag{4}$$

where $1 \le n \le N-1$ is the integer, and $\text{MA}_{N,T}$ is the MA of length N on day T whether calculated simply or exponentially.

Although trial and error is usually the best means for finding the best length, the most popular MA rule as reported in Brock et al. (1992) is the 1 to 200 rule, where the short period is 1 day and the long period is 200 days. Other popular ones are the 1 to 50, 1 to 150, 5 to 200, and the 2 to 200 rules (Gencay and Stengos 1998).

Is there any difference between direction and crossover?

Naturally, it is questioned whether the two mechanisms are the same or not. To respond the question, Equations 5 and 6 for direction and crossover, respectively, are developed and being discussed. Without losing generality, MA in both of the mechanisms is calculated simply, and for crossover, FLMA is considered.

$$\text{MA}_{N,T} - \text{MA}_{N,T-1} = \frac{P_T - P_{T-N}}{N}, \tag{5}$$

$$\text{MA}_{N-n,T} - \text{MA}_{N,T} = \frac{n(P_T + P_{T-1} + \ldots + P_{T-N+n+1}) - (N-n)(P_{T-N+n} + \ldots + P_{T-N+1})}{N(N-n)}. \tag{6}$$

Since the yielded signal depends on positivity or negativity of the equations, for direction the relation between magnitudes of P_T and P_{T-N}, and for crossover the relation between magnitude of $\frac{P_T + P_{T-1} + \ldots + P_{T-N+n+1}}{N-n}$ and $\frac{P_{T-N+n} + \ldots + P_{T-N+1}}{n}$ determines the final signal. Accordingly, the direction signal depends only on the first and last input data, while the crossover signal depends on the average of last $(N-n)$ data and the average of first n data. Hence, it is obvious that two approaches of direction and crossover are not the same and apply different mechanisms.

Evaluation system

To compare different MA-based TA models, there should be an evaluation system according to which they are being assessed. The intended evaluation system of the paper is based on the number of correct and wrong signals and their associated weights. The general method to derive the weights is founded on wrong signals. That is, firstly the weights of wrong signals are calculated then the weights of correct signals by considering a coefficient will be achieved. The wrong signals are sell for buy, buy for sell, hold for buy, buy for hold, hold for sell, and sell for hold that, after considering the fact that whether the stock is in the portfolio or not, become twelve states as shown in Table 1.

Losses that resulted from wrong signals, according to their degree (as can be seen in Table 1), are categorized into five: 'double losing', 'losing', 'neutral−', 'neutral', and

Table 1 Different probable errors of a TA system under different conditions with their results

State	Correct signal	TA signal	Is the stock in the stock trader portfolio?	Result
1	Buy	Sell	Yes	Double losing
2			No	Losing
3	Sell	Buy	Yes	Losing
4			No	Double losing
5	Buy	Hold	Yes	Benefiting
6			No	Losing
7	Hold	Buy	Yes	Neutral
8			No	Neutral–
9	Sell	Hold	Yes	Losing
10			No	Neutral
11	Hold	Sell	Yes	Neutral–
12			No	Neutral

Table 3 Weights of different errors

Error	Correct signal	TA signal	Error weight
Sell for buy	Buy	Sell	80
Buy for sell	Sell	Buy	80
Hold for sell	Sell	Hold	35
Hold for buy	Buy	Hold	30
Buy for hold	Hold	Buy	15
Sell for hold	Hold	Sell	15

two states, the weight of each error should be a function of its corresponding states' weights. In this study, simple averaging has been used to come up with Table 3.

According to Table 3, the negative score (Ers) of a TA on the basis of a particular data set can be calculated by Equation 7:

$$\text{ErS} = 80 \times (\text{nisb} + \text{nibs}) + 35 \times (\text{nihs}) + 30 \times (\text{nihb}) + 15 \times (\text{nibh} + \text{nish}), \tag{7}$$

where nisb is the number of incorrect sell for buy signals; nibs, the number of incorrect buy for sell signals; nihs, the number of incorrect hold for sell signals; nihb, the number of incorrect hold for buy signals; nibh, the number of incorrect buy for hold signals; and nish, the number of incorrect sell for hold signals.

On the basis of the fact that many evaluation systems consider weight of a correct answer three times larger than an incorrect one, this study does the same to calculate the weight of correct buying, selling, and holding signals as 150, 150, and 90, respectively. Hence, the total score of a TA processor is calculated by Equation 8:

$$\text{TSc} = \frac{1}{\text{tns}}\left(150 \times (\text{ncb} + \text{ncs}) + 90 \times \text{nch} - \text{ErS}\right), \tag{8}$$

where tns is the total number of signals.

Benchmark signals

To determine whether the issued signal of TA for a particular day is correct or not, there should be a benchmark signal for that day. The following two parts first

'benefiting'. Double losing happens when the stock trader sell (buy) the stock when its price will increase (decrease) like state 1 (4). Losing happens when the stock trader does not buy (sell) the stock when its price will increase (decrease) like states 2 and 6 (3 and 9). Neutral– happens when the stock trader buy or sell the stock when its price will experience no tangible change in the future like state 8 or 11. Neutral happens when, in spite of wrong signals of the TA processor, the stock trader lose nothing, for example, if the stock is not in the investor portfolio, there will be no difference between sell and hold signals like states 10 and 12. The same thing happens for buy and hold signal, i.e., when the stock trader has the stock and the correct signal is holding like state 7. This can also benefit the investor if the correct signal is buying because according to signal of holding (that is wrong) the investor will keep it and will experience its increase and will benefit, exactly as happens in state 5. The states are arranged in a descending order in Table 2.

To differentiate between different ranks, the assigned weights to them are 100, 60, 20, 10, and 0 for ranks of 1 to 5, respectively. Whereas each error is associated with

Table 2 Ranking of different states of errors according to the magnitude of resulted losses

Rank	State	Correct signal	TA signal	Rank	State	Correct signal	TA signal
1	1	Buy	Sell	3	8	Hold	Buy
	4	Sell	Buy		11	Hold	Sell
2	2	Buy	Sell	4	7	Hold	Buy
	3	Sell	Buy		10	Sell	Hold
	6	Buy	Hold		12	Hold	Sell
	9	Sell	Hold	5	5	Buy	Hold

Table 4 Benchmark signals for trading stocks of Yahoo

Number	Date	Price	$\frac{P_{t+1}-P_t}{P_t}$	Benchmark signal	
				sl = 2%	sl = 5%
1	7/1/2008	20.48	-	Buy	Buy
2	7/2/2008	21.89	6.88%	Sell	Hold
3	7/3/2008	21.35	−2.47%	Buy	Buy
4	7/7/2008	23.4	9.60%	Hold	Hold
5	7/8/2008	23.83	1.84%	Buy	Hold
6	7/9/2008	24.74	3.82%	Sell	Hold
7	7/10/2008	23.76	−3.96%	Sell	Hold
8	7/11/2008	23	−3.20%	Hold	Hold
9	7/14/2008	23.12	0.52%	Sell	Sell
10	7/15/2008	21.79	−5.75%	-	-

describe the implication of each sign and then the methodology of deriving the benchmark signals.

Interpretation

A TA signal does not always mean taking actions; for instance, a buy signal for the stock which is in the portfolio does not necessarily imply increasing its share in the portfolio. In fact, a buy signal means that the stock price will increase in the future, and it is recommended to have the stock; thus, if the stock trader has the stock, the TA processor implies keeping of it. This is similar for a sell signal that recommends not keeping the stock and selling it completely. As a matter of fact, if the investor does not have the stock, a selling signal implies to do nothing. Hence, t uninterrupted buy (sell) signals tell the investor to (not to) have the stock in his/her portfolio for the t-day period. Lastly, a hold signal means there will be no considerable change in the price of the stock in the future.

Methodology

The mechanism of deriving the benchmark signals in this study for running the experiments is shown in Equation 9:

Table 5 Some points about time series of data for running experiments

Item	Index or stock	Notation	Time interval		Number of data
			Start	End	
1	S&P 500	a	1/02/1990	8/14/2008	4,695
2	Nikkei 225	b	1/04/1990	8/14/2008	4,584
3	Egypt CMA Genl	c	7/02/1997	8/14/2008	2,178
4	Yahoo	d	4/12/1996	8/12/2008	3,105
5	China South Air	e	4/16/1998	8/12/2008	2,590
6	Dell	f	1/02/1990	8/12/2008	4,693
7	HP	g	1/02/1990	8/12/2008	4,693

Table 6 Two sets of data and their VAR and SD

Data	1	2	3	4	5	6	7	8	VAR	SD
1	0.01	−0.6	−0.01	1	−0.05	0.8	−2	−1	0.9393	0.9692
2	1,000	1,020	1,050	990	995	1,020	1,080	970	1,253.1	35.4

$$S_t \begin{cases} \text{Buy if } \dfrac{P_{t+1} - P_t}{P_t} > \text{sl} \\ \text{Sell if } \dfrac{P_{t+1} - P_t}{P_t} < -\text{sl where sl} \geq 0, \\ \text{Hold Otherwise} \end{cases} \quad (9)$$

where S_t is the benchmark signal on day t, and sl is the sensitivity level for considering buy or sell opportunities.

Table 4 shows the results of applying Equation 9 to the stock price of Yahoo Company for 10 days between 1 July 2008 and 15 July 2008.

Design and running of experiments
Data

The experiments have been run using data on the three market indices and four individual stocks as shown in Table 5. While it is possible to create MAs from the open, the high, and the low data points as usual this study uses the closing price.

Variability The common measures of literature to quantify variability of a data set are variance (VAR), standard deviation (SD), and so on. However, there is a main problem with them that not consider proportional variability and focus on the absolute magnitude of changes. To see the point better, consider Table 6 in which the variability of two sets of data are calculated by both measures of VAR and SD.

According to Table 6, the measures evaluate the variability of the second set considerably bigger than the first that does not make sense because change percentage of the first set of data from a particular data to the next one is considerably bigger than the second set. To meet the challenge, Equation 10 is devised to quantify the variability of input data:

$$\text{Vol} = \frac{\sum_{i=2}^{n} \left| \frac{P_i - P_{i-1}}{P_{i-1}} \right|}{n - 1}, \quad (10)$$

Table 7 Variability of data that are used in this study according to Equation 10

	S&P 500	Nikkei 225	Egypt CMA Genl	Yahoo	China South Air	Dell	HP
Volatility	0.72%	1.25%	1.09%	3.03%	2.65%	2.38%	1.8%

Table 8 Different amounts of parameters

Parameter	Amounts	Number
Band	0%, 1%, 2%, 3%, 4%, and 5%	6
sl	0.5%, 1%, 1.5%, 2%, 3%, and 5%	6
l	1, 2, . . . , 199, 200	200
(m, n)	$(s, s + i); s = 1, 2, 3, 5, 13, 21, 50, 89, 150, 200, 250; i = 1, 2, . . ., 100$	1,100

Parameters engaged in development of a MA model for running the experiments.

Where P_i is the stock price on day i, and n is the total number of data that is to be analyzed.

After applying Equation 10 to the data in Table 6, volatility of data achieved are 2,643.33% and 4.25% for the first and second data sets, respectively, that are completely sensible. According to Equation 10, Table 7 shows the volatility of data sets used in the study.

Parameters

Table 8 shows different amounts of band, sl, l, and (m, n) where l denotes length of moving average (LMA) in approach of direction and m and n denote lengths of short and long MA in approach of crossover, respectively.

According to StockCharts.com Inc. (1999), some of the more popular lengths of MA include 21, 50, 89, 150, and 200 days as well as 10, 30, and 40 weeks. On the basis of these lengths and the ones that are recommended by other scholars, the amounts of m are decided to be 1, 2, 3, 5, 13, 21, 50, 89, 150, 200, and 250, and for each $m = a$, there are 100 ns of $(a + 1, a + 2, . . ., a + 100)$ that totally becomes 1,100 states.

All combinations of these parameters, beside simple or exponential calculation of MA, deliver a total of 93,600 MA trading rules. To make the calculated TScs applicable for the final inferences, the achieved TScs need some kind of manipulation. The data manipulation for the approach of direction transforms 36 data into 1, while for crossover, 3,600 data into 1. The logic of transformation for both of the approaches is the same and says if a parameter affects the benchmark signals, none of its amount should be ignored, and if it affects the generated signals, the one that maximizes the performance should be considered during

the manipulation process because setting the parameters to generate signals is a trial-and-error process while the other kind of parameter depends on the investor.

Direction

After running the experiments by direction 100800, TScs are achieved. In fact, for each l under a particular technique of calculating MA, there are 36 TScs that are notated by $TSc_{ij} i = 1, . . ., 6; j = 1, . . ., 6,$ where i denotes category of band, and j denotes category of sl. On the basis of what has been discussed before, Equations 11 and 12 work the 36 input data into one. After applying the equations to the 100800 TScs, 2,800 $\bar{\bar{T}}Scs$ are achieved. Table 9 summarizes these results by means of seven indices of 1 to 7.

$$\bar{T}Sc_i = \frac{\sum_{j=1}^{6} TSc_{ij}}{6}, \ i = 1, 2, . . ., 6, \quad (11)$$

$$\bar{\bar{T}}Sc = Max\{\bar{T}Sc_1, . . ., \bar{T}Sc_6\}. \quad (12)$$

Index 1 calculates the maximum difference between $\bar{\bar{T}}Scs$ to determine how much selection of l or technique of calculating MA is important, while index 2 seeks the maximum difference between scores of simple and exponential techniques with similar length to determine the importance of being careful toward selection of the technique. Index 3 calculates the difference between maximum and minimum of $\bar{\bar{T}}Scs$ for the simple technique to highlight the magnitude of effect that LMA has on performance of direction, while index 4 do the same but for exponential technique. Index 5 specifies that between 400 $\bar{\bar{T}}Scs$ of each market index or stock, what proportion

Table 9 Results of applying indices to results of running experiments on the approach of direction

Index	S&P 500	Nikkei 225	Egypt CMA Genl	Yahoo	China South Air	Dell	HP
1	0.31%	1.39%	0.54%	11.76%	5.82%	2.98%	2.41%
2	0.071%	0.073%	0.087%	4.862%	3.337%	1.946%	0.345%
3	0.31%	1.39%	0.54%	11.76%	5.59%	2.98%	2.41%
4	0.31%	1.39%	0.54%	8.16%	5.81%	1.38%	2.30%
5	55%	50%	37.5%	0	90%	35%	27.5%
6	200, 199, 198	199, 200, 197	96, 95, 94	67, 66, 32	136, 140, 135	60, 57, 56	13, 12, 35
7	200, 199, 198	199, 200, 197	96, 95, 94	37, 36, 47	142, 140, 141	199, 192, 198	13, 34, 14

of top forty $\bar{\bar{\bar{T}}}$Scs, belongs to the exponential technique. Then, indices of 6 and 7 concentrate on the LMAs that are associated with the three top $\bar{\bar{\bar{T}}}$Scs for simple and exponential techniques, respectively. Based on the average performance for different amounts of l for the stocks Nikkei 225, Egypt CMA Genl, Yahoo, Dell, and HP, the approach simple is superior, but for the stocks S&P500 and China South Air Ltd., the approach exponential is better.

Crossover

Manipulating the 554,400 outputs of crossover because of an extra length parameter needs more processing. In fact, for each m under a particular technique of calculating MA, there are 3,600 TSCs that are notated by TSc_{ijk}, where $i = 1, 2, \ldots, 100$ denotes category of n, $j = 1, 2, \ldots, 6$ denotes category of band, and $k = 1, 2, \ldots, 6$ denotes category of sl. Equations 13, 14, and 15 work the 3,600 input data into one:

$$\bar{T}Sc_{ij} = \frac{\sum_{k=1}^{6} TSc_{ijk}}{6}, \tag{13}$$

$$\bar{\bar{T}}ScSc_{i} = \text{Max}\{\bar{T}Sc_{i1}, \bar{T}Sc_{i2}, \ldots, \bar{T}Sc_{i6}\}, \tag{14}$$

$$\bar{\bar{\bar{T}}}Sc = \text{Max}\left\{\bar{\bar{T}}Sc_{1}, \bar{\bar{T}}Sc_{2}, \ldots, \bar{\bar{T}}Sc_{100}\right\}, \tag{15}$$

After applying the equations to the 554,400 TSCs, 154 $\bar{\bar{\bar{T}}}$Scs are achieved that prove the priority of the approach exponential to the simple, and also Table 10 summarizes them according to five indices of 1 to 5.

Indices 1 to 4 are the same as Table 9 just l is replaced by m. Index 5 specifies that what proportion of top three $\bar{\bar{\bar{T}}}$Scs of each data set that are calculated simply or exponentially belongs to the exponential technique. Lastly, Table 11 shows the best ns that have the maximum TSc for some particular ms used in the experiments.

Another thing about crossover that should be analyzed is the behavior of TSc for a fixed m while n increases continuously. In the experiment, ms are the same as the previous experiment, while n, if $m = a$, gets 200 amounts

Table 10 Results of applying indices to results of running experiments on the approach of crossover

Index	S&P 500 (%)	Nikkei 225 (%)	Egypt CMA Genl (%)	Yahoo (%)	China South Air (%)	Dell (%)	HP (%)
1	0.34	1.19	0.84	16.45	5.92	3.39	1.18
2	0.12	0.09	0.18	7.02	2.93	1.64	0.27
3	0.33	1.17	0.84	16	4.92	2.4	1.18
4	0.34	1.19	0.84	15.85	5.05	3.21	0.8
5	66	66	33	66	100	66	66

of $\{a + 1, a + 2, \ldots, a + 200\}$. The behavior for all the seven data sets is, to some extent, the same and just as the instance in Figure 1 shows the results for S&P 500 when MA is calculated simply, while Figure 2 shows the results for HP when MA is calculated exponentially for different ms.

Direction versus crossover

After studying two approaches of direction and crossover individually in this section, their performances according to the criterion of processed TSc are compared. The comparison is done by three indices as shown in Table 12. Index 1, by calculating the difference between the best performance of direction and crossover, determines how much choosing between direction and crossover is important. For indices 2 and 3, firstly top five TScs of direction and top five TScs of crossover are combined to form 10 TScs. In this regard, indices 2 and 3 mention the ranks of crossover and direction, respectively.

There are two points about the results of all the experiments of this study. First, the measure of TSc is designed in the way to show the effects of different factors on performance of the TA for a particular and not different sets of data. Thus, it is not a proper measure to determine how much a TA is appropriate or applicable for a particular set of data. Second, all the results whether in the form of table or figure are optimal ones over 6 amounts of band. Consequently, the comments have been given on the basis of the assumption that an appropriate band has been chosen.

Conclusion

In this paper, after developing a measure to evaluate the performance of a TA processor, efficiency of different

Table 11 The ns that produce the maximum TSC with the corresponding m

	Simple							Exponential						
m	a	b	c	d	e	f	g	a	b	c	d	e	f	g
1	2	2	2	2	6	2	2	9	2	2	3	6	2	2
2	4	3	3	9	6	3	3	10	7	3	12	5	3	3
3	6	4	4	5	5	4	4	13	7	4	9	4	7	6
5	10	7	6	10	6	6	6	12	8	7	11	15	6	7
13	15	15	15	21	18	14	14	15	14	14	25	21	14	14
21	22	25	22	33	30	27	22	22	22	32	37	30	27	25
50	57	61	51	66	53	62	51	57	66	51	73	64	55	56
89	93	90	97	103	105	93	91	93	90	97	116	113	100	94
150	198	178	151	168	155	163	154	226	184	151	236	169	190	163
200	226	241	202	206	204	208	203	297	243	202	245	259	239	203
250	297	257	251	261	274	261	257	330	257	251	261	284	267	269

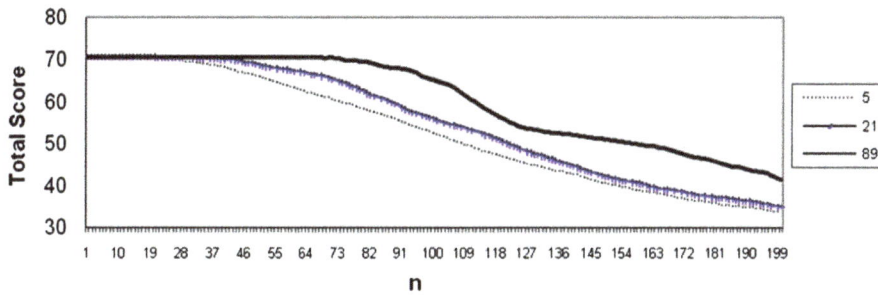

Figure 1 The behavior of crossover on the basis of data from S&P 500. Each line represents a particular *m*.

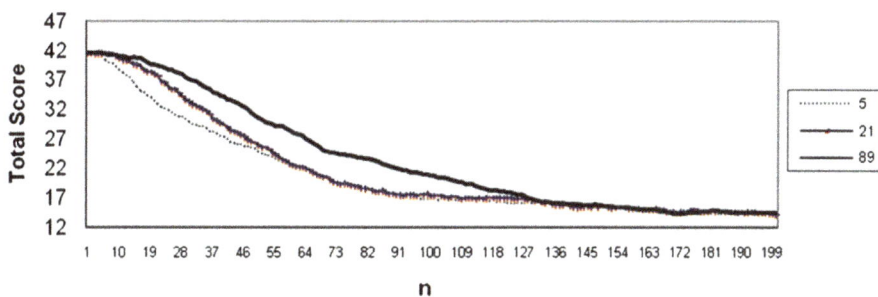

Figure 2 The behavior of crossover on the basis of data from HP. Each line represents a particular *m*.

states of a MA model are tested to get some clues to construct a better TA processor. To our surprise, the results, on the basis of the ranges that have been considered for the parameters, show that some factors in spite of what has been thought only have negligible effects. The factors are as follows:

- In the conditions that other factors are set appropriately, choosing between approaches of direction and crossover, although in most cases, crossover proves to be slightly more efficient.
- In the conditions that other factors are set appropriately, choosing between simple or exponential calculation; although for the approach of direction, simple and for the approach of crossover, exponential prove to be slightly more efficient.

- For the approach of direction, specification of *l* proves to be more important than choosing the technique of calculating MA.
- For the approach of crossover, specification of one of the two parameters when the other has not specified yet. That is, whether *m* or *n* is specified first, the other length (*n* or *m*) can be determined in the way to have acceptable performance.

On the other hand, the factor that affects the performance of crossover considerably is the combination of (*m*, *n*). It is proved that the difference between *m* and *n* should not be much. As a matter of fact, for a fixed *m* if *n* increases, the performance of MA deteriorates. Although there may be some improvements in the performance of MA by increasing *n*, they are venial, and the general trend is descending. Lastly, it is concluded that the more volatile the input data, the more sensitive the performance of TA model to the discussed factors.

As was noted before, the results are independent of the parameter of band. Studying this area to get some points in the selection of band and monitoring the performance of MA for different amounts of it would be a good research area to continue this study. On the other hand, focusing on other evaluation systems of TA models or even optimizing weights of the system that has been presented in this study to present the best sensitivity toward performance of TA is recommended.

Table 12 Results of three indices on comparing direction versus crossover

Index	S&P 500	Nikkei 225	Egypt CMA Genl	Yahoo	China South Air	Dell	HP
1	0.14%	0.31%	0.00%	9.84%	1.92%	1.01%	0.43%
2	1 to 5	1 to 4	1, 2	1 to 10	1 to 7	1, 2, 4, 8	1, 2, 3, 5, 6
3	6 to 10	5 to 10	3 to 10	-	8 to 10	3, 5, 6, 7, 9, 10	4, 7, 8, 9, 10

Author details

[1]Department of Industrial Engineering, Islamic Azad University, Masjed Soleyman Branch, Masjed Soleyman, 61649, Iran. [2]Department of Industrial Engineering, Amirkabir University of Technology, Tehran, 15914, Iran.

Competing interests

The authors declare that they have no competing interests.

Authors' contribution

MJ has made substantial contributions to conception and design of the study alongside data collection and analysis. AMK participated in its design and coordination and helped to draft the manuscript. All authors read and approved the final manuscript.

References

Alexander SS (1964) Price movement in speculative markets: trend or random walks, No. 2. In: Cootner P (ed) The random character of stock market prices. MIT Press, Cambridge

Allen H, Taylor MP (1990) Charts, noise and fundamentals in the London foreign exchange market. Economic Journal 100:49–59

Andrada-Felix J, Fernandez-Rodriguez F (2008) Improving moving average trading rules with boosting and statistical learning methods. Journal of Forecasting 27:433–449

Ausloos M, Ivanova K (2002) Mechanistic approach to generalized technical analysis of share prices and stock market indices. European Physical Journal B 27:177–187

Bessembinder H, Chan K (1995) The profitability of technical trading rules in the Asian stock markets. Journal of Pacific-Basin Finance 3:257–284

Blanchet-Scalliet C, Diop A, Gibson R, Talay D, Tanre E (2007) Technical analysis compared to mathematical models based methods under parameters mis-specification. Journal of Banking & Finance 31:1351–1373

Blume L, Easley D, O'Hara M (1994) Market statistics and technical analysis: the role of volume. Journal of Finance 49:153–181

Brock V, Lakonishok J, LeBaron B (1992) Simple technical trading rules and the stochastic properties of stock returns. Journal of Finance 47(5):1731–1764

Brorsen BW, Irwin SH (1987) Futures funds and price volatility. Review of Futures Markets 7:118–135

Brown D, Jennings R (1989) On technical analysis. Review of Financial Studies 4:527–551

Cesari R, Cremonini D (2003) Benchmarking, portfolio insurance and technical analysis: a Monte Carlo comparison of dynamic strategies of asset allocation. Journal of Economic Dynamics & Control 27:987–1011

Chang PHK, Osler CL (1999) Methodical madness: technical analysis and the irrationality of exchange-rate forecasts. Economic Journal 109:636–661

Dourra H, Siy P (2002) Investment using technical analysis and fuzzy logic. Fuzzy Sets and Systems 127:221–240

Fama EF, Blume M (1966) Filter rules and stock market trading profits. Journal of Business 39:226–241

Gençay R (1996) Nonlinear prediction of security returns with moving average rules. Journal of Forecasting 15:165–174

Gencay R, Stengos T (1998) Moving average rules, volume and the predictability of security returns with feedforward networks. Journal of Forecasting 17:401–414

Gunasekarage A, Power DM (2001) The profitability of moving average trading rules in South Asian stock markets. Emerging Markets Review 2:17–33

Hamilton WD (1922) The stock market barometer. Harper, New York

Hellwig MF (1982) Rational expectations equilibrium with conditioning on past prices: a mean variance example. Journal of Econometric Theory 26:279–312

Hudson R, Dempsey M, Keasey K (1996) A note on the weak form efficiency of capital markets: the application of simple technical trading rules to UK stock prices - 1935 to 1994. Journal of Banking Finance 20:1121–1132

James FE Jr (1968) Monthly moving averages – an effective investment tool? Journal of Financial and Quantitative Analysis 3:315–326

Jensen MC, Benington GA (1970) Random walks and technical theories: some additional evidence. Journal of Finance 25(2):469–482

Knez P, Ready M (1996) Estimating the profits from trading strategies. Review of Financial Studies 9:1121–1164

Leigh W, Purvis R, Ragusa JM (2002) Forecasting the NYSE composite index with technical analysis, pattern recognizer, neural network, and genetic algorithm: a case study in romantic decision support. Decision Support Systems 32:161–174

Levich R, Thomas LR (1993) The significance of technical trading-rules profits in the foreign exchange market: a bootstrap approach. Journal of International Money and Finance 12:451–474

Levy R (1967) Relative strength as a criterion for investment selection. Journal of Finance 22:595–610

Mills TC (1997) Technical analysis and the London stock exchange: testing trading rules using the FT30. International Journal of Finance & Economics 2:319–331

Neely C, Weller P, Dittmar R (1997) Is technical analysis in the foreign exchange market profitable? A genetic programming approach. Journal of Financial and Quantitative Analysis 32(4):405–426

Pruitt SW, White RE (1988) The CRISMA trading system: who says technical analysis can't beat the market? Journal of Portfolio Management 14:55–58

Ratner M, Leal RPC (1999) Test of technical trading strategies in the emerging equity markets of Latin America and Asia. Journal of Banking and Finance 23:1887–1905

Reilly FK, Brown KC (1994) Investment analysis and portfolio management. Dryden Press, Orlando

Rhea R (1932) Dow theory. Barrons, New York

Roberts MC (2005) Technical analysis and genetic programming: constructing and testing a commodity portfolio. The Journal of Futures Markets 25(7):643–660

StockCharts.com Inc (1999) StockCharts.com., September 06 2008. http://www.stockcharts.com

Sullivan R, Timmermann A, White H (1999) Data-snooping, technical trading rule performance and the bootstrap. Journal of Finance 54:1647–1691

Sweeney R (1988) Some new filter rule tests: methods and results. Journal of Financial and Quantitative Analysis 23:285–300

Taylor MP, Allen H (1992) The use of technical analysis in the foreign exchange market. Journal of International Money and Finance 11:304–14

Tian GG, Wan G, Guo M (2002) Market efficiency and the returns to simple technical trading rules: new evidence from U.S. equity market and Chinese equity markets. Asia-Pacific Financial Markets 9:241–258

Treynor JL, Ferguson R (1985) In defense of technical analysis. Journal of Finance 40:757–772

Van Horne JC, Parker GGC (1967) The random-walk theory: an empirical test. Journal of Financial Analysts 23:87–92

Wang JL, Chan SH (2007) Stock market trading rule discovery using pattern recognition and technical analysis. Expert Systems with Applications 33:304–315

Permissions

All chapters in this book were first published in JIEI, by Springer; hereby published with permission under the Creative Commons Attribution License or equivalent. Every chapter published in this book has been scrutinized by our experts. Their significance has been extensively debated. The topics covered herein carry significant findings which will fuel the growth of the discipline. They may even be implemented as practical applications or may be referred to as a beginning point for another development.

The contributors of this book come from diverse backgrounds, making this book a truly international effort. This book will bring forth new frontiers with its revolutionizing research information and detailed analysis of the nascent developments around the world.

We would like to thank all the contributing authors for lending their expertise to make the book truly unique. They have played a crucial role in the development of this book. Without their invaluable contributions this book wouldn't have been possible. They have made vital efforts to compile up to date information on the varied aspects of this subject to make this book a valuable addition to the collection of many professionals and students.

This book was conceptualized with the vision of imparting up-to-date information and advanced data in this field. To ensure the same, a matchless editorial board was set up. Every individual on the board went through rigorous rounds of assessment to prove their worth. After which they invested a large part of their time researching and compiling the most relevant data for our readers.

The editorial board has been involved in producing this book since its inception. They have spent rigorous hours researching and exploring the diverse topics which have resulted in the successful publishing of this book. They have passed on their knowledge of decades through this book. To expedite this challenging task, the publisher supported the team at every step. A small team of assistant editors was also appointed to further simplify the editing procedure and attain best results for the readers.

Apart from the editorial board, the designing team has also invested a significant amount of their time in understanding the subject and creating the most relevant covers. They scrutinized every image to scout for the most suitable representation of the subject and create an appropriate cover for the book.

The publishing team has been an ardent support to the editorial, designing and production team. Their endless efforts to recruit the best for this project, has resulted in the accomplishment of this book. They are a veteran in the field of academics and their pool of knowledge is as vast as their experience in printing. Their expertise and guidance has proved useful at every step. Their uncompromising quality standards have made this book an exceptional effort. Their encouragement from time to time has been an inspiration for everyone.

The publisher and the editorial board hope that this book will prove to be a valuable piece of knowledge for researchers, students, practitioners and scholars across the globe.

List of Contributors

Charan Jeet Singh
Department of Mathematics Guru Nanak Dev University, Amritsar, Punjab, 143005, India

Madhu Jain
Department of Mathematics, Indian Institute of Technology Roorkee, Roorkee, Uttarakhand, 247667, India

Binay Kumar
Department of Mathematics M.L.U. DAV College, Phagwara, Punjab, 144402, India

Mehrab Bahri
Department of Industrial Engineering, Science and Research Branch, Islamic Azad University, Tehran 14778, Iran

Mohammad Jafar Tarokh
Department of Industrial Engineering, K. N. Toosi University of Technology, Tehran 1439955471, Iran

Omid Momen
Karafarin Bank, Tehran, Iran
Amirkabir University of Technology, Tehran, Iran

Alimohammad Kimiagari
Amirkabir University of Technology, Tehran, Iran

Eaman Noorbakhsh
Karafarin Bank, Tehran, Iran
Insead Business School, Paris, France

Ariful Islam
Department of Industrial and Production Engineering, Shahjalal University of Science and Technology, Sylhet 3114, Bangladesh

Des Tedford
Department of Mechanical Engineering, The University of Auckland, Auckland 1142, New Zealand

Mitra Bokaei Hosseini
Department of IT, Faculty of Industrial Engineering, K. N. Toosi University of Technology, Tehran, 1631714191, Iran

Mohammad Jafar Tarokh
Department of IT, Faculty of Industrial Engineering, K. N. Toosi University of Technology, Tehran, 1631714191, Iran

Houshang Taghizadeh
Department of Management, Tabriz Branch, Islamic Azad University, Tabriz, Iran

Ehsan Hafezi
Industrial Engineering - System Management and Productivity at the Non-Governmental and Private Higher Education Institution of ALGHADIR, Tabriz, Iran

Samaneh Noori-Darvish
Department of Industrial Engineering, Allame Mohades Noori University, PC: 46415-451, Noor, Iran

Reza Tavakkoli-Moghaddam
Department of Industrial Engineering, College of Engineering, University of Tehran, PC: 14399-57131, Tehran, Iran

Ali Mohammadi Nasrabadi
Department of Industrial Engineering, K. N. Toosi University of Technology, Tehran, Iran

Mohammad Hossein Hosseinpour
Department of Industrial and Mechanical Engineering, Islamic Azad University, Qazvin Branch, Qazvin, Iran

Sadoullah Ebrahimnejad
Department of Industrial Engineering, Islamic Azad University, Karaj Branch, Alborz, Iran

Sanjay Kumar
International Institute of Technology and Management, Murthal, Haryana 131039, India

Sunil Luthra
National Institute of Technology, Kurukshetra, Haryana 136119, India

Abid Haleem
Faculty of Engineering and Technology, Jamia Mallia Islamia, New Delhi 110025, India

Elahe Faghihinia
Department of Industrial Engineering, Islamic Azad University of Najafabad, Isfahan 8514143131, Iran

Naser Mollaverdi
Department of Industrial Engineering, Isfahan University of Technology, Isfahan 8415683111, Iran

Ghasem Tohidi
Department of Mathematics, Islamic Azad University of Central Tehran Branch, Simaye Iran Ave., Tehran 14676-86831, Iran

Maryam Khodadadi
Department of Mathematics, Islamic Azad University of Central Tehran Branch, Simaye Iran Ave., Tehran 14676-86831, Iran

Hans J Thamhain
Management Department Bentley University, 175 Forest Street Waltham, Massachusetts, USA

Parveen Farooquie
Department of Mechanical Engineering, Aligarh Muslim University (AMU), Aligarh, India

Abdul Gani
Department of Mechanical Engineering, Aligarh Muslim University (AMU), Aligarh, India

Arsalanullah K Zuberi
Department of Mechanical Engineering, Aligarh Muslim University (AMU), Aligarh, India

Imran Hashmi
Department of Mechanical Engineering, Aligarh Muslim University (AMU), Aligarh, India

Seyed Ali Hadighi
National Iranian Oil Company - Mahmoudabad Training Center, Mahmoudabad, Iran Department of Industrial Engineering, Mazandaran University of Science and Technology, Babol, Iran

Navid Sahebjamnia
School of Industrial Engineering, College of Engineering, University of Tehran, Tehran, Iran

Iraj Mahdavi
Department of Industrial Engineering, Mazandaran University of Science and Technology, Babol, Iran

Hadi Asadollahpour
Department of Industrial Engineering, Mazandaran University of Science and Technology, Babol, Iran

Hosna Shafieian
Department of Industrial Engineering, Mazandaran University of Science and Technology, Babol, Iran

Mahdi Ghane
Mechatronics Department, K.N.Toosi University of Technology, Tehran 19697 64499, Iran

Mohammad Jafar Tarokh
Industrial Engineering Department, K.N.Toosi University of Technology, Tehran 19697 64499, Iran

Vinod Kumar Mishra
Department of Computer Science & Engineering, B. T. Kumaon Institute of Technology, Dwarahat, Almora, Uttarakhand 263653, India

Lal Sahab Singh
Department of Mathematics & Statistics, Dr. Ram Manohar Lohia Avadh University, Faizabad, Uttar Pradesh 224001, India

Rakesh Kumar
Department of Mathematics & Statistics, Dr. Ram Manohar Lohia Avadh University, Faizabad, Uttar Pradesh 224001, India

Mir-Bahador Aryanezhad
Department of Industrial Engineering, Iran University of Science and Technology, Tehran 16846-13114, Iran

Nima Fakhim Hashemi
Department of Industrial Engineering, Iran University of Science and Technology, Tehran 16846-13114, Iran

Ahmad Makui
Department of Industrial Engineering, Iran University of Science and Technology, Tehran 16846-13114, Iran

Hasan Javanshir
Department of Industrial Engineering, Islamic Azad University, South Tehran Branch, Tehran 11518-63411, Iran

Milad Jasemi
Department of Industrial Engineering, Islamic Azad University, Masjed Soleyman Branch, Masjed Soleyman, 61649, Iran

Ali M Kimiagari
Department of Industrial Engineering, Amirkabir University of Technology, Tehran, 15914, Iran

www.ingramcontent.com/pod-product-compliance
Lightning Source LLC
Chambersburg PA
CBHW050454200326
41458CB00014B/5181